Quantitative Proteomics

New Developments in Mass Spectrometry

Titles in the Series:
1: Quantitative Proteomics

How to obtain future titles on publication:
A standing order plan is available for this series. A standing order will bring delivery of each new volume immediately on publication.

For further information please contact:
Book Sales Department, Royal Society of Chemistry, Thomas Graham House, Science Park, Milton Road, Cambridge, CB4 0WF, UK
Telephone: +44 (0)1223 420066, Fax: +44 (0)1223 420247
Email: booksales@rsc.org
Visit our website at www.rsc.org/books

Quantitative Proteomics

Edited by

Claire Eyers
Institute of Integrative Biology, University of Liverpool, UK
Email: Claire.Eyers@liverpool.ac.uk

and

Simon J Gaskell
Queen Mary University of London, UK
Email: principal@qmul.ac.uk

New Developments in Mass Spectrometry No. 1

ISBN: 978-1-84973-808-8
ISSN: 2044-253X

A catalogue record for this book is available from the British Library

Published by The Royal Society of Chemistry,
Thomas Graham House, Science Park, Milton Road,
Cambridge CB4 0WF, UK

Registered Charity Number 207890

For further information see our web site at www.rsc.org

Preface

Understanding the roles and regulation of biological macromolecules in controlling cellular function is a classic tenet of biochemistry. Strategies to characterise the amount, stability and modification status of proteins, the macromolecular 'workhorses' of the cell, are of interest both to enhance our basic understanding of how cells function and communicate and also as a marker of when cellular processes have gone awry. The advent of mass spectrometry (MS)-based proteomics in the 1990s enabled protein characterisation on a much larger scale than had previously been possible with studies primarily focussed on qualitative investigations, such as identifying protein binding partners and defining gross changes in cellular composition. The rapid development in more recent years of high specification instrumentation, both for the separation and the analysis of polypeptides, has facilitated significant improvements in the rate of acquisition and depth of coverage of the proteome. Yet, any undergraduate biochemist will be able to tell you that the protein composition of a cell is not static, but changes both as a function of cell cycle and age and, critically, in response to extracellular stimuli; in this manner cells can take advantage of nutrients and protect themselves from unwanted stresses. Taking a snapshot of the protein composition of a system is therefore not nearly as important as being able to monitor quantitative changes in the proteome of that system over time or under variant conditions. However, robust methods for these types of in-depth quantitative experiments have lagged and increasingly sophisticated strategies have needed to be developed in order to assess these quantitative changes with the required degree of accuracy and precision. Sometimes the subtleties of the biological changes can be masked by the quantitative strategy employed; the techniques used therefore need to be selected carefully depending on the question being addressed.

New Developments in Mass Spectrometry No. 1
Quantitative Proteomics
Edited by Claire Eyers and Simon J Gaskell

Published by the Royal Society of Chemistry, www.rsc.org

All our current strategies for quantitative proteomics rely on using peptide surrogates as a read-out for protein amount, be that using isotope-labelled reference peptide standards, or employing label-free strategies that use peptide ion sampling or relative signal intensity as a measure of abundance. These strategies assume that, if we can determine peptide quantity, these numbers will function as a reliable read-out for protein amount. However, while it is currently significantly easier to quantify peptides and there are good reasons to make these assumptions, such inferences do not necessarily hold true when we start to consider changes in the levels of the key peptide surrogates due to post-translational modification. In addition to the oft-considered modification of proteins by the covalent attachment of functional groups, such as phosphate or glycans, which are known to affect protein function and/or stability (arguably the true feature of interest), the activity of native cellular proteases is often altered under disease conditions (*e.g.* cardiovascular disease) and will generate peptides of undefined termini that can then no longer be utilised as quantification standards. Point mutations and single-nucleotide polymorphisms, any of which may affect protein function, may also inadvertently influence such quantification strategies. It is likely therefore that, in the long term, we will move towards protein-level quantification, measuring changes in the amounts of the protein workhorses themselves rather than using a select few of their constituent peptides. However, such quantitative analysis is currently not feasible at the level required: much work will be needed to enable quantification of the proteome directly and the current peptide-based strategies are likely to be generating significant amounts of high quality data for the foreseeable future. These data will continue to be used to enhance what we know about biological systems and their regulation, and they are increasingly used instead of immunochemical-based assays to screen for biomarkers of disease.

Quantitative Proteomics aims to outline the state-of-the-art in mass spectrometry-based quantitative proteomics, describing recent advances and current limitations in the instrumentation used (Chapters 1 and 2), together with the various methods employed for generating high quality data. Strategies describing how stable isotope labelling can be applied for either relative or absolute protein quantification are detailed (Chapters 3–5), as are methods for performing quantitative analysis of proteins in a label-free manner (6–8). The utility of these strategies to understanding cellular protein dynamics are then exemplified with chapters looking at spatial proteomics (Chapter 9), dynamics of protein function as determined by quantifying changes in protein post-translational modification (Chapters 10 and 11) and protein turnover (Chapter 12). Finally, a key application of these techniques to biomarker discovery and validation is presented (Chapter 13 and 14), together with the rapidly developing area of quantitative analysis of protein-based foodstuffs (Chapter 15).

While there will undoubtedly be continued development in the area of quantitative proteomics for many years to come, particularly in the algorithms used for the analysis of the data, the key principles underlying sample preparation, data acquisition and the use of stable isotopes as quantification standards are now firmly established. As the number and type of applications to which MS-based quantitative strategies are applied expands, so new technology will be developed to meet their growing needs. Ultimately, this will further our understanding of the ever-changing proteomes of complex biological systems.

Claire E. Eyers
University of Liverpool

Contents

Technology

Chapter 1 Practical Considerations and Current Limitations in Quantitative Mass Spectrometry-based Proteomics 3
Adam M. Hawkridge

1.1 Introduction 3
1.2 Electrospray Ionization 6
1.2.1 Ionization Mechanisms 7
1.2.2 Ionization Response and Bias 8
1.2.3 Ion Transmission Efficiency 10
1.3 Contemporary LC-MS/MS Instrumentation 12
1.3.1 Liquid Chromatography 12
1.3.2 High-performance Tandem Mass Spectrometers 13
1.3.3 High-performance Characteristics of LC-MS/MS: Cycle Time, Peak Capacity, and Dynamic Range 14
1.4 Optimizing LC-MS/MS for Quantitative Proteomics Using Design of Experiments 16
1.5 Summary 19
Acknowledgments 20
References 20

Chapter 2 High Resolution/Accurate Mass Targeted Proteomics 26
A. Bourmaud, S. Gallien and B. Domon

2.1 Introduction 26

New Developments in Mass Spectrometry No. 1
Quantitative Proteomics
Edited by Claire Eyers and Simon J Gaskell

Published by the Royal Society of Chemistry, www.rsc.org

2.2 LC-MS-based Targeted Proteomics 27
2.2.1 Characteristics of Targeted Experiments 27
2.2.2 Targeted Experiments Using Selected Reaction Monitoring 29
2.3 Targeted Proteomic Experiments Using High Resolution/Accurate Mass Instrumentation 32
2.3.1 Characteristics of High Resolution/Accurate Mass Instruments 32
2.3.2 Quadrupole–Orbitrap Instrument 33
2.4 Quantification Performed on Precursor Ions 33
2.4.1 Trapping Capabilities 35
2.4.2 Quantification in Selected Ion Monitoring 36
2.4.3 Parameters Used in Selected Ion Monitoring 38
2.5 Quantification Performed on Product Ions 39
2.5.1 Comparison of SIM and PRM Modes 39
2.5.2 Quantification in Parallel Reaction Monitoring 41
2.5.3 Parameters Used in Parallel Reaction Monitoring 43
2.6 Conclusion 45
Acknowledgments 46
References 46

Label-based Protein Quantification

Chapter 3 Making Sense Out of the Proteome: the Utility of iTRAQ and TMT 51
Narciso Couto, Caroline A. Evans, Jagroop Pandhal, Wen Qiu, Trong K. Pham, Josselin Noirel and Phillip C. Wright

3.1 Introduction—Mass Spectrometry-based Quantitative Proteomics, an Overview 51
3.1.1 Labelling Approaches—Metabolic Labelling 53
3.1.2 Chemical Labelling 53
3.2 Isobaric Tagging—iTRAQ and TMT Labelling 55
3.2.1 Structure of iTRAQ and TMT Reagents 55
3.2.2 iTRAQ and TMT Protocol Overview 55
3.2.3 iTRAQ and TMT Variants 58
3.3 Technical Limitations of iTRAQ/TMT 59
3.4 Mass Spectrometry Platforms 62
3.5 Downstream Analysis of Proteomic Data 63
3.6 iTRAQ and TMT Relative Quantification Applied to the Analysis of Post-translational Modifications (PTMs) 64
3.6.1 Phosphoproteomic Analysis Using Isobaric Tagging 65

3.6.2 Glycoproteomics Analysis Using Isobaric Tagging 66
3.7 Bio-engineering and Biomedical Applications of Isobaric Tagging Technology 67
3.7.1 Application to Biological Engineering and Systems Research 67
3.7.2 Applications to Medical Research 68
3.8 Conclusions 70
Acknowledgements 71
References 71

Chapter 4 Getting Absolute: Determining Absolute Protein Quantities *via* Selected Reaction Monitoring Mass Spectrometry 80
Christina Ludwig and Ruedi Aebersold

4.1 Introduction 80
4.1.1 Why Do We Need Absolute Protein Quantification? 80
4.1.2 Established Technologies for Absolute Protein Quantification 82
4.2 Targeted Mass Spectrometry 83
4.2.1 Principles of SRM 84
4.2.2 The SRM Assay 85
4.2.3 Why is SRM Suited for Absolute Protein Quantification? 87
4.2.4 Next-generation Targeted Mass Spectrometry 87
4.3 Absolute Protein Quantification Combined with SRM 88
4.3.1 Stable Isotope Labeling 88
4.3.2 Absolute Label-free Protein Quantity Estimation 94
4.4 Challenges of Absolute Protein Quantification 96
4.4.1 The Challenge of Complete Protein Extraction 96
4.4.2 The Challenge of Specific and Complete Protein Digestion 97
4.4.3 The Challenge of Optimal Peptide Selection 98
4.4.4 The Challenge of Translating Label-free MS Intensities into Absolute Protein Quantities 100
4.5 Conclusions and Further Perspectives 101
Acknowledgements 103
References 103

Chapter 5 Proteomics Standards with Controllable Trueness—Absolute Quantification of Peptides, Phosphopeptides and Proteins Using ICP- and ESI-MS 110
Anna Konopka, Christina Wild, Martin E. Boehm and Wolf D. Lehmann

5.1 Introduction 110
5.2 μLC-ICP-MS as an Absolute Quantification Method for Peptides 113
5.2.1 Phosphorus-quantified Phosphopeptide Standards 113
5.2.2 Phosphorus-quantified but Phosphorus-free Peptide Standards 116
5.2.3 Peptide/Phosphopeptide Ratio Standards 118
5.3 ICP-MS as an Absolute Quantification Method for Proteins 118
5.3.1 Selenium-quantified Protein Standards 119
5.3.2 Selenium-quantified but Selenium-free Protein Standards 123
5.4 Summary and Outlook 125
Acknowledgements 125
References 126

Label-free Protein Quantification

Chapter 6 Overview and Implementation of Mass Spectrometry-Based Label-Free Quantitative Proteomics 131
Erik J. Soderblom, J. Will Thompson and M. Arthur Moseley

6.1 Introduction 131
6.2 Label-free LC-MS Quantification Strategies 134
6.2.1 Spectral Counting and Derived Indices for Label-free Quantification 134
6.2.2 Ion Intensity-based Label-free Quantification 137
6.3 Software Packages for Label-free Quantitative Proteomics 143
6.4 Label-free Experimental Design Considerations 145
6.4.1 General Sample Handling Guidelines for Label-free Quantitative Experiments 145
6.4.2 LC-MS System Qualification 148
6.4.3 Quality Control and Data Outlier Determination 149
6.5 Conclusions 150
Acknowledgment 150
References 151

Chapter 7 MS1 Label-free Quantification Using Ion Intensity Chromatograms in Skyline (Research and Clinical Applications) **154**
Birgit Schilling, Brendan X. MacLean, Alexandria D'Souza, Matthew J. Rardin, Nicholas J. Shulman, Michael J. MacCoss and Bradford W. Gibson

7.1 Introduction—Label-free Quantification Using MS1 Filtering 154
7.2 Applications, Improved Features and Applied Statistical Processing 157
7.2.1 Improved and Expanded Features for Quantification Using Skyline MS1 Filtering 158
7.2.2 Development of Statistical Tools for Processing MS1 Filtering Data Sets 161
7.3 Research and Clinical Applications 162
7.3.1 MS1-based Label-free Quantification Tools to Assess and Optimize Sample Preparation Workflows 162
7.3.2 MS1-based Label-free Quantification to Investigate Subtype-Specific Breast Cancer 164
7.3.3 Laboratory Research Pipelines Connecting MS1-based Label-free Quantification with Targeted, Data-independent Quantification Methods 168
7.4 Conclusions and Future Outlook 171
Acknowledgements 172
References 172

Chapter 8 Label-free Quantification of Proteins Using Data-Independent Acquisition **175**
Yishai Levin

8.1 Introduction 175
8.1.1 Fundamentals 177
8.2 Identifying Proteins 178
8.3 Quantifying Proteins 178
8.4 Ion Mobility and MS^E 179
8.5 Performance 180
8.6 Conclusions 183
Acknowledgements 183
References 183

Dynamic Protein Quantification

Chapter 9 Spatial Proteomics: Practical Considerations for Data Acquisition and Analysis in Protein Subcellular Localisation Studies 187
Andy Christoforou, Claire Mulvey, Lisa M. Breckels, Laurent Gatto and Kathryn S. Lilley

9.1 Introduction 187
9.2 Traditional Approaches for Characterising Protein Localisation 188
9.2.1 Microscopy-driven Analysis 188
9.2.2 *In Silico* Methods and Analyses 189
9.2.3 Quantitative Mass Spectrometry: A Complementary Technology 190
9.3 Subcellular Fractionation 190
9.3.1 Separative Centrifugation 190
9.3.2 Differential Permeabilisation 191
9.3.3 Immunocapture and Affinity Purification 191
9.3.4 Zone Electrophoresis 192
9.3.5 Selecting a Suitable Fractionation Workflow 192
9.4 Mass Spectrometric Analysis 193
9.4.1 Organelle Purification and Cataloguing 193
9.4.2 Subtractive Proteomics 195
9.4.3 Analytical Fractionation 198
9.4.4 Protein Correlation Profiling (PCP) 198
9.4.5 Localisation of Organelle Proteins by Isotope Tagging (LOPIT) 199
9.5 Data Analysis 202
9.6 Validation of Localisation Studies 205
9.7 Conclusions 207
References 207

Chapter 10 Quantitative Analyses of Phosphotyrosine Cellular Signaling in Disease 211
Hannah Johnson

10.1 Introduction 211
10.1.1 Phosphotyrosine Cellular Signaling 212
10.1.2 Deregulation of Tyrosine Phosphorylation in Disease 213
10.2 Enrichment Techniques for Phosphotyrosine Profiling 214
10.2.1 Phosphotyrosine-specific Antibodies as Enrichment Tools 214

10.2.2 SH2 Domains as Phosphotyrosine Profiling Tools 215
10.2.3 Tandem Mass Spectrometry Phosphorylation Site Localization 216
10.3 Quantification of Tyrosine Phosphorylation 218
10.3.1 Relative Quantification by Mass Spectrometry 218
10.3.2 Absolute Quantification by Mass Spectrometry 223
10.3.3 Label-free Quantification by Mass Spectrometry 223
10.3.4 Quantification of Phosphotyrosine Signaling at the Single-cell Level 224
10.4 Functional Analyses 226
10.4.1 Computational Modeling 226
10.4.2 Bioinformatics Tools 227
10.5 Future Perspectives: Quantitative Proteomics to Identify Novel Therapeutic Targets 228
References 228

Chapter 11 Next Generation Proteomics: PTMs in Space and Time 233
Dalila Bensadek, Armel Nicolas and Angus I. Lamond

11.1 Introduction 233
11.2 Next Generation Proteomics: Addressing the Challenges in Applying Global Proteomics to Cell Biology 234
11.2.1 Sample Preparation: Pre-fractionation 236
11.3 Quantitative Proteomics 240
11.3.1 Metabolic Labelling 240
11.3.2 Chemical Labelling 241
11.3.3 Label-free Quantification 243
11.4 Post-translational Modifications (PTMs) 246
11.4.1 PTM Enrichment 247
11.5 Data Analysis and Big Data 250
11.6 Concluding Remarks 251
References 253

Chapter 12 Experimental and Analytical Approaches to the Quantification of Protein Turnover on a Proteome-wide Scale 257
Amy J. Claydon, Dean E. Hammond and Robert J. Beynon

12.1 The Significance of Proteome Dynamics 257
12.2 Labelling Strategies for Turnover Studies 258
12.2.1 Label Choice 259

12.2.2 Control of Precursor Enrichment and the Significance of Precursor RIA 260
12.2.3 Sampling Times and Frequency 261
12.3 Solutions for the Analysis of Proteome Dynamics Data 262
12.3.1 Measurement of Precursor RIA 262
12.3.2 Analytical Approaches to the Measurement of Protein Turnover Rates 264
12.3.3 Acquisition of Precursor Peptide Intensities for Labelled and Unlabelled Variants 265
12.3.4 Manual Analysis 266
12.3.5 Software Solutions for the Calculation of Protein Turnover Rates on a Proteome-wide Scale 266
12.3.6 Open Source Solutions for Turnover Analysis 271
12.3.7 Optimal Software for the Analysis of Proteome Turnover Data 273
Acknowledgments 275
References 275

Applications of Quantitative Proteomics

Chapter 13 Protein Quantification by MRM for Biomarker Validation 279
L. Staunton, T. Clancy, C. Tonry, B. Hernández, S. Ademowo, M. Dharsee, K. Evans, A. C. Parnell, R. W. Watson, K. A. Tasken and S. R. Pennington

13.1 Introduction 279
13.1.1 Biomarker Panels 281
13.1.2 Protein Biomarker Discovery and Development 283
13.1.3 Clinical Samples 286
13.2 From Candidate Biomarker Discovery to Verification 287
13.2.1 Assembly of Candidate Protein Biomarker Panels 288
13.3 Integrative Bioinformatics for Biomarker Prioritisation 289
13.3.1 Gene Ontology as a Tool for Biomarker Discovery 290
13.3.2 Data- and Text-mining Approaches to Prioritise Biomarkers 291

13.3.3 Network and Pathway Approaches to Rank Candidates 291
13.3.4 Towards Next Generation Integrative Bioinformatics Approaches 293
13.4 Statistical Methods for Analysing MRM Data: Random forests and Support Vector Machines 294
13.4.1 Random Forests 295
13.4.2 Support Vector Machines (SVM) 297
13.5 Measurement of Protein Biomarker Panels—From Panels to Protein Signatures 299
13.5.1 MRM Development: Peptide Selection 299
13.5.2 MRM Assays for Large Scale Analysis 299
13.5.3 Methods for MRM Quantification 300
13.5.4 Quality Control 301
13.5.5 Skyline Software 303
13.5.6 Integrated Quality Control 305
13.5.7 Standardisation within the Lab and Inter-lab 306
13.6 Conclusion 306
Appendix 308
Acknowledgements 308
References 309

Chapter 14 MRM-based Protein Quantification with Labeled Standards for Biomarker Discovery, Verification, and Validation in Human Plasma 316
Andrew J. Percy, Andrew G. Chambers, Carol E. Parker and Christoph H. Borchers

14.1 Introduction 316
14.2 MRM for Biomarker Verification and Validation 320
14.3 MRM for Biomarker Discovery 322
14.4 Conclusions 323
Acknowledgements 325
References 325

Chapter 15 Mass Spectrometry-based Quantification of Proteins and Peptides in Food 329
Phil E. Johnson, Justin T. Marsh and E.N. Clare Mills

15.1 Introduction 329
15.2 Challenges for Food Proteomics 330
15.2.1 Food Proteome Annotation 330
15.2.2 Processing-induced Modifications of Food Proteins 331

15.2.3 Food Matrix Effects and Protein Extraction 332
15.2.4 Protease Digestion in Mass Spectrometry Analysis of Proteins 334
15.2.5 Target Selection for Analysis of Food Proteins 334
15.2.6 Calibrants and Reference Materials 336
15.3 Applications of Proteomic Profiling and Quantitative Proteomics in Food Analysis 337
15.3.1 Detection of Allergens in Foods 337
15.3.2 Proteomic Profiling for Food Product Quality 344
15.4 Conclusions 346
References 346

Subject Index **349**

TECHNOLOGY

CHAPTER 1

Practical Considerations and Current Limitations in Quantitative Mass Spectrometry-based Proteomics

ADAM M. HAWKRIDGE

Departments of Pharmaceutics & Pharmacotherapy and Outcomes Sciences, Virginia Commonwealth University School of Pharmacy, Richmond, VA 23298, USA
Email: amhawkridge@vcu.edu

1.1 Introduction

Mass spectrometry (MS)-based proteomics has become a prominent technology platform for quantitatively studying protein expression, modification, interaction, and degradation.[1–4] Although more established protein quantification methods using antibodies, gel electrophoresis, and radiochemical labelling remain important in biological research, they do not provide the level of molecular specificity and breadth of unbiased proteome coverage achieved by MS-based approaches. Among the many types and configurations of mass spectrometer, reverse phase liquid chromatography coupled to tandem mass spectrometry (LC-MS/MS) is perhaps the most commonly used for MS-based proteomics studies. LC-MS/MS can be configured to accommodate many types of quantitative experiments that span discovery (*i.e.*, global shotgun proteomics) to targeted protein quantification (protein

New Developments in Mass Spectrometry No. 1
Quantitative Proteomics
Edited by Claire Eyers and Simon J Gaskell

Published by the Royal Society of Chemistry, www.rsc.org

cleavage-isotope dilution mass spectrometry, PC-IDMS). However, optimizing the performance of an LC-MS/MS system for quantitative proteomics measurements can be a daunting challenge when considering the type of sample, choice of quantitative strategies (*e.g.*, isobaric tagging, SILAC, label-free, PC-IDMS, *etc.*), extensive number of instrument parameters that can be adjusted, and the rapidly changing technology landscape. In order to optimize LC-MS/MS for detection limits, proteome coverage, precision, and accuracy, it is important to have a fundamental understanding of the inter-related effects of instrument settings and experimental conditions on data quality.

LC-MS/MS is not an inherently quantitative technique due in large part to bias in the electrospray ionization (ESI) response.[5] Co-eluting species in a typical LC-MS/MS experiment must compete for a finite amount of charge in the ESI plume before reaching the mass spectrometer. For LC-MS/MS-based proteomics, the competing species are typically proteolytic peptides derived from intact proteins, each of which has unique physical (*e.g.*, molecular weight) and chemical (*e.g.*, hydrophobicity, isoelectric point) properties that give rise to different ESI responses. These physico-chemical properties also affect additional LC-MS/MS performance characteristics, including chromatographic, transmission, and fragmentation efficiency, further complicating LC-MS/MS quantification. Compounding these challenges is the pre-analytical variability introduced during sample preparation, including digestion efficiency, differential degradation rates of proteins and peptides, and preparation-induced modifications (*e.g.*, methionine oxidation), that can impact quantitative accuracy. All of these factors must be considered in quantitative LC-MS/MS-based proteomics, which covers a broad spectrum of experimental workflows. These workflows can be categorized as either relative (Figure 1.1A) or absolute (Figure 1.1B).

There are several relative quantification methods developed for LC-MS/MS-based proteomics (Figure 1.1A), including isobaric tagging (*e.g.*, iCAT[6],

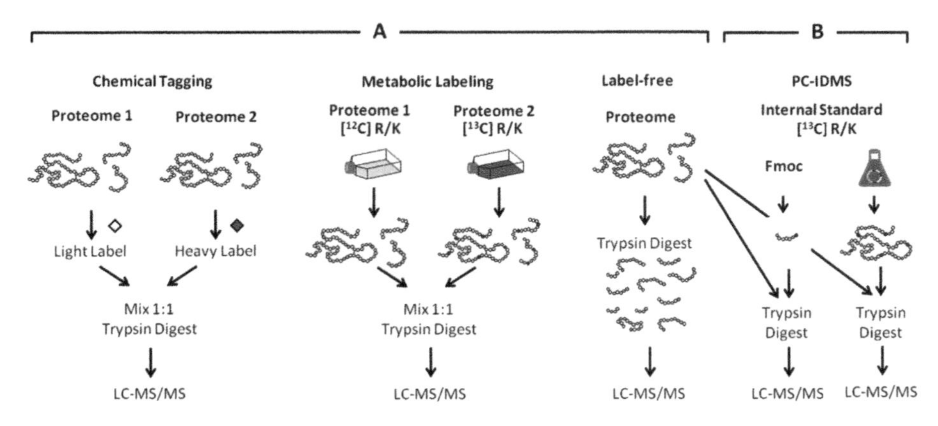

Figure 1.1 Quantitative proteomics strategies for global protein expression (A) and targeted protein quantification (B).

iTRAQ[7], and TMT[8]; see Chapter 3 for further information), *in vitro* metabolic labeling (*e.g.*, SILAC[9], ^{15}N[10]; see Chapters 10 and 11 for further information), label-free (see Chapter 6 for further information),[11–15] and $^{16}O/^{18}O$ labeling.[16] Label-free LC-MS/MS quantification requires a high degree of LC run-to-run reproducibility[13,17] to achieve good precision as the data-dependent MS/MS settings impact accuracy, dynamic range, and proteome coverage.[18] Isobaric tagging improves run-to-run precision but can suffer from imprecision at the sample preparation stage and is subject to reduced accuracy.[19] Although the multiplexing capability of isobaric tagging is a major strength, the dynamic range of measured protein abundance suffers with increasing numbers of tags. Digesting a protein sample in ^{18}O water introduces a shift of 2 Da in the proteolytic peptide relative to a protein sample digested in ^{16}O water. This global approach is attractive for clinical proteome samples (*e.g.*, plasma and tissues) that cannot be labelled *in vitro* but it suffers from overlapping isotopic distributions, which make identification and quantification difficult.[20] *In vitro* labeling, particularly stable isotope labeling of amino acids in cell culture (SILAC), provides excellent quantitative precision for global protein expression studies accounting for pre-analytical variability during sample preparation and is less sensitive to run-to-run LC reproducibility. However, because SILAC generates peptide pairs (2-plex labeling) or triplets (3-plex labeling), proteome coverage could, in principle, be reduced under the same LC-MS/MS method compared with label-free or isobaric tagging.[21,22] This is a consequence of the instrument spending twice or three times as much time performing MS/MS on differentially labeled peptides of the same primary sequence and the added charge to AGC-based instruments, which limit the number of ions that can be detected.

Absolute quantification methods for LC-MS/MS-based proteomics (*i.e.*, PC-IDMS) require internal standards, such as stable isotope-labeled proteins or peptides with known concentrations (Figure 1.1B; see Chapter 4 for further information).[23–29] The accuracy and precision of the LC-MS/MS method is largely dependent on the quality and stability of the internal standard. Ideally, a purified stable isotope-labeled form of the intact protein is most desirable as it would, in principle, generate the measured (*i.e.*, signature) tryptic peptide at the same rate and to the same extent as the endogenous protein target.[30] However, stable isotope-labeled proteins are sometimes challenging to express and purify, making them more expensive and less common. Another strategy, termed QconCAT,[31–34] uses a synthetic gene inserted into an expression vector that contains a series of the desired tryptic peptides for subsequent stable isotope labeling and enzymatic digestion. This strategy has great potential for multiplexed assays that require reproducible production of targeted peptides. Although the digestion efficiency of the concatenated stable isotope-labeled peptides may not be the same relative to their endogenous targeted proteins, thus impacting accuracy, the peptide production rate should be similar irrespective of sample, providing a high level of precision. The most common type of PC-IDMS method uses stable isotope-labeled synthetic peptides generated by solid phase synthesis techniques at a lower cost and with high purity. However, studies have shown that the quantitative accuracy of synthetic stable isotope-labeled tryptic peptides is highly variable depending on the rates of formation and stability of the tryptic peptides.[35,36] Thus, the strengths and weaknesses of each approach must be carefully considered when developing targeted LC-MS/MS assays. Optimization of LC-MS/MS conditions for PC-IDMS largely focuses on limits of detection, limits of quantification, and linear dynamic range while concurrently minimizing analysis time. Triple quadrupole LC-MS/MS is typically the instrument of choice for these types of measurements but new hybrid high resolution/high mass accuracy instruments, such as the quadrupole-Orbitrap, have recently been introduced that could potentially shift this paradigm.

This chapter will focus on some of the fundamental and practical aspects of LC-MS/MS that impact quantitative proteomics measurements. The first section covers ESI mechanism(s), bias, and transmission. Ion transmission spans the ESI plume, MS inlet, intermediate pressure region, and then through the ion optics in the ultra-low pressure region of the mass spectrometer. Once these areas have been reviewed, the collective LC-MS/MS platform will be discussed in the context of peak capacity, duty cycle, and dynamic range. The final section will review experiment design (*i.e.*, design of experiments, DoE) and highlight recent interest in this approach for optimizing LC-MS/MS. DoE offers perhaps the most universal empirical method for optimizing LC-MS/MS for complex quantitative proteomics measurements where experimental goals and outcomes must be balanced with a rapidly changing technology landscape.

1.2 Electrospray Ionization

Electrospray ionization (ESI) was introduced by Fenn in 1984[37] and has transformed the fields of separation science and mass spectrometry with no foreseeable limit to its analytical utility. Wilm *et al.*[38,39] introduced nanoelectrospray (nESI) in the mid-1990s, which significantly improved the sensitivity of ESI and allowed for the characterization of low concentration tryptic peptides. The original nESI emitter configuration involved a high voltage applied to a metalized tapered fused silica tip with no applied back pressure. However, this configuration has been replaced with a non-metalized fused silica nESI emitter for LC-MS/MS applications, which is illustrated in Figure 1.2A, with the key dimensions and experimental parameters labeled. A standard nESI emitter is 360 μm OD fused silica capillary pulled to a 10–30 μm ID tapered tip,[40,41] which is typically positioned approximately 1–5 mm from the entrance of the mass spectrometer. A flow rate of 200–500 nL min^{-1} is applied with either a syringe or LC pump, and a high voltage (1500–2500 V) is applied to an anode at the liquid/emitter interface. Despite the evolution and established use of nESI for quantitative proteomics, fundamental questions remain regarding the ionization mechanism(s), bias, dynamic range, and transmission efficiency of gas-phase ions into the mass

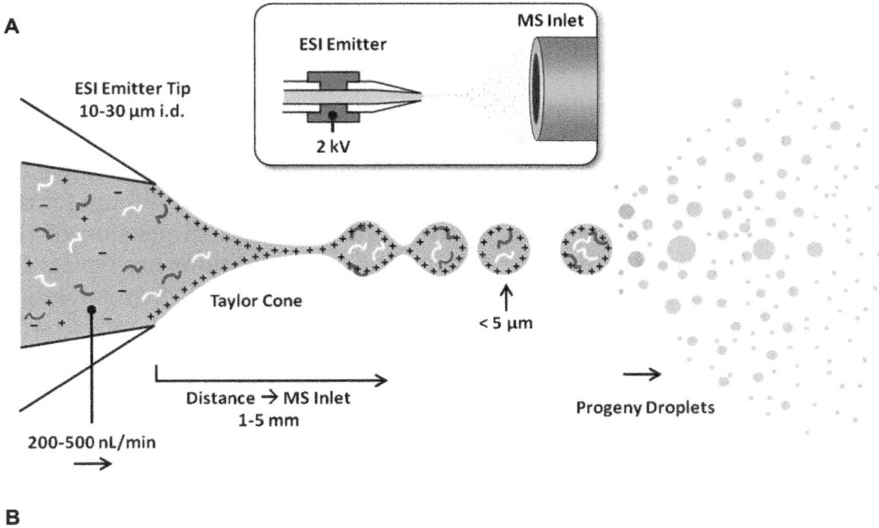

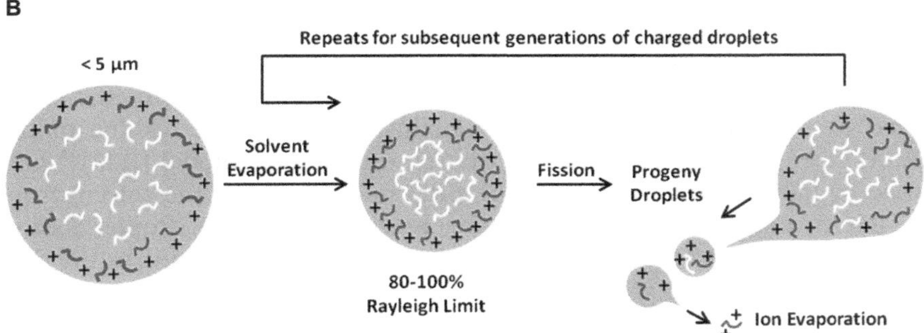

Figure 1.2 Overview of the formation of the Taylor cone from a standard nanoflow electrospray ionization source (A) and the subsequent peptide ionization *via* the ion evaporation model (B). Hydrophilic peptides (white/light gray) reside mainly in the core of the droplet, whereas more surface active/hydrophobic peptides (dark gray) migrate to the surface where they are more readily ionized.

spectrometer. A comprehensive discussion of these topics is provided in a seminal book by Cole *et al.*[42] and several well written review articles.[43–46]

1.2.1 Ionization Mechanisms

The initial stages of ESI are illustrated in Figure 1.2A whereby the positive high voltage applied to the solvent creates an excess of protons at the emitter tip to form a Taylor cone.[39,47] The combination of solvent flow and charge repulsion (*i.e.*, maximizing surface area) create charged droplets that emit from the tip of the Taylor cone. The size of the droplet varies depending on the solvent composition, flow rate, and emitter tip diameter.[38,39,43,48,49] Wilm *et al.*[38] estimated a parent droplet diameter of ~200 nm emitted from a 1 μm metalized emitter tip at 20–40 nL min^{-1}. They had previously shown that the parent droplet was proportional to two thirds the power of the flow rate.[39] Smith *et al.*[49] used Doppler interferometry to study droplet dynamics emitted from a 50 μm metal capillary tip at 1–2 μL min^{-1} for different solvent compositions. They measured parent droplet diameters of water between 10–40 μm. Based on these two studies and the fact that the geometry and voltage junctions are completely different, a conservative estimate for the parent droplet diameter shown in Figure 1.2A is <5 μm for the experimental conditions shown. Regardless of the exact dimension of the parent droplet, the experimental dimensions and conditions are typical for nESI in LC-MS/MS quantitative proteomics measurements. In general, larger diameter droplets (>5 μm) generated at higher flow rates (>1 μL min^{-1}) with larger diameter emitter tips produce smaller charge/volume ratios, requiring sheath and countercurrent gas flow to desolvate the droplets and drive ionization of analytes. Smaller diameter droplets (<5 μm) generated in nESI (<1 μL min^{-1}) have a higher charge/volume ratio, which require less desolvation to drive ionization. With the exception of fast LC separations at μL min^{-1} flow rates for targeted PC-IDMS protein quantification, most LC-MS/MS proteomics work employs nESI because of the enhanced ionization efficiency and, thus, improved analysis/quantification.

The formation of ionized gas-phase analytes from the ESI process is believed to occur by one of two mechanisms: 1. the charge residue model (CRM)[50] proposed by Dole; and 2. the ion evaporation model (IEM)[51] proposed by Iribarne and Thomson. The CRM maintains that the solvent from the parent charged droplet evaporates causing successive droplet fission events until all solvent has evaporated and residual charge ionizes the desolvated analyte. The IEM (Figure 1.2B) maintains that, as solvent evaporates from the parent droplet, the surface charge density increases creating an environment for charged yet solvated analytes to readily escape ('evaporate') once fission occurs at 80–100% of the Rayleigh limit.[43] Although the fundamental ionization mechanism(s) for ESI is still actively studied,[48,52–58] it is generally accepted that the IEM provides the most likely ionization mechanism for peptides and small proteins.

1.2.2 Ionization Response and Bias

The IEM is generally accepted as the most likely ionization mechanism in shotgun proteomics due to the inherent ionization bias observed in complex mixtures. For example, the LC-MS analysis of a pure protein that had been enzymatically digested would not result in equivalent intensities for each individual peptide. We can begin to estimate the ESI response for peptides based on eqn 1.1 published by Fenn.[5]

$$N_{iz} = 3Ae^{\left(-\Delta G^0_{iz}/RT\right)}(N_i/r)e^{\left(\Delta rzQ/4\pi\varepsilon_0 RTr^2\right)} \tag{1.1}$$

The ESI response (or ion flux), N_{iz}, of an ion (i) with z charges is a function of a proportionality constant (A) that relates bulk concentration to surface activity for a given analyte, the free energy of solvation ($\Delta G^0{}_{iz}$), the gas constant (R), the temperature (T), the moles of analyte (N_i), the radius of the droplet (r), the distance the ions must travel to become desolvated (Δr), the excess charge (Q), and the gas permittivity constant (ε_0). Although, in principle, all of the experimental factors from eqn 1.1 can be adjusted to optimize ESI response, the most practical include the droplet size (r), temperature (T), desolvation distance (Δr), and concentration (N_i).

The importance of droplet size (r) on ionization efficiency was already discussed in the context of ESI *vs.* nESI. Because r is inversely related to the ESI response, smaller droplets are best for shotgun proteomics measurements. The trade-off for nESI and low flow rates is sample throughput as equilibration of the LC column takes longer. Temperature is another factor that can be controlled to tailor ESI response, particularly at higher flow rates where desolvation of large droplets is more critical.[59] Heated nESI sources have also been developed[60,61] to study protein complex thermodynamics and level the ESI response factors for target analytes. A thorough study of the effect of nESI emitter temperature on quantitative proteomics datasets may reveal significant advantages, particularly for label-free proteomics where lower responding peptides may increase in abundance and subsequently be selected for sequencing. The desolvation distance (Δr) for a set r and T can be adjusted by changing the nESI emitter tip-to-capillary inlet distance or capillary length.[41,62] Geromanos *et al.*[41] found an emitter-capillary distances of 1 mm to be optimal for peptide signal intensity, and results from Page *et al.*[62] showed a two-fold decrease in peak intensity of reserpine when positioning the nESI emitter tip from 2 mm to 5 mm. It is important to note, however, that these distances are highly source dependent, thus the reason for giving a conservative range (1–5 mm) in Figure 1.2A. Finally, analyte concentration can be increased (N_i) to improve peptide signal intensity but saturation begins to occur above ~1–10 pmoles (the amount of a unique proteolytic peptide trapped on a column) for contemporary nLC-MS/MS systems. The primary drivers for ESI response bias include the inter-related partition constant (A) and $\Delta G^0{}_{iz}$, both of which are based on the physico-chemical properties of the proteolytic peptides.

Peptide ionization bias in the ESI process has been attributed to several physico-chemical properties, including molecular weight, basicity, hydrophobicity, 3D structure, and solubility. Collectively, these metrics are being used to understand the ESI response bias observed in nLC-MS/MS shotgun proteomics studies. Early fundamental work by Enke *et al.* [63–65] showed that more hydrophobic amino acid residues preferentially ionize relative to more hydrophilic residues. Similar findings were observed by Muddiman *et al.* for intact proteins[66] and DNA oligonucleotides[67], all of which corroborate the seminal study by Fenn[5] with quaternary alkyl amines. Researchers have built on these earlier fundamental studies of simple systems and developed interpretative and predictive informatics tools to dissect the ESI response of complex proteomics datasets.[68–72] Hydrophobicity, basicity, and molecular weight were found to be the top ranked physico-chemical properties that predict ESI response. The development and evolution of these bioinformatic tools will likely facilitate more efficient development (*i.e.*, less trial and error) of targeted PC-IDMS assays and potentially expand our fundamental understanding of peptide ESI response in complex mixtures.

1.2.3 Ion Transmission Efficiency

The transmission of ESI-generated ions is an inefficient process relative to the number of ions generated. Modern LC-MS systems can routinely achieve femtomole to attomole detection limits for targeted PC-IDMS studies. However, it has been shown experimentally that the ionization efficiency of ESI (and nESI), which approaches unity, far exceeds the transmission efficiency, which is estimated to be less than 1%.[38,39] Figure 1.3A illustrates the ionization efficiency relative to the distance between the nESI emitter tip and the heated capillary inlet. As the charged droplets approach the MS inlet, the population of ions increases exponentially with successive fission events occurring as illustrated in Figure 1.2B. Furthermore, as discussed in the previous section, there is inherent bias during ionization with the better ionizing peptide generating a higher ESI response (Figure 1.2B and Figure 1.3A). Thus, when considering that contemporary LC-MS systems can routinely provide low femtomole detection limits for tryptic peptides, it becomes clear that there is enormous opportunity to improve detection limits with higher transmission efficiencies.

There have been several approaches to improve the ion transmission efficiency both at the MS inlet (Figure 1.3A) and within the MS inlet capillary/orifice (Figure 1.3B). These include fabricating multiple MS inlets,[73–75] flared capillaries,[76,77] atmospheric pressure separation (*e.g.*, FAIMS),[78–87] and air amplifiers.[88,89] Each study has demonstrated improvements from two- to twenty-fold depending on the analyte. The amount of ions that make it into the MS inlet also depends on the nESI emitter-to-capillary distance.[62] Too far and the majority of ions are lost at the face of the capillary inlet. Too close and ions do not have sufficient time and distance to form from the charge droplets (see section 1.2.2). Furthermore, Page *et al.*[62] showed that significantly more current is lost to the inner wall of the capillary (Figure 1.3B), reducing ion formation and/or transmission. The MS inlet capillary or orifice illustrated in Figure 1.3B has been increasingly shortened from ~6 inch-long glass capillaries to ~1 inch metal inlets that vary from capillaries to skimmer cones (*e.g.*, Micromass Z-spray). The length, inner diameter, and temperature of the inlet dictate the conductance[90,91] and the pumping speed necessary in the intermediate pressure regime (Figure 1.3C). For commercial instruments where the physical dimensions are not easily shortened to improve transmission efficiency,[90] the only experimental parameters that can be adjusted are the applied voltage, temperature, and the proximity of the nESI emitter to the MS inlet. Temperature is adjusted such that it is sufficiently high to effectively desolvate charged droplets (*i.e.*, promote peptide ionization) yet low enough to facilitate effective conductance into

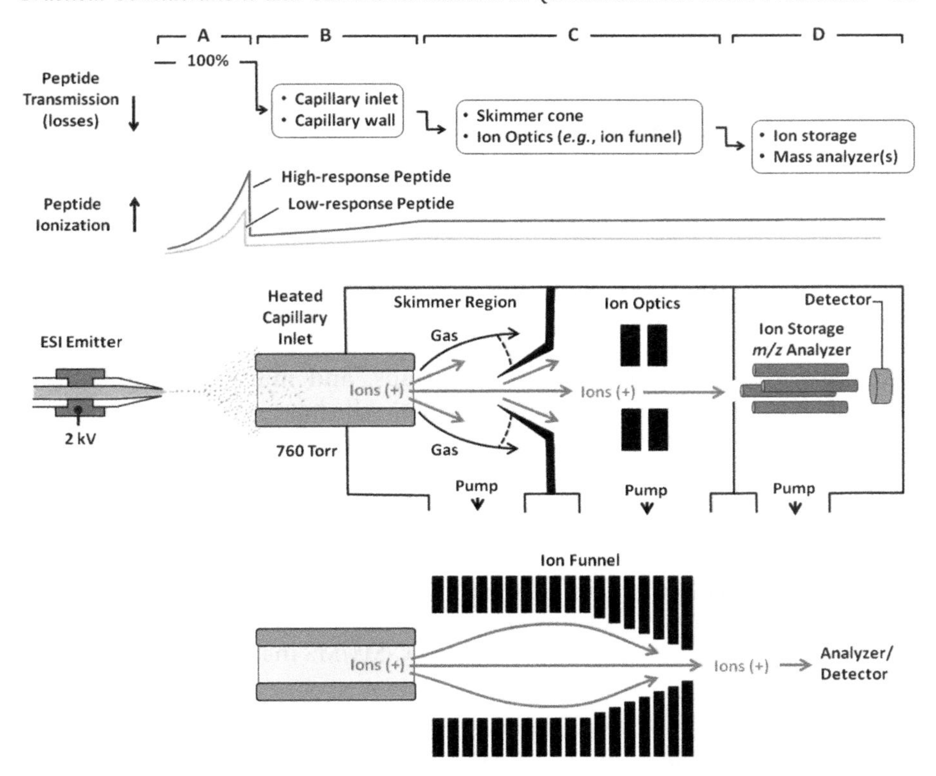

Figure 1.3 Schematic representation of peptide ionization and transmission through different regions of a mass spectrometer. Peptide ionization increases exponentially from the emitter tip to the MS inlet without any transmission losses. Note the more surface active peptide (dark gray) has higher ionization efficiency than the more hydrophilic peptide (white/light gray) in reference to Figure 1.2B (A). Once the ions reach the MS inlet, transmission drops significantly due to losses on the face of the capillary and the inner wall of the capillary. Ions continue to be generated inside the capillary (B). Ion formation is assumed to be negligible after the exit of the MS inlet but ion transmission is impacted by the transition from atmospheric pressure to the reduced pressure region. Ion losses can occur on the skimmer cone due to the expansion of gases from the exit orifice of the inlet, but ion funnel technology has reduced these losses (C). Ion losses are highly variable during storage and analysis (D).

the intermediate pressure region. Increasing the MS inlet temperature can result in a drop in pressure in the intermediate pressure region, which could in turn negatively impact ion transmission through the inner capillary wall. For example, ion transmission is sensitive to the position of the Mach disc relative to the skimmer cone, the former being sensitive to the amount of gas exiting the capillary. The illustration in Figure 1.3C gives a generic overview of both older ESI sources (skimmer cone) and newer ESI sources that incorporate ion funnel technology. The ion funnel developed in the Smith lab[73,74,92–101] has replaced some of the skimmer cone-based sources in commercial LC-MS/MS platforms in an effort to improve ion transmission in the high/intermediate pressure regions immediately following the exit orifice of the mass spectrometer inlet. Although these technologies have been proven effective in custom-built instruments, it is difficult to quantify how they impact commercial instruments with restricted access to instrument controls and an inability to modify or measure losses at each stage of ion formation and transport. The same can be said for understanding losses during ion storage and mass analysis because each contemporary instrument platform from a given manufacturer is unique. Quadrupoles, linear ion traps, time-of-flight analyzers, Orbitraps, and multiple combinations thereof (*i.e.*, hybrid MS/MS instruments) all have different operating principles and performance characteristics that make independent cross-instrument platform comparisons of ion transmission challenging. The reality of quantitative LC-MS/MS-based proteomics is that the user has increasingly less control over how commercial instruments function and limited ability to modify the instrument in an effort to improve performance. To keep pace with the rate of instrument improvements and maximize performance, section 1.4 of this chapter provides a discussion of a systematic and empirical method for optimizing LC-MS/MS instruments, regardless of instrument manufacturer.

1.3 Contemporary LC-MS/MS Instrumentation

1.3.1 Liquid Chromatography

Chromatography is essential prior to MS-based proteomics analysis for complex protein purification, pre-fractionation, and separation. Reverse-phase (RP) high-performance liquid chromatography (HPLC) is synonymously used as the 'LC' component in LC-MS/MS because it provides excellent separation efficiency of tryptic peptides and the mobile phase compositions are compatible with ESI. The demands of MS-based proteomics have driven innovation in LC technology, particularly with regard to nanoflow systems. The benefits of lower flow rates have already been discussed, yet producing stable and reproducible nanoflow gradients on capillary reverse phase columns is not trivial. Early efforts by researchers to achieve nanoflow gradients involved splitting the flow from a standard μL min^{-1} or mL min^{-1} HPLC system prior to the capillary column. Run-to-run reproducibility was challenging because back pressure on the analytical capillary column would increase with successive sample injections. Furthermore, the flow rates would change with different gradient conditions as back pressure was highest at the beginning (high aqueous) relative to the end (high organic) of the gradient. The changes in flow rate affected the nESI conditions, which in turn produced variable electrospray responses and poor inter-LC-MS/MS run reproducibility.

Splitless nanoflow LC systems were introduced around ten years ago and have become the standard for high-performance LC-MS/MS-based quantitative proteomics due to their superior flow rate stability and retention time reproducibility.[102] Flow rate stability is critical for generating uniformly sized charged droplets to ensure reproducible ionization efficiency across multiple LC-MS/MS runs. Retention times are also important, particularly for label-free quantification and scheduled SRMs. Average retention time reproducibility for peptides in complex shotgun proteomic mixtures is <0.5% RSD for some of the more widely used commercial nLC systems.[103] The technological advances in nLC systems related to retention time reproducibility have also been extended into the microflow rates. This is becoming increasingly important with the introduction of hybrid LC-MS/MS instruments that can perform both global 'discovery' and targeted quantitative proteomics studies, the former typically employing nLC, whereas the latter uses μLC. Furthermore, these developments have been integrated into ultra-high pressure (UHPLC) systems that operate at pressures above 10 000 psi. UHPLC, in general, provides higher peak capacity (and therefore improves detection limits) and the possibility of faster LC-MS/MS run times compared with HPLC.[104–106] Versatile splitless nano-microflow HPLC and UHPLC systems have been critical for improved precision and accuracy in quantitative LC-MS/MS. Current trends suggest more integrated nLC-nESI interfaces in the future with chip-based columns, reduced dead volume connections, and better control over column and ESI conditions (*e.g.*, column temperature, high voltage placement). Collectively, these advances will serve to make quantitative proteomics, particularly label-free quantitative studies, more routine and reproducible.

1.3.2 High-performance Tandem Mass Spectrometers

Contemporary global protein expression studies use high-performance tandem mass spectrometers, such as quadrupole time-of-flights (Q-TOF) and linear ion trap/quadrupole Orbitraps (LTQ-Orbitrap/Q-Exactive), to provide two essential types of data: 1. the accurate mass of the intact tryptic peptide (<5 ppm); and 2. the product ion spectra of the tryptic peptide. These two pieces of data provide a means to confidently identify the peptide relative to the relevant protein database. In principle, the faster the instrument can generate these two forms of data of high quality, the greater the number of proteins that can be identified. Protein identification is a blunt metric for assessing LC-MS/MS performance, particularly in the context of quantitative proteomics given the wide ranging goals one can set. However, the rapid pace of mass spectrometric instrumentation development over the past 10–15 years is astounding. Consider that, from 2001 to 2011, reports of comprehensive proteome coverage for yeast increased 2-fold from 1484[107] to 2990[108] proteins with essentially identical analysis times (28 h *vs.* 24 h, respectively), yet the number of identified peptides increased 2.5-fold from 5540[107] to 13 682,[108] respectively. The increased proteome coverage is primarily attributed to improvements in high-performance MS/MS technology, which produces high mass accuracy/high resolving power MS and MS/MS data, achieve faster duty cycles, and benefit from improved ion transmission.

Triple quadrupole LC-MS/MS systems have been mainstay instruments for the targeted quantification of small molecules. These systems have become increasingly used for targeted protein quantification using protein cleavage isotope dilution (PC-IDMS).[23,24,27,28] Triple quadrupole LC-MS/MS systems do not provide high resolving power or high mass accuracy data, yet they are capable of rapidly measuring analytes in complex mixtures with high specificity over four to five orders of magnitude in concentration. The specificity is achieved with the rapid scan speeds for the mass selective quadrupoles. For a single precursor tryptic peptide ion, multiple signature product ions can be measured *via* rapid mass selective transitions to quantify and qualify the targeted peptide, which is representative of the protein concentration. Transition scan speeds are typically 10–50 Hz depending on the complexity of the sample, the number of peptides, the number of transitions necessary to uniquely define a peptide, and the concentration of the targeted peptide(s). Thus, monitoring a peptide with five transitions for 0.1 s each, one could collect 40 data points over a 20 s peak width. For two co-eluting peptides (*i.e.*, target peptide and SIL internal standard) that require five transitions each at 0.1 s, one could obtain 20 data points for both peptides over the same 20 s peak width.

A potential shift away from the use of triple quadrupole LC-MS/MS systems is the introduction of hybrid, high resolving power and high mass accuracy instruments (*e.g.*, Triple-TOF[109] and Q-Exactive[110]). The newer hybrid systems do not have the mass selection/transition scan speeds found with classic triple quadrupole instruments (*e.g.*, 10–100 Hz), but the high resolving power full scan of precursor product ions provides potentially all the information needed to offset this shortcoming. The evolution of high-performance LC-MS/MS instrumentation will continue to challenge the existing paradigms for how quantitative proteomics experiments are designed, the type(s) of information that can be collected, and the need, or lack thereof, for distinct instrument platforms to make different types of quantitative measurements.

1.3.3 High-performance Characteristics of LC-MS/MS: Cycle Time, Peak Capacity, and Dynamic Range

'High-performance' LC-MS/MS has been used as a general term in this Chapter to describe many different types of instruments. Here, the term is given some figures of merit that can be used to better define LC-MS/MS performance. Two main types of instruments have been discussed in this Chapter: 1. high-performance LC-MS/MS for global protein expression level studies; and 2. high-performance LC-MS/MS for targeted protein quantification. However, it should be re-emphasized that, as LC-MS/MS instrumentation continues to evolve, the distinctions in application for one type of instrument *versus* the other continue to blur. Three metrics are discussed in the context of quantitative proteomics: duty cycle, peak capacity, and dynamic range.

LC-MS/MS duty cycle can be defined differently depending on the type of instrument, experiment, and desired data. The basic definition of LC-MS/MS duty cycle is given by eqn 1.2.

$$\text{Duty Cycle}_{\text{LC-MS/MS}} = \frac{T_{\text{ion}}}{T_{\text{cycle}}}, \tag{1.2}$$

where T_{ion} is the amount of time the instrument spends measuring an ion and the cycle time (T_{cycle}) is the total amount of time between each measurement. If we consider a typical global proteomics LC-MS/MS data-dependent acquisition (DDA) experiment, a high resolution/high mass accuracy full scan of the eluting proteolytic peptides provides a list of peptide *m/z* values for subsequent mass selection, dissociation, and detection. A typical full scan (300–2000 *m/z*) can provide up to ~30 unique peptides that can be subjected to MS/MS. A standard DDA is set-up so that the top ten most abundant peptides are selected for MS/MS analysis ('top ten'). Thus, the full scan + ten MS/MS scans + the signal processing time represents one cycle. Consider a Q-Exactive operating in a typical 'top ten' mode[110] with a 256 ms high-resolution scan and 10 × 64 ms MS/MS scans to give a total of 1.06 s: the actual time for this full cycle is 1.22 s if one accounts for the signal processing overhead. Thus, the duty cycle for the Q-Exactive DDA, as defined by eqn 1.2, is 1.06/1.22 = 0.87. In other words, the instrument spends 87% of its time detecting and fragmenting ions, which is remarkably efficient. However, this can be misleading if the goal is to identify/quantify as many peptides as possible, or if only specific precursor ions are of interest. For global LC-MS/MS studies, peak capacity is a potentially more useful measure of instrument performance.

The peak capacity for a single LC-MS/MS run was estimated by Wolters *et al.*[111] using eqn 1.3.

$$n_{\text{c}} = \left(\frac{L}{4\sigma}\right) \times \left(\frac{\#\ \text{MS/MS}}{\text{Cycle Time}} \times 4\sigma\right) \tag{1.3}$$

The terms are the total time the peptides have to elute (L), the average peak width (4σ), the number of MS/MS events in a cycle (# MS/MS), and the time for one DDA MS/MS cycle (Cycle time). If we assume a 90 min LC gradient, average peak width of 25 s, # MS/MS events = 10 (top ten), and a cycle time of 1.5 s; the maximum theoretical peak capacity is equal to 7.8×10^6:

$$7.8 \times 10^6 = \left(\frac{5400\ \text{secs}}{25\ \text{secs}}\right) \times \left(\frac{10}{1.5} \times 25\ \text{secs}\right)$$

Of course, not all MS/MS events result in useful product ion spectra enabling peptide identification due to a lack of signal or poor quality mass spectra. Furthermore, there are several practical limitations encountered during a shotgun DDA that impact the theoretical peak capacity. Abundant peptides can dominate the full scan mass spectrum limiting the detection and subsequent analysis of lower abundant peptides. The abundance of the peptide may be indicative of its true expression level (*e.g.*, albumin in plasma) and/or it may be the result of higher ionization efficiency during the ESI process. To counter this effect, pre-fractionation of the sample or adjustment of the DDA LC-MS/MS method may be necessary. Shifting to UHPLC in theory improves peak capacity for the chromatography, but it increases the burden on the MS/MS. If we substitute a peak width of 10 s into eqn 1.3, the theoretical peak capacity decreases by half. However, narrower peak widths could potentially open elution windows that give peptides with lower surface affinities a better chance to effectively compete for charges on the droplet surface for subsequent detection and sequencing.

The abundances of proteins in mammalian cells can range from 1 to 10^7 copies per cell[112,113] and, in plasma, circulating levels of proteins can exceed 10^{12}.[114] Global LC-MS/MS protein expression studies in cell lines can provide excellent proteome coverage with recent publications reporting $\sim$50% coverage of all predicted proteins for human cell lines.[112,113] Detecting quantitative changes for each protein across this range of abundances requires extensive fractionation, significant instrument time, and one or more quantitative strategies. For this discussion, three quantitative strategies will be considered, including SILAC, isobaric tagging, and label-free approaches. The magnitude of detectable protein expression changes are defined by the linear dynamic range for each approach. The linear dynamic range for SILAC and isobaric tagging is around two orders of magnitude, whereas label-free quantification is around three orders of magnitude.[115] However, the linear dynamic range can be extended in some cases depending on the complexity of the sample, LC-MS/MS platform, and LC-MS/MS settings. Targeted protein quantification studies by PC-IDMS LC-MS/MS have reported four to five orders of magnitude of linear dynamic range.[115] The upper limit for the linear dynamic range occurs when the concentration (N_i, eqn 1.1) begins to approach and exceed the charges available per droplet. Thus, there reaches a point where more sample loaded onto the column becomes deleterious to quantitative measurements.

1.4 Optimizing LC-MS/MS for Quantitative Proteomics Using Design of Experiments

The diversity of MS-based quantitative proteomics experiments is immense, making a single set of optimal LC-MS/MS conditions impractical. Goals for each experiment can be significantly different and involve different LC-MS/MS platforms with distinct operating parameters. Furthermore, the technical expertise and established workflows for individual laboratories will be different resulting in a high degree of inter-laboratory variability.[116] Given these factors, combined with the rapidly changing landscape of LC-MS/MS technology that is increasingly 'blackbox', it becomes necessary for each laboratory to efficiently and empirically develop optimized strategies for a broad range of quantitative proteomics experiments.

Design of experiments (DoE) was developed in the early part of the 20th century by Sir Ronald Fisher and has been extensively used to efficiently increase productivity in a wide range of fields, including manufacturing, agriculture, engineering, and the basic sciences.[117,118] In DoE, an experimental framework is constructed that incorporates the principles of replication, randomization, and blocking (*i.e.*, comparisons of parts of the overall experiment that are expected to be more homogenous) to assess the statistical significance of experimental conditions (*i.e.*, factors) on the desired outcome(s) (*i.e.*, response). This framework allows for the statistical evaluation of responses (*e.g.*, proteome coverage, limits of detection) from multiple factors (*e.g.*, ESI voltage, capillary temperature, CID energy, number of MS/MS events, *etc.*) that both reduce the time to optimize the LC-MS/MS system and also detect factors (or combinations of factors) that interact with one another. Many types of factorial designs have been developed and described in detail and fall outside the scope of this Chapter.[119,120] The main types discussed here are two-level full factorial design (FullFD) and fractional factorial design (FracFD), where only two levels are considered. These experiments (commonly referred to as two-level screens) make the following assumptions: 1. factors are fixed; 2. the responses between the fixed factors are linear; and 3. responses are normally distributed.

DoE has been used for optimizing MS-based measurements,[121–124] but it has not been widely reported by researchers in academia compared with industrial researchers, who are not always permitted to publish in peer-reviewed literature. An informative tutorial by Riter *et al.*[125] makes this point and showed how DoE could be used to efficiently optimize LC-MS/MS for bottom-up proteomics. This approach was recently extended to newer LC-MS/MS systems using total yeast digests. Andrews *et al.*[124] used a combination of FullFD and FracFD to optimize a nanoLC-LTQ-Orbitrap system for qualitatively and quantitatively studying a total yeast digest. The authors identified statistically significant main effects, such as capillary temperature, ionization time, monoisotopic precursor selection, number (#) of MS/MS events, and tube lens voltage, which collectively improved proteome coverage by ~60%. The same group used DoE again to optimize a nanoLC-Q-Exactive with a total yeast digest and empirically derived the critical factors that influenced protein identification.[126] Clearly, there is great potential for DoE to empirically explore and determine the best LC-MS/MS settings for virtually any type of quantitative proteomics study.

Two-level FullFD provides the most comprehensive assessment of factors, and combinations of higher-order interactions, that yield statistically significant response(s). Figure 1.4A shows a FullFD design for testing the response of four factors. However, this level of detail comes at the expense of efficiency, where the number of experiments necessary to complete a

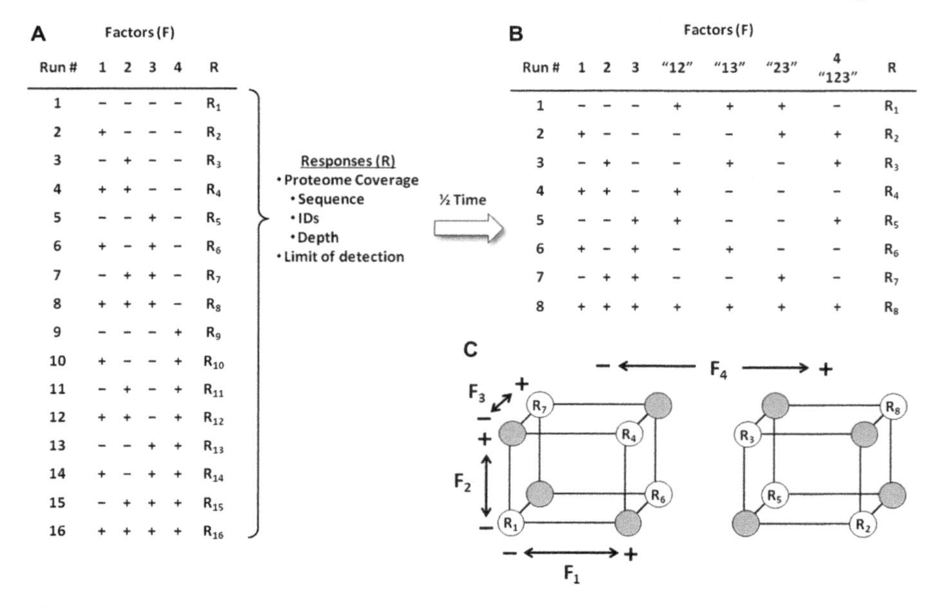

Figure 1.4 Generic two-level full factorial design (A) and fractional factorial design (B) for four factors. The + and – represent high and low values for the factors (1–4) and the responses can be any number of outputs depending on the goals of the quantitative LC-MS/MS experiment. The fractional factorials design (B) combines factors (*e.g.*, "12", "123") to reduce the number of experiments by half. A 3D representation of this fractional factorial design is illustrated (C).

two-level FullFD for k factors is defined as 2^k experiments for a single replicate. If we assume each LC-MS/MS experiment (*i.e.*, Run #) takes 90 min, it will take 24 h to complete a single replicate of all four experimental conditions. This is not an impractical interval of time to evaluate four experimental factors, even if done in triplicate over three days. However, it would not be unreasonable to identify a greater number of experimental factors that would impact the optimization of a typical quantitative LC-MS/MS experiment.[124] If we consider a ten-factor FullFD study, a single replicate would occupy almost two months of instrument time assuming a 90 min LC gradient. Thus, more efficient and practical factorial designs are needed to explore the full range of factors one would encounter in an LC-MS/MS workflow, particularly for global protein expression studies or multi-peptide SRM-based assays.

Two-level FracFD are commonly used in place of FullFD when the number of experimental factors is large and the time to optimize an experiment or process is limited. Figure 1.4B illustrates a common FracFD for the same four factors as in Figure 1.4A, which takes half the time. The gain in efficiency, however, is typically at the expense of the extent of confounding (*i.e.*, an inability to differentiate contributions from individual or combined factors on a response), which is a measure of the FracFD's resolution.

A lower resolution requires fewer experiments, but results in greater confounding, where a response cannot be attributed to a single factor or higher-order factor interactions. For example, a two-level FracFD of ten factors with a resolution of three requires 16 LC-MS/MS runs to complete a single replicate. This FracFD allows all main factors to be assessed without confounding, but some main factor–two factor interactions and all two factor–two factor interactions are confounded. For the same study, FracFD can be designed at a resolution of four that requires 32 LC-MS/MS runs for a single replicate. Such a design can assess responses for main factors and main factor–two factor interactions, but two factor–two factor interactions are confounded. If we compare the necessary LC-MS/MS analysis time for a two-level, ten-factor FullFD *vs.* a FracFD (resolution three) *vs.* a FracFD (resolution four), the necessary LC-MS/MS analysis time is decreased from two plus months to one day and two days, respectively.

DoE represents one of the most underutilized approaches for efficiently optimizing LC-MS/MS systems for complex quantitative proteomics studies. Previous efforts to optimize LC-MS/MS systems typically study the effects of one factor at a time, which is inefficient and insensitive to significant interactions that impact proteome coverage, quantitative precision and accuracy, and limits of detection. By applying some general knowledge of LC-MS/MS instrumentation, one can focus on 5–15 critical factors and develop a two-level FracFD experiment to screen for the main effects and higher-order interactions. Additional FracFD or FullFD with more than two levels can be developed to further optimize a system, particularly for targeted assays, where subtle changes are likely more important. As more researchers become aware of the power of DoE, it is likely to become a major factor in developing LC-MS/MS methods for quantitative proteomics.

1.5 Summary

Quantitative LC-MS/MS-based proteomics has evolved at a tremendous rate over the past 10–15 years. For example, instrument manufacturers introduce new products about every two years, making it very difficult to evaluate instrument performance and optimize for complex proteome analysis. Despite the frenetic pace of new instrument releases, there remain significant opportunities to improve the utility and quality of quantitative LC-MS/MS measurements. Fundamental studies focusing on the mechanism of ESI promise to yield important fundamental information about improving ionization efficiency for low-responding peptides and potentially leveling the response so as to improve proteome coverage. Ion transmission remains a significant barrier to measuring low-abundant peptides and small populations of cells. The trend toward faster MS/MS scan speeds will continue to expand our ability to measure more proteins at greater proteome depth and promises to effectively exploit the higher peak capacity of UHPLC. DoE is an underutilized strategy for optimizing complex LC-MS/MS measurements that should find more widespread use in the proteomics community.

The growth and acceptance of a broad class of quantitative proteomics approaches is a testament to the power and untapped potential of LC-MS/MS technology.

Acknowledgments

AMH gratefully acknowledges funding support from NIH (K25CA128666) and Virginia Commonwealth University.

References

1. R. Aebersold and M. Mann, *Nature*, 2003, **422**(6928), 198–207.
2. S. E. Ong, L. J. Foster and M. Mann, *Methods*, 2003, **29**(2), 124–130.
3. J. Cox and M. Mann, *Annu. Rev. Biochem.*, 2011, **80**, 273–299.
4. R. Aebersold, H. Lee, D. Han, M. Wright, H. L. Zhou, T. Griffin, S. Purvine and D. Goodlett, *Faseb J*, 2002, **16**(4), A12–A13.
5. J. B. Fenn, *J. Am. Soc. Mass Spectrom.*, 1993, **4**(7), 524–535.
6. S. P. Gygi, B. Rist, S. A. Gerber, F. Turecek, M. H. Gelb and R. Aebersold, *Nat. Biotechnol.*, 1999, **17**(10), 994–999.
7. P. L. Ross, Y. N. Huang, J. N. Marchese, B. Williamson, K. Parker, S. Hattan, N. Khainovski, S. Pillai, S. Dey, S. Daniels, S. Purkayastha, P. Juhasz, S. Martin, M. Bartlet-Jones, F. He, A. Jacobson and D. J. Pappin, *Mol. Cell. Proteomics*, 2004, **3**(12), 1154–1169.
8. A. Thompson, J. Schafer, K. Kuhn, S. Kienle, J. Schwarz, G. Schmidt, T. Neumann, R. Johnstone, A. K. Mohammed and C. Hamon, *Anal. Chem.*, 2003, **75**(8), 1895–1904.
9. S. E. Ong, B. Blagoev, I. Kratchmarova, D. B. Kristensen, H. Steen, A. Pandey and M. Mann, *Mol. Cell. Proteomics*, 2002, **1**(5), 376–386.
10. Y. Oda, K. Huang, F. R. Cross, D. Cowburn and B. T. Chait, *Proc. Natl. Acad. Sci. USA*, 1999, **96**(12), 6591–6596.
11. P. V. Bondarenko, D. Chelius and T. A. Shaler, *Anal. Chem.*, 2002, **74**(18), 4741–4749.
12. M. P. Washburn, R. Ulaszek, C. Deciu, D. M. Schieltz and J. R. Yates, 3rd, *Anal. Chem.*, 2002, **74**(7), 1650–1657.
13. M. P. Washburn, R. R. Ulaszek and J. R. Yates, 3rd, *Anal. Chem.*, 2003, **75**(19), 5054–5061.
14. H. B. Liu, R. G. Sadygov and J. R. Yates, *Anal. Chem.*, 2004, **76**(14), 4193–4201.
15. Y. Ishihama, Y. Oda, T. Tabata, T. Sato, T. Nagasu, J. Rappsilber and M. Mann, *Mol. Cell. Proteomics*, 2005, **4**(9), 1265–1272.
16. X. D. Yao, C. Afonso and C. Fenselau, *J. Proteome Res.*, 2003, **2**(2), 147–152.
17. K. L. Johnson, C. J. Mason, D. C. Muddiman and J. E. Eckel, *Anal. Chem.*, 2004, **76**(17), 5097–5103.
18. Y. Zhang, Z. Wen, M. P. Washburn and L. Florens, *Anal. Chem.*, 2009, **81**(5), 6317–6326.

19. S. Y. Ow, M. Salim, J. Noirel, C. Evans, I. Rehman and P. C. Wright, *J. Proteome Res.*, 2009, **8**(11), 5347–5355.
20. K. L. Johnson and D. C. Muddiman, *J. Am. Soc. Mass Spectrom.*, 2004, **15**(4), 437–445.
21. T. S. Collier, S. M. Randall, P. Sarkar, B. M. Rao, R. A. Dean and D. C. Muddiman, *Rapid Commun. Mass Spectrom.*, 2011, **25**(17), 2524–2532.
22. T. S. Collier, P. Sarkar, W. L. Franck, B. M. Rao, R. A. Dean and D. C. Muddiman, *Anal. Chem.*, 2010, **82**(20), 8696–8702.
23. D. R. Barnidge, M. K. Goodmanson, G. G. Klee and D. C. Muddiman, *J. Proteome Res.*, 2004, **3**(3), 644–652.
24. D. R. Barnidge, G. D. Hall, J. L. Stocker and D. C. Muddiman, *J. Proteome Res.*, 2004, **3**(3), 658–661.
25. J. R. Barr, V. L. Maggio, D. G. Patterson, G. R. Cooper, L. O. Henderson, W. E. Turner, S. J. Smith, W. H. Hannon, L. L. Needham and E. J. Sampson, *Clin. Chem.*, 1996, **42**(10), 1676–1682.
26. C. Dass, J. J. Kusmierz and D. M. Desiderio, *Biol. Mass Spectrom.*, 1991, **20**(3), 130–138.
27. S. A. Gerber, J. Rush, O. Stemman, M. W. Kirschner and S. P. Gygi, *Proc. Nat. Acad. Sci. USA*, 2003, **100**(12), 6940–6945.
28. E. Kuhn, J. Wu, J. Karl, H. Liao, W. Zolg and B. Guild, *Proteomics*, 2004, **4**(4), 1175–1186.
29. N. L. Anderson, N. G. Anderson, L. R. Haines, D. B. Hardie, R. W. Olafson and T. W. Pearson, *J. Proteome Res.*, 2004, **3**(2), 235–244.
30. S. Hanke, H. Besir, D. Oesterhelt and M. Mann, *J. Proteome Res.*, 2008, **7**(3), 1118–1130.
31. K. M. Carroll, F. Lanucara and C. E. Eyers, *Methods Enzymol.*, 2011, **500**, 113–131.
32. J. M. Pratt, D. M. Simpson, M. K. Doherty, J. Rivers, S. J. Gaskell and R. J. Beynon, *Nat. Protoc.*, 2006, **1**(2), 1029–1043.
33. J. Rivers, D. M. Simpson, D. H. Robertson, S. J. Gaskell and R. J. Beynon, *Mol. Cell. Proteomics*, 2007, **6**(8), 1416–1427.
34. R. J. Beynon, M. K. Doherty, J. M. Pratt and S. J. Gaskell, *Nat. Methods*, 2005, **2**(8), 587–589.
35. P. Brownridge and R. J. Beynon, *Methods*, 2011, **54**(4), 351–360.
36. C. M. Shuford, R. R. Sederoff, V. L. Chiang and D. C. Muddiman, *Mol. Cell. Proteomics*, 2012, **11**(9), 814–823.
37. M. Yamashita and J. B. Fenn, *J. Phys. Chem.*, 1984, **88**(20), 4451–4459.
38. M. Wilm and M. Mann, *Anal. Chem.*, 1996, **68**(1), 1–8.
39. M. S. Wilm and M. Mann, *Int. J. Mass Spectrom.*, 1994, **136**(2–3), 167–180.
40. J. C. Hannis and D. C. Muddiman, *Rapid Commun. Mass Spectrom.*, 1998, **12**(8), 443–448.
41. S. Geromanos, G. Freckleton and P. Tempst, *Anal. Chem.*, 2000, **72**(4), 777–790.
42. R. B. Cole, ed. *Electrospray Ionization Mass Spectrometry*, John Wiley & Sons, Inc.: USA, 1997, 577.

43. P. Kebarle and U. H. Verkerk, *Mass Spectrom. Rev.*, 2009, **28**(6), 898–917.
44. N. B. Cech and C. G. N. Enke, *Mass Spectrom. Rev.*, 2001, **20**(6), 362–387.
45. P. Kebarle and M. Peschke, *Anal. Chim. Acta*, 2000, **406**(1), 11–35.
46. P. Kebarle and L. Tang, *Anal. Chem.*, 1993, **65**(22), A972–A986.
47. A. T. Blades, M. G. Ikonomou and P. Kebarle, *Anal. Chem.*, 1991, **63**(19), 2109–2114.
48. J. N. Smith, R. C. Flagan and J. L. Beauchamp, *J. Phys. Chem. A*, 2002, **106**(42), 9957–9967.
49. R. D. Smith, J. A. Loo, C. G. Edmonds, C. J. Barinaga and H. R. Udseth, *Anal. Chem.*, 1990, **62**(9), 882–899.
50. M. Dole, L. L. Mack and R. L. Hines, *J. Chem. Phys.*, 1968, **49**(5), 2240–2249.
51. J. V. Iribarne and B. A. Thomson, *J. Chem. Phys.*, 1976, **64**(6), 2287–2294.
52. E. Ahadi and L. Konermann, *J. Am. Chem. Soc.*, 2011, **133**(24), 9354–9363.
53. L. Konermann, E. Ahadi, A. D. Rodriguez and S. Vahidi, *Anal. Chem.*, 2013, **85**(1), 2–9.
54. E. Ahadi and L. Konermann, *J. Am. Chem. Soc.*, 2010, **132**(32), 11270–11277.
55. E. Ahadi and L. Konermann, *J. Phys. Chem. B*, 2009, **113**(20), 7071–7080.
56. S. Consta, *J. Phys. Chem. B*, 2010, **114**(16), 5263–5268.
57. R. L. Grimm and J. L. Beauchamp, *Anal. Chem.*, 2002, **74**(24), 6291–6297.
58. R. L. Grimm and J. L. Beauchamp, *J. Phys. Chem. A*, 2010, **114**(3), 1411–1419.
59. M. G. Ikonomou and P. Kebarle, *J. Am. Soc. Mass Spectrom.*, 1994, **5**(9), 791–799.
60. J. L. Benesch, F. Sobott and C. V. Robinson, *Anal. Chem.*, 2003, **75**(10), 2208–2214.
61. J. L. Frahm, D. C. Muddiman and M. J. Burke, *J. Am. Soc. Mass Spectrom.*, 2005, **16**(5), 772–778.
62. J. S. Page, R. T. Kelly, K. Tang and R. D. Smith, *J. Am. Soc. Mass Spectrom.*, 2007, **18**(9), 1582–1590.
63. N. B. Cech and C. G. Enke, *Anal. Chem.*, 2000, **72**(13), 2717–2723.
64. N. B. Cech and C. G. Enke, *Anal. Chem.*, 2001, **73**(19), 4632–4639.
65. N. B. Cech, J. R. Krone and C. G. Enke, *Anal. Chem.*, 2001, **73**(2), 208–213.
66. E. F. Gordon and D. C. Muddiman, *Rapid Commun. Mass Spectrom.*, 1999, **13**(3), 164–171.
67. A. P. Null, A. I. Nepomuceno and D. C. Muddiman, *Anal. Chem.*, 2003, **75**(6), 1331–1339.
68. P. Mallick, M. Schirle, S. S. Chen, M. R. Flory, H. Lee, D. Martin, J. Ranish, B. Raught, R. Schmitt, T. Werner, B. Kuster and R. Aebersold, *Nat. Biotechnol.*, 2007, **25**(1), 125–131.
69. W. S. Sanders, S. M. Bridges, F. M. McCarthy, B. Nanduri and S. C. Burgess, *BMC Bioinformatics*, 2007, **8**(Suppl 7), S23.

70. H. Tang, R. J. Arnold, P. Alves, Z. Xun, D. E. Clemmer, M. V. Novotny, J. P. Reilly and P. Radivojac, *Bioinformatics*, 2006, **22**(14), e481–e488.
71. V. A. Fusaro, D. R. Mani, J. P. Mesirov and S. A. Carr, *Nat. Biotechnol.*, 2009, **27**(2), 190–198.
72. C. E. Eyers, C. Lawless, D. C. Wedge, K. W. Lau, S. J. Gaskell and S. J. Hubbard, *Mol. Cell. Proteomics*, 2011, **10**(11), M110003384.
73. T. Kim, K. Tang, H. R. Udseth and R. D. Smith, *Anal. Chem.*, 2001, **73**(17), 4162–4170.
74. T. Kim, H. R. Udseth and R. D. N. Smith, *Anal. Chem.*, 2000, **72**(20), 5014–5019.
75. R. T. Kelly, J. S. Page, K. Tang and R. D. Smith, *Anal. Chem.*, 2007, **79**(11), 4192–4198.
76. R. B. Dixon and D. C. Muddiman, *Rapid Commun. Mass Spectrom.*, 2007, **21**(19), 3207–3212.
77. S. Wu, K. Zhang, N. K. Kaiser, J. E. Bruce, D. C. Prior and G. A. Anderson, *J. Am. Soc. Mass Spectrom.*, 2006, **17**(6), 772–779.
78. R. W. Purves and R. Guevremont, *Anal. Chem.*, 1999, **71**(13), 2346–2357.
79. J. D. Canterbury, X. Yi, M. R. Hoopmann and M. J. MacCoss, *Anal. Chem.*, 2008, **80**(18), 6888–6897.
80. J. Saba, E. Bonneil, C. Pomies, K. Eng and P. Thibault, *J. Proteome Res.*, 2009, **8**(7), 3355–3366.
81. G. Bridon, E. Bonneil, T. Muratore-Schroeder, O. Caron-Lizotte and P. Thibault, *J. Proteome Res.*, 2012, **11**(2), 927–940.
82. K. E. Swearingen, M. R. Hoopmann, R. S. Johnson, R. A. Saleem, J. D. Aitchison and R. L. Moritz, *Mol. Cell. Proteomics*, 2012, **11**(4), M111014985.
83. A. J. Creese, N. J. Shimwell, K. P. Larkins, J. K. Heath and H. J. Cooper, *J. Am. Soc. Mass Spectrom.*, 2013, **24**(3), 431–443.
84. D. A. Barnett, B. Ells, R. Guevremont and R. W. Purves, *J. Am. Soc. Mass Spectrom.*, 2002, **13**(11), 1282–1291.
85. S. T. Wu, Y. Q. Xia and M. Jemal, *Rapid Commun. Mass Spectrom.*, 2007, **21**(22), 3667–3676.
86. Y. Q. Xia, S. T. Wu and M. Jemal, *Anal. Chem.*, 2008, **80**(18), 7137–7143.
87. R. W. Purves, *Anal. Bioanal. Chem.*, 2013, **405**(1), 35–42.
88. L. Zhou, B. F. Yue, D. V. Dearden, E. D. Lee, A. L. Rockwood and M. L. Lee, *Anal. Chem.*, 2003, **75**(21), 5978–5983.
89. A. M. Hawkridge, L. Zhou, M. L. Lee and D. C. Muddiman, *Anal. Chem.*, 2004, **76**(14), 4118–4122.
90. J. S. Page, I. Marginean, E. S. Baker, R. T. Kelly, K. Tang and R. D. Smith, *J. Am. Soc. Mass Spectrom.*, 2009, **20**(12), 2265–2272.
91. B. W. Lin and J. Sunner, *J. Am. Soc. Mass Spectrom.*, 1994, **5**(10), 873–885.
92. S. A. Shaffer, K. Q. Tang, G. A. Anderson, D. C. Prior, H. R. Udseth and R. D. Smith, *Rapid Commun. Mass Spectrom.*, 1997, **11**(16), 1813–1817.
93. S. A. Shaffer, D. C. Prior, G. A. Anderson, H. R. Udseth and R. D. Smith, *Anal. Chem.*, 1998, **70**(19), 4111–4119.

94. S. A. Shaffer, A. Tolmachev, D. C. Prior, G. A. Anderson, H. R. Udseth and R. D. Smith, *Anal. Chem.*, 1999, **71**(15), 2957–2964.
95. B. H. Clowers, Y. M. Ibrahim, D. C. Prior, W. F. Danielson, 3rd, M. E. Belov and R. D. Smith, *Anal. Chem.*, 2008, **80**(3), 612–623.
96. Y. Ibrahim, M. E. Belov, A. V. Tolmachev, D. C. Prior and R. D. Smith, *Anal. Chem.*, 2007, **79**(20), 7845–7852.
97. Y. Ibrahim, K. Tang, A. V. Tolmachev, A. A. Shvartsburg and R. D. Smith, *J. Am. Soc. Mass Spectrom.*, 2006, **17**(9), 1299–1305.
98. Y. M. Ibrahim, M. E. Belov, A. V. Liyu and R. D. Smith, *Anal. Chem.*, 2008, **80**(14), 5367–5376.
99. J. S. Page, B. Bogdanov, A. N. Vilkov, D. C. Prior, M. A. Buschbach, K. Tang and R. D. Smith, *J. Am. Soc. Mass Spectrom.*, 2005, **16**(2), 244–253.
100. J. S. Page, A. V. Tolmachev, K. Tang and R. D. Smith, *J. Am. Soc. Mass Spectrom.*, 2006, **17**(4), 586–592.
101. K. Tang, A. V. Tolmachev, E. Nikolaev, R. Zhang, M. E. Belov, H. R. Udseth and R. D. Smith, *Anal. Chem.*, 2002, **74**(20), 5431–5437.
102. C. J. Mason, K. L. Johnson and D. C. Muddiman, *J. Biomol. Tech.*, 2005, **16**(4), 414–422.
103. A. Liu, J. S. Cobb, J. L. Johnson, Q. Wang and J. N. Agar, *J. Chrom. Sci.*, 2013, DOI: 10.1093/chromsci/bms255.
104. J. E. MacNair, K. C. Lewis and J. W. Jorgenson, *Anal. Chem.*, 1997, **69**(6), 983–989.
105. N. Wu and A. M. Clausen, *J. Sep. Sci.*, 2007, **30**(8), 1167–1182.
106. N. Wu, J. A. Lippert and M. L. Lee, *J. Chrom. A*, 2001, **911**(1), 1–12.
107. M. P. Washburn, D. Wolters and J. R. Yates, 3rd, *Nat. Biotechnol.*, 2001, **19**(3), 242–247.
108. S. S. Thakur, T. Geiger, B. Chatterjee, P. Bandilla, F. Frohlich, J. Cox and M. Mann, *Mol. Cell. Proteomics*, 2011, **10**(8), M110003699.
109. G. L. Andrews, B. L. Simons, J. B. Young, A. M. Hawkridge and D. C. Muddiman, *Anal. Chem.*, 2011, **83**(13), 5442–5446.
110. A. Michalski, E. Damoc, J. P. Hauschild, O. Lange, A. Wieghaus, A. Makarov, N. Nagaraj, J. Cox, M. Mann and S. Horning, *Mol. Cell. Proteomics*, 2011, **10**(9), M111011015.
111. D. A. Wolters, M. P. Washburn and J. R. Yates, 3rd, *Anal. Chem.*, 2001, **73**(23), 5683–5690.
112. M. Beck, A. Schmidt, J. Malmstroem, M. Claassen, A. Ori, A. Szymborska, F. Herzog, O. Rinner, J. Ellenberg and R. Aebersold, *Mol. Syst. Biol.*, 2011, **7**, 1–8.
113. N. Nagaraj, J. R. Wisniewski, T. Geiger, J. Cox, M. Kircher, J. Kelso, S. Paabo and M. Mann, *Mol. Syst. Biol.*, 2011, **7**, 1–8.
114. N. L. Anderson and N. G. Anderson, *Mol. Cell. Proteomics*, 2002, **1**(11), 845–867.
115. M. Bantscheff, M. Schirle, G. Sweetman, J. Rick and B. Kuster, *Anal. Bioanal. Chem.*, 2007, **389**(4), 1017–1031.

116. A. G. Paulovich, D. Billheimer, A. J. Ham, L. Vega-Montoto, P. A. Rudnick, D. L. Tabb, P. Wang, R. K. Blackman, D. M. Bunk, H. L. Cardasis, K. R. Clauser, C. R. Kinsinger, B. Schilling, T. J. Tegeler, A. M. Variyath, M. Wang, J. R. Whiteaker, L. J. Zimmerman, D. Fenyo, S. A. Carr, S. J. Fisher, B. W. Gibson, M. Mesri, T. A. Neubert, F. E. Regnier, H. Rodriguez, C. Spiegelman, S. E. Stein, P. Tempst and D. C. Liebler, *Mol. Cell. Proteomics*, 2010, **9**(2), 242–254.
117. R. A. Fisher, *The Design of Experiments*, Macmillan, Edinburgh, 1935, 252.
118. R. A. Fisher, *Statistical Methods for Research Workers*, Oliver and Boyd, Edinburgh, 1925, 239.
119. G. E. Box, J. S. Hunter, W. G. Hunter, *Statistics for Experimenters: Design, Innovation, and Discovery*, John Wiley & Sons, Hoboken, NJ, 2nd edn, 2005, 639.
120. D. C. Montgomery, *Design and Analysis of Experiments*, John Wiley & Sons, USA, 4th edn, 1997.
121. S. K. Huang, S. B. Glancy and W. F. Stansbury, *Rapid Commun. Mass Spectrom.*, 1993, **7**(8), 722–724.
122. C. Seto, K. P. Bateman and B. Gunter, *J. Am. Soc. Mass Spectrom.*, 2002, **13**(1), 2–9.
123. S. J. Bruce, I. Tavazzi, V. Parisod, S. Rezzi, S. Kochhar and P. A. Guy, *Anal. Chem.*, 2009, **81**(9), 3285–3296.
124. G. L. Andrews, R. A. Dean, A. M. Hawkridge and D. C. Muddiman, *J. Am. Soc. Mass Spectrom.*, 2011, **22**(4), 773–783.
125. L. S. Riter, O. Vitek, K. M. Gooding, B. D. Hodge and R. K. Julian, Jr., *J. Mass Spectrom.*, 2005, **40**(5), 565–579.
126. S. M. Randall, H. L. Cardasis and D. C. Muddiman, *J. Am. Soc. Mass Spectrom.*, 2013, **24**, 1501–1512.

CHAPTER 2

High Resolution/Accurate Mass Targeted Proteomics

A. BOURMAUD,[†] S. GALLIEN[†] AND B. DOMON*

Luxembourg Clinical Proteomics Center (LCP), CRP-Santé, Strassen, Luxembourg
*Email: bdomon@crp-sante.lu

2.1 Introduction

Over the past two decades, mass spectrometry (MS)-based approaches have emerged as powerful techniques to study proteomes. The proteins constituting the proteome are enzymatically digested and the resulting peptides separated by liquid chromatography (LC) and analyzed by single-stage and tandem MS. Two different MS-based strategies can be applied.[1] First, the shotgun or discovery proteomics method based on data-dependent acquisition is widely used to identify thousands of proteins in a biological sample under non-supervised conditions.[2] Data-dependent acquisition consists in performing an initial survey scan in single-stage MS mode, which enables the selection of the precursor ions to be fragmented, and in turn, the generation of tandem mass spectrometry (MS/MS) spectra; peptide identification is then performed by a database search. Although very successful in establishing the profiles of proteomes, as illustrated by a wealth of publications,[3,4] the so-called "shotgun" method presents limitations for

[†]These authors contributed equally to this work.

New Developments in Mass Spectrometry No. 1
Quantitative Proteomics
Edited by Claire Eyers and Simon J Gaskell

Published by the Royal Society of Chemistry, www.rsc.org

quantitative studies. This is mainly due to its restricted sensitivity and reproducibility resulting from the nature of its data acquisition process. These shortcomings have prompted the development of alternative strategies, such as "targeted proteomics", which provide more precise quantitative results. This method generates data with high sensitivity owing to the systematic analysis of the peptides used as surrogates for a preselected group of proteins to answer a specific biological or clinical question, ignoring other proteins.[5] Triple quadrupole instruments operated in selected reaction monitoring (SRM) mode are commonly used in such hypothesis-driven workflows due to their high sensitivity and reproducibility of measurements, allowing the systematic detection of low abundance analytes in complex samples. Its application to proteomics has resulted in significant achievements benefiting from continuous technology development.[6] However, the analytical performance of SRM measurements are currently reaching a plateau, reflecting the performance shortcomings of triple quadrupole instruments, especially during the analysis of high complexity samples, such as bodily fluids. Thus, new technologies and methodologies based on high resolution (HR) instruments provide alternatives for performing targeted proteomics, especially when high selectivity is required.

2.2 LC-MS-based Targeted Proteomics

In contrast to liquid chromatography coupled to mass spectrometry (LC-MS) analysis of small molecules, proteomics applications require the prior selection of peptides to be used as surrogates for the proteins of interest.

2.2.1 Characteristics of Targeted Experiments

The workflow of a targeted proteomics experiment is more complex and typically relies on two main steps. First, from the proteins of interest defining a biological question, surrogate peptides are selected based on their proteotypicity, *i.e.*, the uniqueness of their amino acid sequence and their consistent observation in LC-MS(/MS) analyses. In addition, to avoid erroneous results, *e.g.*, due to proteolytic events or the presence of splice variants, several surrogate peptides distributed across the full sequence should ideally be selected.[7] In a second phase, the acquisition method is designed by defining for each peptide the attributes required for their LC-MS/MS analysis (*i.e.*, mass-to-charge ratio of ions selected as surrogates of targeted peptides and predicted chromatographic retention time windows) and the instrumental parameters promoting their detection. Once the analyses have been completed, the actual quantification of the peptides is performed using their corresponding ion traces. Quantification is usually carried out using stable isotope dilution. This approach consists of the analysis of endogenous peptides present in the biological sample, together with their corresponding synthetic isotopically labeled counterparts. In this case, the labeled peptide is used as an internal standard correcting possible bias in inter-sample measurements.

In a targeted acquisition method, 100% of measurement time is devoted to the analytes of interest, translating into better analytical performance, but also in interdependence between the number of peptides measured in one run and the time allocated to measure each peptide. Thus, an advanced management of acquisition time is required to measure the full set of peptides while maintaining a high performance level. As the benefit of increasing the time spent measuring each analyte is obtained at the expense of the number of analytes measured in a single LC-MS(/MS) experiment, and *vice versa*, a trade-off is required between these two factors. In practice two main types of targeted experiments can be considered, each with a specific purpose and factors adjusted accordingly. The first type of experiment, or screening mode, aims at establishing the detection and the relative quantification of a large number of peptides. The second one is the "*true*" quantification mode, which is used for the precise and sensitive measurement of fewer analytes. In contrast to screening experiments, in which the number of peptides is the priority, quantification experiments require higher sensitivity and selectivity obtained with a detrimental effect on the number of peptides analyzed (Figure 2.1).[8]

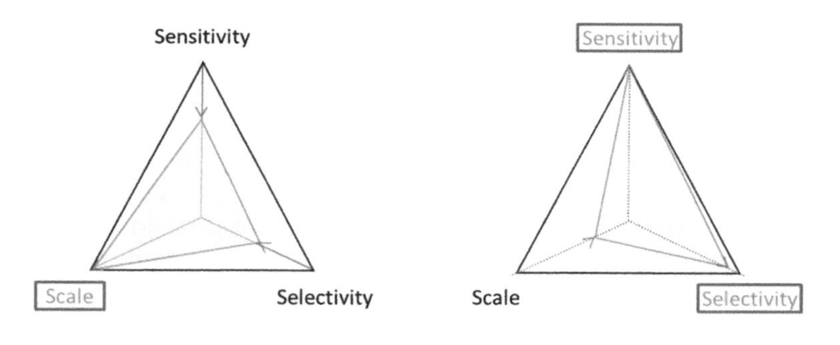

Figure 2.1 Trade-off in quantitative proteomics approaches. The interdependence between sensitivity, selectivity and scale has prompted to consider two distinct methods. The first one, the screening mode, is intended to assess detectability and to determine the relative concentrations of a large number of peptides at the expense of sensitivity and selectivity. The second method, called "true" quantification mode, aims at the precise quantification of a limited set of analytes with high sensitivity and selectivity using internal standards.
Adapted from ref. 8.

2.2.2 Targeted Experiments Using Selected Reaction Monitoring

SRM performed on a triple quadrupole instrument is nowadays commonly employed to conduct targeted proteomics experiments. This technology exploits the capabilities of the triple quadrupole to accurately measure the signal of ions over a wide dynamic range with exquisite sensitivity and selectivity. During SRM analysis, the first and third quadrupoles serve as mass filters while the second quadrupole is used as a collision cell (Figure 2.2). The first quadrupole isolates a predefined precursor ion within a selection window to undergo fragmentation in the second quadrupole, followed by the isolation of a specific product ion by the third quadrupole to be transmitted to the detector.[9]

An SRM experiment thus consists in measuring sequentially a series of transitions (pairs of precursor/product ions) specific for the targeted peptides. This singularity requires the implementation of a particular hypothesis-driven workflow depicted in Figure 2.3.[10] The selection of target peptides associated with the proteins of interest in the biological context is followed by the definition of a set of transitions maximizing the sensitivity and the selectivity of measurement. The validation of the transitions is performed by an evaluation of the interferences for each transition in the actual biological matrix. The selected transitions are optimized according to the collision energy applied during fragmentation in the collision cell.

SRM analysis on triple quadrupole mass spectrometers was developed more than two decades ago for the quantification of small molecules.[11] Recently, it has emerged as the reference method for targeted quantitative proteomics due to its high sensitivity, selectivity and throughput. However, the initial application of SRM analysis to proteomics suffered from one main limitation originating from its direct transfer from small molecules to peptides. In contrast to small molecule analysis, targeted proteomics usually requires not only the measurement of large panels of analytes, but also

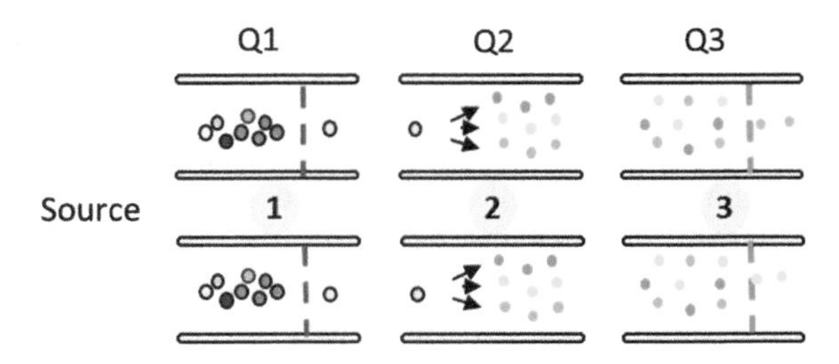

Figure 2.2 Principle of the selected reaction monitoring (SRM) acquisition mode performed on a triple quadrupole mass spectrometer. The precursor ion is selected by the first quadrupole (Q1) prior to undergoing fragmentation in the collision cell (Q2). A single product ion is then selected in the third quadrupole for detection (Q3). Adapted from ref. 8.

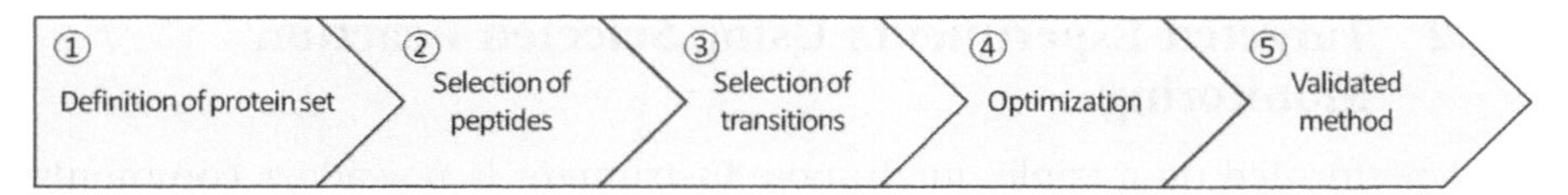

Figure 2.3 Workflow of a targeted proteomics experiment using the SRM technique. (1) A list of candidates is defined, *i.e.*, a set of proteins corresponding to a biological question; (2) a series of peptides is selected and used as surrogates for the proteins of interest; (3) for each peptide a set of transitions is defined to maximize sensitivity and selectivity of measurements; (4) the transitions are validated through the evaluation of the presence of interferences in the biological matrix. Optionally, the collision energy applied during the fragmentation step can be optimized to further enhance the sensitivity; (5) a validated method allows true quantification experiments by LC-MS/MS.

multiple transitions per analytes to ensure the specificity of the measurements. In LC-MS experiments, the cycle time, driven by the number of transitions measured for each peptide and the dwell time of each transition, is adjusted to collect enough data points (eight to ten) across the elution profile. Increasing the overall number of transitions monitored over the full LC-MS analysis decreases the individual dwell times and penalizes the sensitivity. This initial limitation prompted the development of a scheduled approach, the so-called time scheduled SRM mode, which relies on monitoring transitions only during the expected elution time of the peptide. This acquisition mode includes a segmentation of the analytical time (typically 2–4 min) during which a subset of peptides is measured. Time scheduled SRM allows a higher number of peptides to be monitored in a single LC-MS analysis without a decrease in quantification performance.[12] To further improve the selectivity and the scale of SRM experiments, the concept of intelligent SRM (iSRM) was developed.[13] This acquisition mode combines the continuous monitoring of a very limited set of primary transitions (typically two or three transitions per peptide) used for quantification purposes, which are supplemented with the punctual monitoring of secondary transitions (typically five or six transitions), triggered by the signal of primary transitions. These secondary transitions are exploited to generate composite MS/MS spectra used to confirm the identity of the targeted peptides. The iSRM technique improves the specificity of measurements without compromising the dwell time of the primary transitions. It was exploited to perform large-scale experiments as illustrated with the measurement of nearly 1000 peptides in a single LC-MS run.[13] An extension of the iSRM data acquisition technique was proposed, which relies on the dynamic adjustment of the time scheduled monitoring windows to further increase the number of peptides monitored during a single LC-MS analysis.[14] In spite of all these improvements in acquisition, SRM performance (*i.e.*, sensitivity, selectivity and scale) presents limitations inherent to the low resolution of the quadrupole analyzer.[10,15] The resolving power is not always sufficient to reliably discriminate the targeted analytes from their complex background, commonly encountered in biological samples. The two stages of mass filtering employed in SRM acquisition partially overcome the issue but do not systematically eliminate the presence of interferences. In many instances the co-isolation of precursor ions and interferences in the first quadrupole yields nearly isobaric product ions, which cannot be resolved by the third quadrupole, thus leading to convolution of the signals. The interferences can be detrimental for peptide ions present in low abundance in a complex background.[16] The impact of the biochemical background on the selectivity of SRM measurements is illustrated in Figure 2.4 with the traces of five transitions of a peptide spiked in various amounts into two samples, one of low and the second of high complexity. The detection, and thus the quantification, of peptides present in low concentrations in an aqueous solution (Figure 2.4A and B) or in high concentrations in a complex sample (Figure 2.4E) are straightforward. However, the low concentrations of peptides in a complex sample are difficult to measure due to interferences from the matrix (Figure 2.4C and D). This demonstrates the effect of sample complexity on analytical performance and stresses the necessity to systematically evaluate the selectivity of measurements in the matrix of interest to avoid erroneous quantification results.

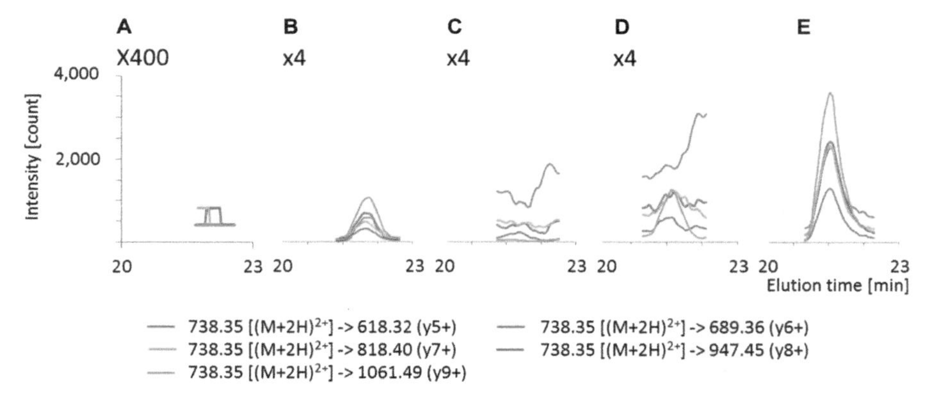

Figure 2.4 Matrix interferences as a function of the sample complexity in SRM analysis. The signals of five SRM transitions were monitored in samples of low and high complexity. (A) The signal of the transitions in a blank sample of low complexity indicated electronic noise. (B) The signal of the transitions of the peptide of interest spiked at a concentration of 50 amol in a sample of low complexity is free from interferences. (C) Interferences are observed in the blank sample of high complexity (500 ng of urine protein digest). (D) The signals of the transitions of the peptide of interest spiked at a concentration of 50 amol are interfered in the highly complex sample. The spurious signals contribute to the signal of the targeted peptide. In this example only one transition can be exploited for quantification as it exhibits a sufficient signal-to-noise ratio. (E) The peptide spiked in a higher amount (concentration of 500 amol) in a complex background does not show interferences. Adapted from ref. 16.

2.3 Targeted Proteomic Experiments Using High Resolution/Accurate Mass Instrumentation

The proteomic analysis of biological samples usually suffers from the intrinsic mismatch between their complexity and the LC-MS peak capacity. The problem is made more severe by the low resolving power of conventional quadrupole mass filters, typically operated at unit mass resolution. Mass analyzers with increased resolution and efficient ion transmission, such as quadrupole instruments equipped with hyperbolic rods, theoretically increase the MS peak capacity by improving the discrimination between ions covering the regular *m*/*z* range. In most favourable cases, quadrupole isolation windows can be set at 0.2 *m*/*z* units, which is not sufficient to discriminate target ions from interferences as predicted by simulations.[7,17] This limitation, together with a loss of sensitivity, explains the restricted number of applications exploiting "enhanced resolution" SRM.[18,19] As an alternative, high resolution mass analyzers, such as time-of-flight (ToF) or orbitrap, enable the separation of ions of close *m*/*z* values both in MS or MS/MS spectra. Initially hampered by the limited speed and sensitivity of high resolution/accurate mass (HR/AM) hybrid mass spectrometers, the latest generation of quadrupole–ToF mass spectrometers and the quadrupole–orbitrap mass spectrometer can support such approaches owing to their high acquisition rates. The implementation of HR/AM technology, which has boosted shotgun proteomics, is now impacting targeted methods.

2.3.1 Characteristics of High Resolution/Accurate Mass Instruments

Quadrupole–ToF and quadrupole–orbitrap instruments offer HR/AM measurement capability using either one or two analysis stages. They rely on different physical principles that are distinct from a triple quadrupole instrument. The triple quadrupole mass spectrometer continuously measures a flux of ions produced in the source. Such a beam-based technique is advantageous for quantification as the measured ion flux is directly proportional to the amounts of analytes present in the sample. In contrast, trap-based instruments are operated in a batch mode. The selected ions are intermediately accumulated in the trapping device for a given time period, which is beneficial for the analysis of low abundant analytes. In this case, quantitative measurements require a step to recalculate the amount of analytes in the sample based on trapping time. The accuracy of quantification on this type of instrument is thus related to the trapping parameters. As long as the trap is operated below its capacity limit, this device provides exquisite sensitivity through its ability to accumulate ions over an extended period of time. The quadrupole–ToF instrument combines an in-beam quadrupole with an orthogonal ToF analyzer. In this case, all product ions are transferred into the ToF as packets in a pulsed mode. This adds constraints to the duty cycle and, in turn, penalizes the sensitivity as only a fraction of the "ion beam" is analyzed.

2.3.2 Quadrupole–Orbitrap Instrument

The hybrid quadrupole–orbitrap mass spectrometer (*Q-Exactive*, Thermo Scientific) presents a unique configuration combining a quadrupole mass filter with an orbitrap mass analyzer.[20] Compared with previous generations of orbitrap-based instruments, it benefits from improved ion transmission in the C-trap and the collision cell (called the HCD-cell), which both offer multiplexing capabilities. The quadrupole at the front-end allows precise selection of the precursor ions and thus overcomes the restricted dynamic range of trapping devices, which otherwise hampers the detection of low abundant analytes in complex samples. The ability to isolate selectively well-defined populations of precursor ions by selecting narrow *m/z* ranges associated with specific accumulation times in the trapping devices can, in turn, significantly improve the sensitivity of the measurements. Furthermore, the multiplexing capabilities of both the C-trap and the HCD-cell further extend the number of peptides that can be monitored. In the MS/MS mode several precursor ions are sequentially fragmented while their product ions are temporarily stored in the HCD-cell to be eventually measured simultaneously in the orbitrap in a single scan. The measurements are performed in high resolution mode (up to 140 000 at *m/z* 200), which also provides high mass accuracy.

These features enable new ways of performing targeted proteomics, either in selected ion monitoring (SIM), relying on single-stage MS, or in parallel reaction monitoring (PRM), based on tandem MS. Both modes, requiring a minimal set of instrument control parameters, lead to a simplification of the analytical workflow compared with that commonly used for SRM analysis (Figure 2.5). While predefining the target precursor ion *m/z* value that corresponds to the peptide of interest is still necessary (optionally with its associated predicted chromatographic retention time window), no *a priori* selection of the transitions to be monitored is required in PRM.[21]

2.4 Quantification Performed on Precursor Ions

The single-stage mass spectra acquired on the quadrupole–orbitrap mass spectrometer can be either a full scan or a partial scan by selecting a narrow mass range, typically using 2–8 *m/z* units in SIM mode. The conventional full scan mode relies on the acquisition of data over the full *m/z* range (200 to 1600 for regular tryptic peptides). In this mode, the quadrupole mass filter operates in "rf-only" mode to transfer all ions within a wide *m/z* range to the C-trap, where they are accumulated according to a predefined target value (automatic gain control, AGC, value), and a maximal fill time. The ions are eventually transferred to the orbitrap mass analyser, where a full HR

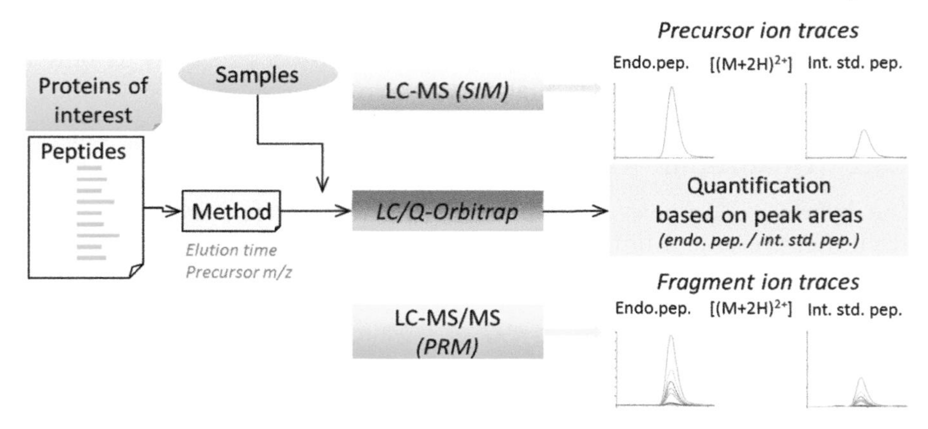

Figure 2.5 Targeted workflow employing a quadrupole–orbitrap mass spectrometer. From a set of proteins of interest defined in the context of a biological question, a list of peptides used as surrogates is established. The measurement of the targeted peptides necessitates only a few key instrument control parameters, *i.e.*, the elution time and the precursor *m*/*z*. Quantification is conducted using the quadrupole–orbitrap instrument either in selected ion monitoring (SIM) mode, relying on single-stage mass spectrometry, or in parallel reaction monitoring (PRM) mode, relying on tandem mass spectrometry.
Adapted from ref. 21.

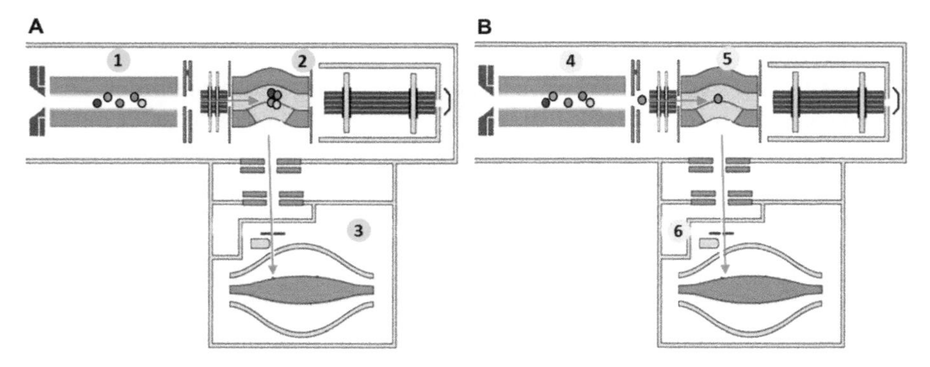

Figure 2.6 Principle of single-stage mass spectrometry acquisition on a quadrupole–orbitrap instrument operated in full scan and SIM modes. (A) In full scan acquisition, the ions covering a wide *m*/*z* range are transferred (1) to the C-trap for accumulation (2). Then, ions are eventually transferred and detected in the orbitrap (3). (B) In SIM mode, ions included in a restricted *m*/*z* range are isolated by the quadrupole (4), followed by their accumulation in the C-trap (5), and eventually their analysis in the orbitrap (6).
Adapted from ref. 21.

spectrum is recorded (Figure 2.6A). In SIM mode, the pre-definition of target ions of interest is required for targeted proteomics. Narrow *m*/*z* ranges (typically 2–3 *m*/*z* units), including target ions, are isolated by the quadrupole and transferred to the C-trap. Similar to the full scan acquisition, following accumulation, ions are transmitted to the orbitrap for mass analysis (Figure 2.6B).

2.4.1 Trapping Capabilities

The sampling of ions over a wide *m*/*z* range, like the full scan mode, results in trapping both the targeted peptides and some background. The complexity of the sample directly impacts the quantification performance, as exemplified in Figure 2.7, with the analysis of a series of eight synthetic isotopologous peptides. These peptides, having the same amino acid sequence (LVALVR) but different ^{15}N and ^{13}C labeling patterns, were mixed together in various relative amounts with a three-fold dilution factor between each point. This allowed the entire dilution series to be measured in one LC-MS run, either neat or added to a yeast protein digest. The analyses were performed in full scan and SIM modes at 70 000 resolution (*m*/*z* 200) with a maximum fill time of 250 ms and a targeted AGC set at 1E6 for full scan and 5E5 for SIM analysis. With a low background, both acquisition

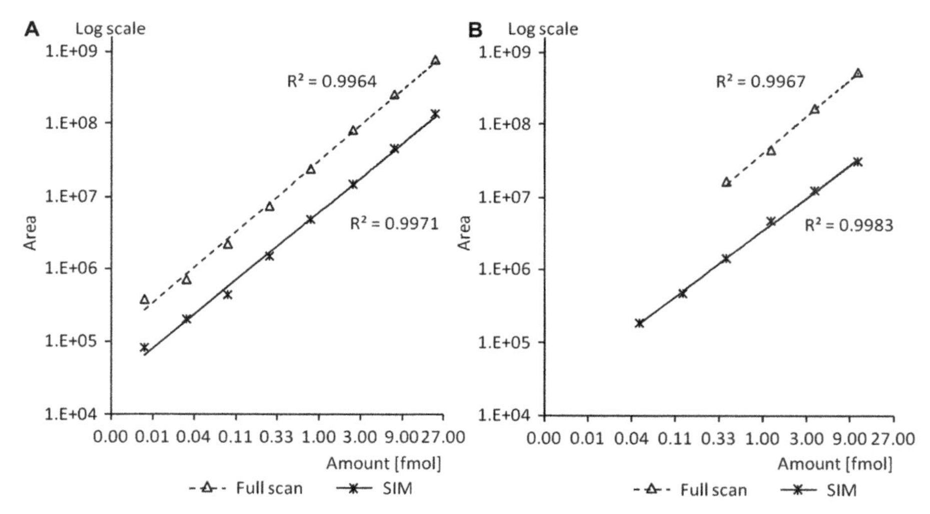

Figure 2.7 Comparison of the analytical performance of full scan and SIM modes for samples of various complexities. A dilution series of eight synthetic isotopologous peptides, containing the same amino acid sequence (LVALVR) with different labeling based on the ^{15}N and ^{13}C incorporation, was measured in one LC-MS analysis. The eight isotopologous peptides were prepared in various amounts, from 10 amol to 21.9 fmol in water and from 5 amol to 10.9 fmol in 500 ng yeast digest. (A) The linearity ranges of the dilution curves obtained for one single LC-MS analysis in low matrix level are equivalent. (B) In a complex biological background, the linearity range of the dilution curve starts from 45 amol in SIM mode and from 405 amol in full scan mode. Adapted from ref. 21.

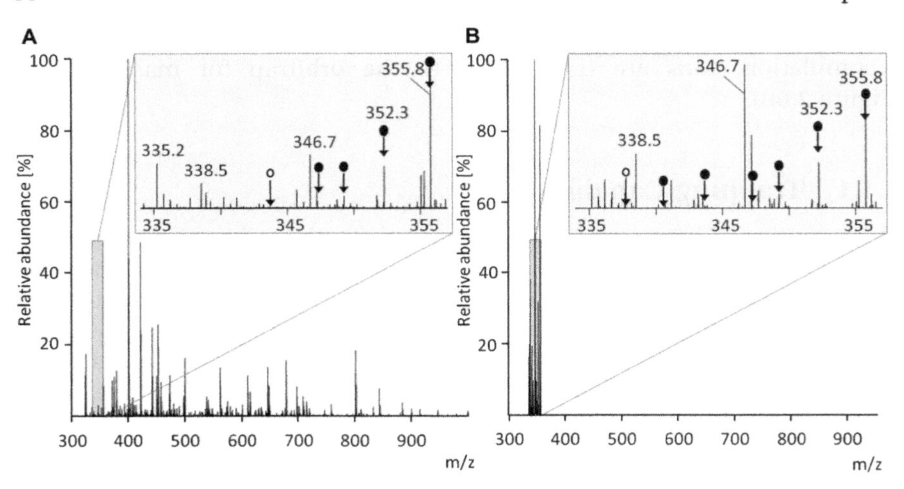

Figure 2.8 Comparison of the analytical performances of full scan (A) and SIM (B) modes in a high matrix background. The mass spectra correspond to the analysis of a dilution series of eight isotopologous peptides with the same amino acid sequence (LVALVR) and different ^{15}N and ^{13}C incorporations. Various amounts (between 5 amol to 10.9 fmol) were spiked in 500 ng yeast digest. The mass spectra at the apex of the elution profile are shown; the precursor ions, highlighted with dark circles, were detected and open circles depict those out of the linearity range. Adapted from ref. 21.

methods exhibited similar results (Figure 2.7A) and the full set of isotopically labeled peptides was detected down to concentrations in the low attomol range. In contrast, in the presence of a significant matrix background, both modes of analysis showed reduced performances, but the SIM technique exhibited a ten-fold higher sensitivity as compared to a conventional full scan analysis (Figure 2.7B). The SIM mode leverages the quadrupole mass filter to extend the accumulation time of ions in the C-trap in order to increase their signal-to-noise ratio and *in fine* the sensitivity. These results demonstrate the performance gain by enriching low abundant components using a narrow isolation window in conjunction with a longer fill time (Figure 2.8). Typically, accumulation during 250 ms was used in SIM mode for the ions at the lowest concentrations, compared to a value as low as 1 ms in full scan mode. The narrow isolation window reduces the co-isolation of background ions and, thus, dramatically increases the fill times for low abundant components to reach ion counts set by the target AGC value.

2.4.2 Quantification in Selected Ion Monitoring

The single-stage analyses performed in SIM mode allow sensitive quantification in complex samples with an extension of the dynamic range. When benchmarked with the SRM technique applied to complex samples, it exhibits similar performance as illustrated in Figure 2.9. The dilution series of

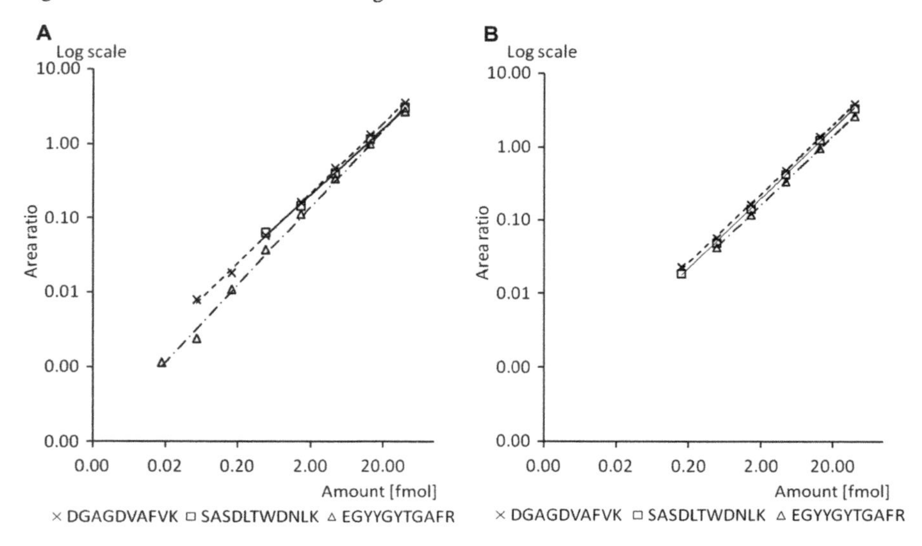

Figure 2.9 SIM and SRM quantification measurements of endogenous human transferrin in urine samples. A dilution series of three isotopically labeled peptides corresponding to the protein was carried out to establish area under the curve ratios for each pair of endogenous/isotopically labeled peptides. (A) Linear ranges of the dilution curves in SIM analysis. (B) Linear ranges of the dilution curves in SRM analysis. Adapted from ref. 21.

Table 2.1 Comparison of results determined with SIM and SRM techniques for the quantification of endogenous human transferrin in a urine matrix. A dilution series of three isotopically labeled peptides corresponding to transferrin was measured in a urine sample. Adapted from ref. 21.

	DGAGDVAFVK		*SASDLTWDNLK*		*EGYYGYTGAFR*	
	SIM	*SRM*	*SIM*	*SRM*	*SIM*	*SRM*
LOQ [fmol]	0.05	0.16	0.49	0.16	0.02	0.49
Amount of endogenous peptide [fmol]	10.0	9.4	11.5	10.7	12.7	13.6

three isotopically labeled peptides representing human transferrin were measured in urine samples using both techniques. Two of these peptides (DGAGDVAFVK̲ and EGYYGYTGAFR̲) actually exhibited a lower LOQ in SIM (Figure 2.9A), whereas the third peptide (SASDLTWDNLK̲) yielded better results in SRM (Figure 2.9B). The actual amount of endogenous peptides determined from the dilution curves showed little difference (<10%) between the two techniques (Table 2.1). This example depicts the results typically obtained during the evaluation of a wider set of 28 peptides, with LOQ values in a similar range for both techniques, usually with better results in SIM mode at low concentration. The HR/AM measurements performed on biological samples using single-stage MS analysis yielded results comparable to those obtained in SRM.

2.4.3 Parameters Used in Selected Ion Monitoring

The high resolution SIM technique requires that the *m*/*z* and elution time for each peptide is defined prior to the analysis. In addition, the optimization of instrument parameters, such as the maximal fill time and the orbitrap resolving power, impacts the quantitative results. As mentioned, the sensitivity of measurements is increased using longer fill times. By maximizing this parameter, remarkable performances were observed, as illustrated with the sensitivity test performed using an isotopologous peptides mixture (Figure 2.10). In this experiment, one partially labeled synthetic peptide (LVALVR) was spiked at extremely low concentrations

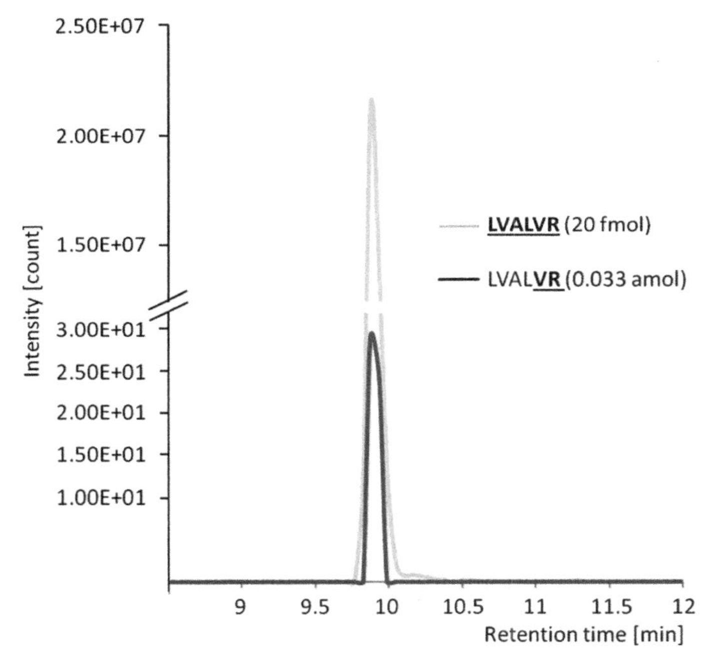

Figure 2.10 Ultimate sensitivity test of the quadrupole–orbitrap instrument operated in SIM mode. A mixture of the synthetic peptide LVALVR (*m*/*z* 343.745, $[M + 2H]^{2+}$) spiked in tiny amounts (0.033 amol μL^{-1}) into an aqueous solution containing the isotopologous peptide LVALVR (*m*/*z* 355.773, $[M + 2H]^{2+}$; at a concentration of 20 fmol μL^{-1}) was measured in SIM mode with maximum fill time set to 3 s. The extracted ion chromatograms demonstrate an apparent intra-scan dynamic range exceeding five orders of magnitude.
Adapted from ref. 21.

(0.9 amol μL^{-1}; 0.3 amol μL^{-1}; 0.1 amol μL^{-1}; 0.033 amol μL^{-1}) into an aqueous solution containing a second fully labeled synthetic peptide (LVALVR) in a larger amount (20 fmol μL^{-1}), which was used as a "carrier". The SIM experiment using fill times up to 3 s allowed the detection of all these amounts of peptide LVALVR. This unprecedented detection level was obtained in a low chemical background, which is not representative of a biological sample; nonetheless, it demonstrates the exquisite sensitivity achievable in SIM mode in the sub-attomol range (33 zmol) using extended trapping times.

The actual limit of detection in more complex samples is usually limited by the presence of interferences co-isolated by the quadrupole, thus reducing the fill time. The use of a narrower isolation window excluding the interference overcomes the problem; however, it comes at the expense of ion transmission efficiency, as mentioned earlier. The increase of the orbitrap resolving power better differentiates the targeted ion from the interferences. The operation at the current maximal resolution of 140 000 (defined at *m*/*z* 200) results in a longer transient time and, in turn, in a reduction of the number of peptides measured during the acquisition cycle.

2.5 Quantification Performed on Product Ions

The selectivity of the measurement is improved if the analysis is not performed on precursor ions but on product ions. The analysis is carried out in the so-called parallel reaction monitoring (PRM) mode, which is alike to selected reaction monitoring, but with the systematic analysis of all MS/MS product ions. In a PRM experiment, ions within a narrow *m*/*z* range centered on the target precursor ion are isolated by the quadrupole and transferred *via* the C-trap into the HCD-cell to undergo fragmentation. The resulting product ions are transferred back to the C-trap and into the orbitrap to record a full tandem mass spectrum (Figure 2.11). Whereas, in SRM experiments, predefined product ions are monitored for a given peptide in a series of sequential acquisitions; in PRM mode, all product ions are monitored simultaneously. Thus, no *a priori* knowledge of the fragmentation behaviour is needed, which dramatically simplifies the experimental design.

2.5.1 Comparison of SIM and PRM Modes

The overall sensitivity (expressed in ion count) in PRM mode is intrinsically lower than in SIM mode due to the fact that the signal of a single precursor ion is split across several product ions during the fragmentation process. Ideally, the sum of all the product ion signals should equal the signal of the precursor ion. Due to unfragmented precursor ions, multiple collisions generating low *m*/*z* products, and non-perfect sampling and transmission of ions, a lower overall signal is usually observed. An example comparing results of SIM and PRM modes is depicted in Figure 2.12 for the isotopically

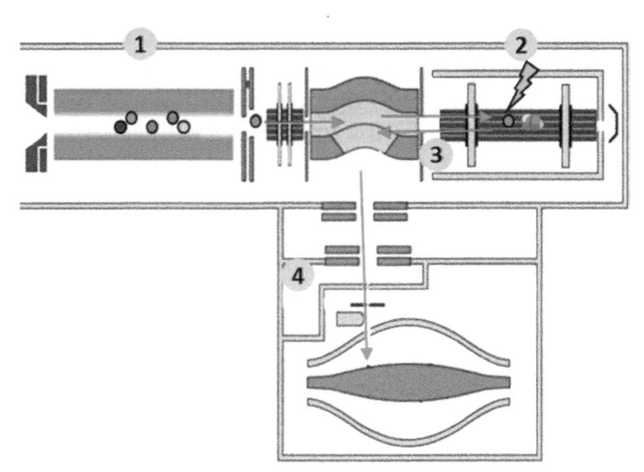

Figure 2.11 Principle of tandem mass spectrometry measurements performed on a quadrupole–orbitrap instrument. Precursor ions are isolated in a narrow *m*/*z* range by the first quadrupole (1), and are then transferred *via* the C-trap into the HDC collision cell to undergo fragmentation (2). All product ions are transferred back into the C-trap (3) and measured in the orbitrap (4).
Adapted from ref. 21.

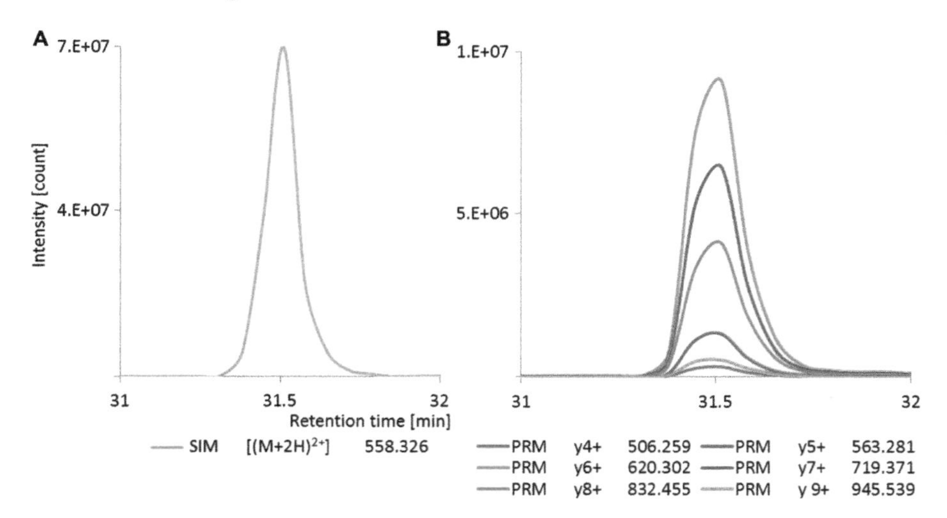

Figure 2.12 Results obtained in SIM (A) and PRM (B) modes for the peptide GLILVGGYGTR̲ in aqueous solution. The fragmentation of the precursor ion (*m*/*z* 558.326, $[M + 2H]^{2+}$) results in several product ions with signals of lower intensities due to the partial fragmentation of the precursor ion, the extensive fragmentation generating low *m*/*z* products, or the non-perfect collection and transmission of ions.

labeled peptide (GLILVGGYGTR̲). The signal of the precursor ion in SIM mode exhibits higher intensity than the corresponding product ions measured by PRM.

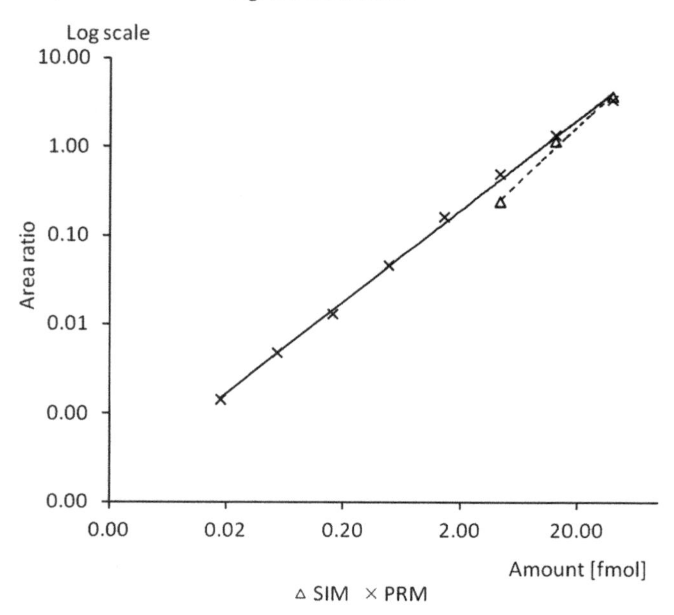

Figure 2.13 Comparison of the selectivity of measurements based on dilution curves established by SIM and PRM analyses in a complex background. A dilution series of the peptide SDLAVPSELALLK (from 2 amol μL^{-1} to 40 fmol μL^{-1}) was prepared in 1 µg μL^{-1} of urine digest. The second analytical stage improves the selectivity of the measurements, as reflected by lower LOQ values for PRM analyses, 20 amol, compared to 4 fmol in SIM mode.
Adapted from ref. 21.

However, the lower intrinsic signal intensity of the PRM mode is usually compensated by a better signal-to-noise ratio, especially for analyses performed in complex biological samples. The additional fragmentation stage dramatically increases the selectivity of the measurements, as shown in Figure 2.13. A dilution series of a peptide (SDLAVPSELALLK) spiked in various amounts into a urine protein digest was measured using the quadrupole–orbitrap instrument operated in either SIM or PRM mode. The increased selectivity observed in the PRM mode, translated in the detection of the peptide at lower amounts, was not observed in the SIM analysis. Thus, tandem MS measurements using the HR/AM capabilities exhibit a higher degree of selectivity, which improves detection and quantification in complex samples. PRM measurements performed on the quadrupole–orbitrap mass spectrometer represent an alternative to SRM analyses when high selectivity is required.

2.5.2 Quantification in Parallel Reaction Monitoring

The analyses in PRM and SRM modes were also compared for a dilution series of isotopically labeled peptides spiked into a urine protein digest

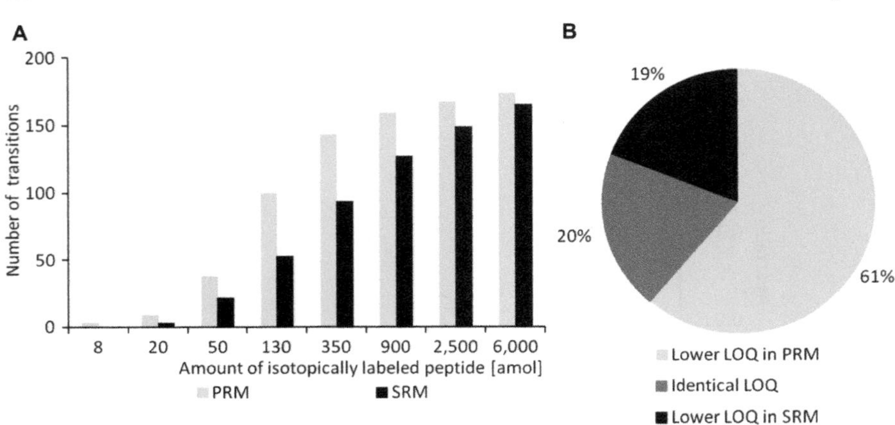

Figure 2.14 Performances of PRM and SRM techniques applied to the analyses of 35 isotopically labeled peptides (175 transitions evaluated) spiked into a biological sample (urine) at different concentrations. (A) Comparison of the number of transitions allowing peptide quantification in PRM and SRM analyzed with different amounts of isotopically labeled peptides. (B) Comparison of the LOQ values obtained in PRM and SRM experiments.
Adapted from ref. 10.

and the results indicated superiority of PRM measurements in terms of selectivity. Across a broad range of concentrations, the number of transitions that did not exhibit apparent interferences was consistently higher in PRM mode as compared to SRM (Figure 2.14). In fact, 60% of transitions exhibited lower LOQs in PRM mode, while 20% of the transitions had lower LOQs in SRM, and another 20% showed similar results. This is illustrated in Figure 2.15 on a selected case with two transitions of the peptide SDALVPSELALLK exhibiting more interference when measured with the low resolution of SRM, resulting in LOQs at 2 and 6 fmol in SRM, compared to 20 and 130 amol in PRM mode. The misaligned elution profiles of the two SRM transitions clearly indicate the presence of interfering ions. The PRM analyses with ion chromatographic traces extracted with a mass tolerance of 10 ppm at *m*/*z* 878.54 and 977.61, respectively, exhibited two perfectly co-eluting profiles. By relaxing the mass tolerance from 10 ppm to 700 ppm to extract the ion traces in order to mimic the low resolution of a quadrupole, the chromatographic profiles exhibited a pattern similar to the SRM analyses, indicative of interferences. This was corroborated by the in-depth inspection of the HR MS/MS data, in which the interfering ions at *m*/*z* 878.45 and 977.52 could be observed. The accurate mass measurement of product ions with HR adds a new dimension to quantitative proteomics, clearly improving the quality and reliability of the results.

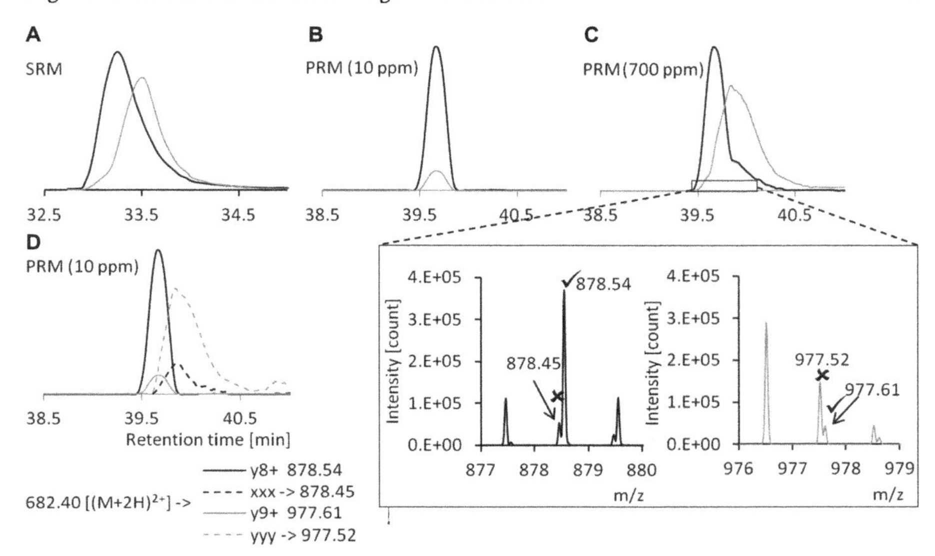

Figure 2.15 Interferences observed in low resolution SRM experiments. (A) The traces of two transitions of the peptide SDALVPSELALLK measured in SRM indicated the presence of interferences. (B) The extraction of PRM data with a tight tolerance (10 ppm) showed two co-eluting transitions. (C) A relaxed mass tolerance (700 ppm) mimicking the resolution of a SRM analysis. The averaged MS/MS spectrum obtained over the 39.5–40 min elution time window confirmed the presence of ions nearly isobaric (labeled ✘) with product ions of interest (labeled ✔). (D) The traces corresponding to these interfering ions are extracted with tight tolerance (10 ppm) from PRM data.
Adapted from ref. 10.

2.5.3 Parameters Used in Parallel Reaction Monitoring

PRM analyses performed on a quadrupole–orbitrap instrument fully leverage the trapping and multiplexing capabilities of the HCD-cell, which enhance performance, but also require the optimization of specific parameters. These include parameters critical to the mass window selection and duration of this selection and additional parameters for the generation, transfer and analysis of product ions. The quadrupole isolation window and the fill time impact on the dynamic range of PRM analyses is, in a way, similar to the one described for SIM analyses. The overall sensitivity is affected by the efficiency of the fragmentation process, controlled by the collision energy; thus, the optimization of this parameter improves sensitivity. As in SRM mode, the fine tuning of PRM increases the signal of specific transitions, leading to better limits of detection. Figure 2.16 exemplifies the effect of collision energy optimization for five product ions of the peptides HFTYLR and NLLSVAYK monitored at different collision energies. The *pseudo*-breakdown curves displayed for both peptides indicate different optimal collision

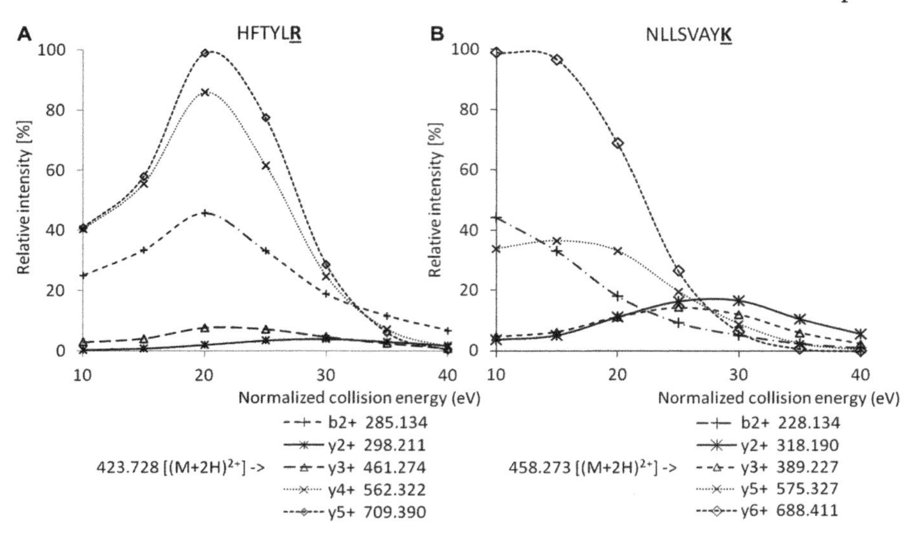

Figure 2.16 Influence of the collision energy on the PRM signal of product ions. The *pseudo*-breakdown curves of five transitions of peptides HFTYLR and NLLSVAYK indicated differences in the optimal collision energy. (A) For the first peptide, the optimal collision energy was 20 eV for four product ions, whereas the other one exhibited higher intensity at 30 eV. (B) The second peptide showed dominating products over a 10 to 30 eV range.

energies for the different products. The signals of high mass y-ions and b-ions are maximized at a lower collision energy than that of low mass y-ions. Considering the peptide HFTYLR, the optimal collision condition for four transitions, *i.e.*, y3, y4, y5, and b2, was observed at 20 eV compared to 30 eV for the last transition y2. Optimal fragmentation of the peptide NLLSVAYK, showed distinct values: with an optimum for y6 and b2 at 10 eV, for y5 at 15 eV and for low mass y3 and y2 ions at 25 and 30 eV, respectively. In practice, at the peptide level, the optimal collision energies were 20 eV and 25 eV for HFTYLR and NLLSVAYK, respectively. The adjustment of this collision energy, at the individual peptide level, is needed to maximize the sensitivity of PRM analyses. By contrast, with SRM, the use of different collision energies for several transitions from the same precursor has limited practical use due to their simultaneous measurement.

The high resolving power of the orbitrap mass analyzer allows discrimination between the product ions of the analytes and possible interferences. The quadrupole isolation width has a major impact on the level of interference in the analyses. The ions co-isolated with the precursor ion of interest are not always separated by HR and identified based on accurate mass in single-stage MS. In tandem MS, as long as the isolation window is maintained sufficiently narrow, the co-fragmentation of the precursor ion and the interferences is more likely to yield distinct product ions, increasing

Table 2.2 Comparison of the proportion of product ions showing interferences in PRM acquisition of 122 isotopically labeled peptides spiked into urine as a function of the quadrupole isolation window. A set of 2172 product ions from 244 peptides was evaluated using isolation windows from 2 to 24 *m/z* units. Adapted from ref. 10.

	Quadrupole isolation window [m/z units]				
	24	*16*	*8*	*4*	*2*
Fraction of product ions with interferences	46	24	16	10	9

the selectivity. But the isolation of relatively wide *m/z* ranges in PRM mode, in order to isolate, simultaneously, the endogenous and the isotopically labeled peptides, is prone to compromise the selectivity of the measurements. This is documented in Table 2.2; the analysis of hundreds of isotopically labeled peptides spiked in a urine protein digest was performed using variable isolation windows between 2 and 24 *m/z* units. Narrowing the isolation window dramatically reduces the number of product ions exhibiting interference and thus ensures better selectivity of measurement. On the other hand, a very broad isolation window width (24 *m/z* units) has very limited selectivity as 46% of the signals showed interferences.

2.6 Conclusion

An overview of the analytical capabilities of the quadrupole–orbitrap instrument used for targeted proteomics experiments was presented in this chapter. The quantification methods were based on high resolution/accurate mass measurements using both selected ion monitoring and parallel reaction monitoring, exploiting the unique capabilities of this instrument. The results were compared to those obtained on a triple quadrupole instrument operated in selected reaction monitoring mode and demonstrated the benefits of high resolution for quantitative proteomics applications.

The quantification in SIM mode takes advantage of the quadrupole capability to isolate precursor ions in narrow *m/z* ranges in conjunction with long fill times of the C-trap. This overcomes the dynamic range limitation commonly observed with conventional trapping devices. For complex samples, the SIM mode clearly outperforms the full scan mode on the quadrupole–orbitrap instrument, while comparable results to those obtained using SRM on a triple quadrupole instrument are observed. The PRM mode provides an additional level of selectivity by performing the quantification on product ions, which yields superior results.

The analytical performance in both SIM and PRM modes is maximal when specific acquisition parameters, such as the precursor isolation window and the collision energy, are fully optimized. The number of peptides monitored during a single LC-MS analysis is dramatically increased by applying the multiplexing capability of the quadrupole–orbitrap instrument. Efforts to fully automate such acquisition methods, including the development of data processing software, are presently on-going and will allow routine large-scale multiplexed experiments.

Acknowledgments

The Luxembourg Fonds National de la Recherche (FNR) is acknowledged for funding (PEARL and CORE grants). This project was supported by Thermo Scientific (Bremen). We are grateful to Catharina Crone, Andreas Huhmer, Markus Kellmann, Thomas Moehring, and Yue Xuan (Thermo Scientific) for helpful discussion.

References

1. A. Doerr, *Nat. Methods*, 2013, **10**, 23.
2. B. Domon and R. Aebersold, *Nat. Biotechnol.*, 2010, **28**, 710–721.
3. R. Aebersold and M. Mann, *Nature*, 2003, **422**, 198–207.
4. B. Domon and R. Aebersold, *Science*, 2006, **312**, 212–217.
5. V. Marx, *Nat. Methods*, 2013, **10**, 19–22.
6. M. A. Gillette and S. A. Carr, *Nat. Methods*, 2013, **10**, 28–34.
7. K. Helsens, M. Mueller, N. Hulstaert and L. Martens, *Proteomics*, 2012, **12**, 1142–1146.
8. S. Gallien, E. Duriez and B. Domon, *J. Mass Spectrom.*, 2011, **46**, 298–312.
9. V. Lange, P. Picotti, B. Domon and R. Aebersold, *Mol. Syst. Biol.*, 2008, **4**, doi: 10.1038/msb.2008.61.
10. S. Gallien, E. Duriez, K. Demeure and B. Domon, *J. Proteomics*, 2013, **81**, 148–158.
11. S. H. Hoke, K. L. Morand, K. D. Greis, T. R. Baker, K. L. Harbol and R. L. M. Dobson, *Int. J. Mass Spectrom.*, 2001, **212**, 135–196.
12. J. Stahl-Zeng, V. Lange, R. Ossola, K. Eckhardt, W. Krek, R. Aebersold and B. Domon, *Mol. Cell. Proteomics*, 2007, **6**, 1809–1817.
13. R. Kiyonami, A. Schoen, A. Prakash, S. Peterman, V. Zabrouskov, P. Picotti, R. Aebersold, A. Huhmer and B. Domon, *Mol. Cell. Proteomics*, 2011, **10**, M110002931.
14. S. Gallien, S. Peterman, R. Kiyonami, J. Souady, E. Duriez, A. Schoen and B. Domon, *Proteomics*, 2012, **12**, 1122–1133.
15. J. Sherman, M. J. McKay, K. Ashman and M. P. Molloy, *Proteomics*, 2009, **9**, 1120–1123.
16. B. Domon, *Proteom. Clin. Appl.*, 2012, **6**, 609–614.
17. H. Rost, L. Malmstrom and R. Aebersold, *Mol. Cell. Proteomics*, 2012, **11**, 540–549.
18. H. Gallart-Ayala, E. Moyano and M. T. Galceran, *J. Chromatogr. A*, 2008, **1208**, 182–188.

19. A. Martinez-Villalba, E. Moyano, C. P. Martins and M. T. Galceran, *Anal. Bioanal. Chem.*, 2010, **397**, 2893–2901.
20. A. Michalski, E. Damoc, J. P. Hauschild, O. Lange, A. Wieghaus, A. Makarov, N. Nagaraj, J. Cox, M. Mann and S. Horning, *Mol. Cell. Proteomics*, 2011, **10**, M111011015.
21. S. Gallien, E. Duriez, C. Crone, M. Kellmann, T. Moehring and B. Domon, *Mol. Cell. Proteomics*, 2012, **11**, 1709–1723.

LABEL-BASED PROTEIN QUANTIFICATION

CHAPTER 3

Making Sense Out of the Proteome: the Utility of iTRAQ and TMT

NARCISO COUTO, CAROLINE A. EVANS,* JAGROOP PANDHAL, WEN QIU, TRONG K. PHAM, JOSSELIN NOIREL AND PHILLIP C. WRIGHT

ChELSI Institute, Department of Chemical and Biological Engineering, University of Sheffield, Mappin Street, Sheffield, S1 3JD, UK
*Email: caroline.evans@sheffield.ac.uk

3.1 Introduction—Mass Spectrometry-based Quantitative Proteomics, an Overview

Analysis of the proteome yields information on how cells function at the molecular level, for example, by monitoring protein expression profiles or protein–protein interactions. Information can be derived at the qualitative (identification and characterisation of proteins) and quantitative (relative and absolute) levels. Mass spectrometry (MS) has become an indispensible bio-analytical tool for the identification and quantification of proteins for the study of cell culture systems, biological fluids and tissues. MS generates data, qualitative in nature, for protein identification; it is not, however, intrinsically quantitative: when peptides of known concentrations are analysed by MS, there is poor correlation between a peptide's abundance and its intensity in the mass spectrum. A number of factors are responsible for this: peptides have different physico-chemical properties (size, charge

New Developments in Mass Spectrometry No. 1
Quantitative Proteomics
Edited by Claire Eyers and Simon J Gaskell

Published by the Royal Society of Chemistry, www.rsc.org

state, hydrophobicity and gas-phase basicity), which affect their ionisation efficiency and thus 'flyability'.[1,2] Consequently, new approaches had to be developed to provide biologists with quantitative information.

Some quantitative MS approaches are best suited for targeted measurements of specific proteins, while others are best suited for large-scale, global studies. Some methods derive the quantitative information in MS mode; other methods make use of tandem mass spectrometry (MS/MS) for simultaneous identification and quantification. Regardless, quantification is attained by integrating signal, either peak intensity or peak area.

There are two main quantitative proteomics approaches: 'label-based' and 'label-free'. Label-based methods employ stable isotopes for labelling proteins or peptides: light and heavy atoms are chemically combined in compounds so as to encode the origin of a protein or a peptide. The most commonly used heavy isotopes are ^{2}H, ^{13}C, ^{15}N and/or ^{18}O. These isotopes can be incorporated into the protein or the peptide either *in vivo*, *via* incorporation of isotopically labelled nutrients during cell growth, or *in vitro*, *via* enzymatic or chemical modification of the proteins or peptides.

Label-free methods are based on either fragment counts or extracted ion currents (see Chapter 6 for further information). Label-free experiments require chromatographic and signal reproducibility since the comparison of peptide abundance is performed across different liquid chromatography (LC) MS runs. At the MS stage, reproducible ionisation efficiency and ion transmission, or flyability, is required.[3]

Labeling methods are used for either absolute or relative quantification. Typically, mass difference isotopes are used in absolute quantification strategies where a heavy version of a standard is spiked in the real sample at a known concentration. Peptides, either alone (AQUA)[4] or from concatenated proteins (QConCAT),[5] as well as truncated proteins (PrEST)[6] and full length protein standards (protein standard absolute quantification, PSAQ™),[7] have been used as standards in MS (see Chapter 4 for further information).

In terms of popularity, isobaric tags (predominantly isobaric tags for relative and absolute quantification (iTRAQ), as well as tandem mass tags (TMT)), metabolic labelling by 'stable isotope labelling by amino acids in cell culture' (SILAC) and label-free methods are the major workflows employed in relative quantification for proteomics discovery.[8–11] A bibliometric survey indicates that isobaric tagging and label-free strategies are both used more than SILAC for relative quantitative proteomics, probably due to their universal applicability to different sample types (Figure 3.1).

Current label-free quantitative approaches have gained ground largely from developments in LC systems and, in particular, ultra high performance liquid chromatography (UHPLC), which provides high-resolution HPLC peptide separation with high reproducibility.[12–14] There have also been improvements in MS resolution and the introduction of bioinformatics tools to process data. The label-free route also dispenses with the cost and handling of chemical reagents, thus reducing sample processing (relative to iTRAQ and TMT) prior to LC-MS analysis. However, there is a requirement to

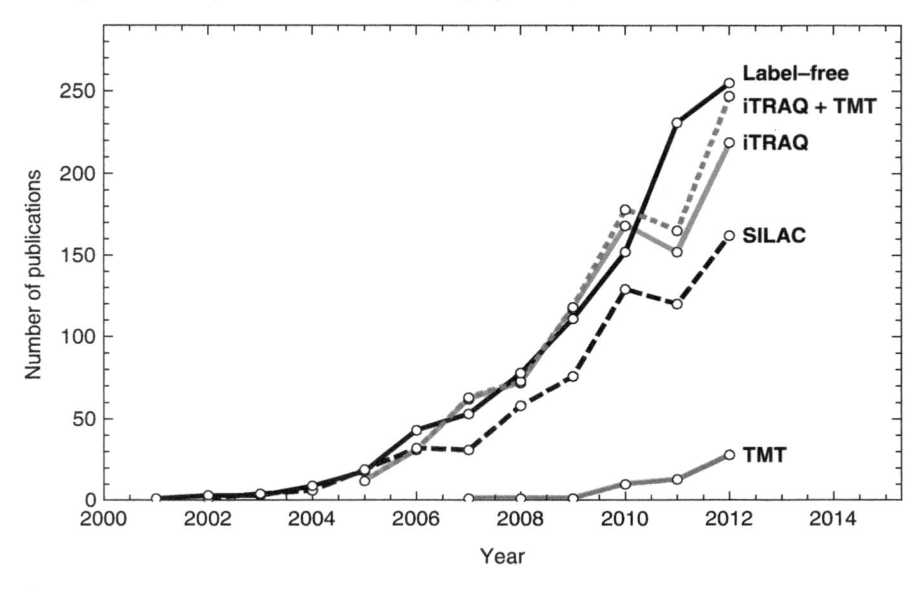

Figure 3.1 Bibliometric evaluation of quantitative-based proteomics publications. Data from searching Web of Science (accessed 1 July 2013). The following strings were used in this search: 'iTRAQ proteom*', 'SILAC proteom*', 'TMT proteom*', 'SILAC proteom*' and '("label-free" OR "label free" OR labelfree) AND proteom*'. The numbers of publications are plotted year on year since 2000–2012 (*y* and *x* axes, respectively).

run more technical replicates (typically 3–5/sample) in order to evaluate the precision of the data.

3.1.1 Labelling Approaches—Metabolic Labelling

Metabolic labelling can be performed either by the incorporation of ^{15}N (in the form of labelled ^{15}N ammonium salts) as the sole source of nitrogen or by SILAC *in vivo*.[15,16] Such labelling confers the ability to distinguish the origin of a protein due to the distinct mass of the labels. The SILAC method requires complete incorporation of heavy amino acids. The main limitation of SILAC is that the analysis of tissues and biological fluids is not directly possible since the labelling step requires the sample to metabolically incorporate the label. A great strength of SILAC, when it is applicable, is that label incorporation takes place at the earliest possible stage; consequently, technical errors associated with downstream processing (such as lysis, subcellular fractionation and/or protein purification) are minimised.

3.1.2 Chemical Labelling

Chemical labelling has broader biological applicability than metabolic labelling since the label is introduced after sample preparation rather than during cell growth. It is therefore applicable to samples such as cell lysates, tissues or biological fluids. Proteins and peptides possess several functional groups susceptible to derivatisation. While both proteins and peptides can be derivatised, labelling of peptides is the most popular approach. Functional groups present in peptides susceptible to labelling are the *C*- and *N*-termini, the ε-amino groups of lysine residues and the carboxylic groups present on the side chains of aspartic acid and glutamic acid residues.

Non-isobaric tagging can be achieved enzymatically, for example, differential ^{18}O labelling of carboxylic acid groups at the peptide *C*-termini, whereby ^{16}O atoms are replaced with ^{18}O atoms by enzyme-catalyzed oxygen-exchange in the presence of $H_2{}^{18}O$.[17–19] Usually high spectral complexity is obtained due to some variability in ^{18}O incorporation; therefore, extraction of quantification information is difficult. For that reason, high specificity chemical labelling (in particular, directed to free amine residues) is often preferentially applied.

Differential labelling of primary amines can be achieved by acetylation using acetic anhydride ($^{1}H_6$ *versus* $^{2}H_6$)[20] or *via* dimethyl labelling (reductive amination) with formaldehyde ($^{1}H_2$ *versus* $^{2}H_2$).[21,22] Moreover, using *N*-acetoxysuccinimide ($^{1}H_3$ *versus* $^{2}H_3$) or nicotinyl-*N*-hydroxysuccinimide ($^{1}H_4$ *versus* $^{2}H_4$), specific labelling of the *N*-terminal amines is achieved without affecting the ε-amino groups of the lysine residues *via* acetylation.[23–25] The well-known non-isobaric, isotope-coded affinity tags (ICATs) were designed to specifically label cysteine residues.[26] Originally, ICAT used deuterium coding. However, non-deuterated peptides and their deuterated counterparts do not elute at exactly the same time during reverse-phase chromatography and, therefore, quantification is not accurate.[27] In order to eliminate the chromatographic shift caused by the deuterated tag, the heavy ICAT reagent was modified to incorporate the ^{13}C isotope instead. Although highly specific, cysteine is a relatively rare amino acid in proteins and some residues can be alkylated *in vivo*. This strategy is, therefore, not ideal for high throughput proteomics workflows[28,29] and, for these reasons, ICAT has now fallen largely out of fashion.

In general, non-isobaric tags, whether introduced metabolically (like SILAC), enzymatically or chemically, generate supernumerary peaks for each and every labelled peptide. This causes a dramatic increase in the spectral complexity and a collateral reduction in the sensitivity of the analysis. This is particularly problematic for complex biological samples. Isobaric tags were initially developed to provide an alternative quantification method. Essentially, since they have the same mass, isobaric-tagged peptides are indistinguishable in the MS survey scan. Relative quantification is achieved using diagnostic signature ions in the low mass region of each MS/MS spectrum produced following fragmentation of the intact peptide and its associated chemical tag. This *m*/*z* region is usually free of other ions and therefore of peaks that could significantly interfere with the accuracy of quantification. Two sets of isobaric tags (iTRAQ and TMT) have become popular because of their multiplexing capabilities. iTRAQ allows up to eight samples to be analysed (iTRAQ 8-plex), while TMT can be used to simultaneously analyse up to six samples (6-plex). Thanks to multiplexing, biological and technical replicates can be easily introduced to facilitate the statistical analysis of the results.[30,31]

3.2 Isobaric Tagging—iTRAQ and TMT Labelling

3.2.1 Structure of iTRAQ and TMT Reagents

Isobaric tag reagents contain a reactive group, a balance (iTRAQ)/normaliser (TMT) group and a reporter group. The reactive group is typically an *N*-hydroxysuccinimide ester (NHS-ester), which is known for its high reactivity to primary amines. The reaction is highly specific and efficient and irreversibly labels free amines found at the *N*-terminus and the amine group of the lysine side chain in proteins and peptides. The reaction results in the dissociation of the reactive group, leaving the peptides labelled with the isobaric groups (balance/normaliser and reporter). Balance and reporter groups have strategically placed different isotopes ($^{1}H/^{2}H$, $^{12}C/^{13}C$ and $^{14}N/^{15}N$) to generate reagents whose masses are identical.

Isobaric tags were designed to generate a strong signal (low mass reporter ions) at the MS/MS level for relative quantification. Simultaneously, the other MS/MS product ions are amino acid sequence informative and used for identification. Due to the isobaric nature of the differently labelled precursor peptide ions, an increase in MS signal is observed as the isobaric-tagged peptides from all the samples are co-isolated and used for sequence identification (providing an advantage over MS level quantification methods). This results in signal amplification and the sensitivity of the MS/MS analysis accordingly increases.

3.2.2 iTRAQ and TMT Protocol Overview

Both iTRAQ and TMT workflows share similarities. Although they have a different structure, iTRAQ and TMT reagents have the same reactive group, the NHS-ester, which is highly specific and efficient toward free amines (Figure 3.2).

For iTRAQ and TMT labelling, biological samples should be prepared in a compatible buffer. An essential requirement is that the buffer is amine-free to prevent any interference with peptide labelling. Triethyl ammonium bicarbonate (TEAB), 4-(2-hydroxyethyl)-1-piperazineethanesulfonic acid (HEPES), 3-(*N*-morpholino)propanesulfonic acid (MOPS) and phosphate buffer are widely used for this reason. The optimal pH lies in the range 7.5–8.5, which is also compatible with proteolytic digestion using trypsin.

Before labelling, all biological samples (including replicates) should contain a comparable amount of starting material; the protein content extracted from each sample needs to be assayed. Once the amount of protein in each sample is known, the same amount of protein needs to be reduced and alkylated and enzymatically digested. Typically, disulfide crosslinks of cysteine residues in proteins are reduced to sulfhydryl groups either using D,L-dithiothreitol (DTT) or tris-(2-carboxyethyl)phosphine (TCEP), followed

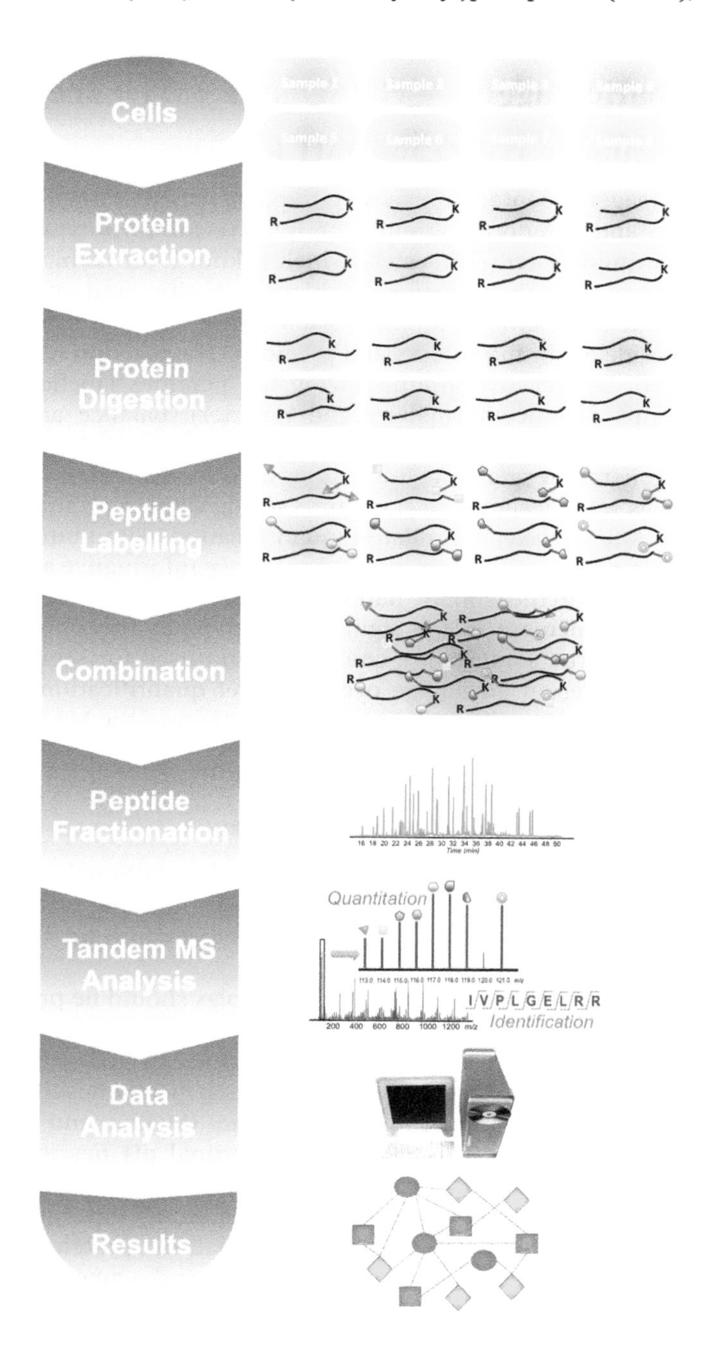

by sulfhydryl group alkylation using *S*-methyl methanethiosulfonate (MMTS) or iodoacetamide. Subsequent enzymatic proteolysis is usually performed in the presence of trypsin; however, other proteolytic enzymes have been used. Lys-C,[32–34] as well as elastase proteases, were used with TMT labelling.[35,36]

Peptides derived from each sample are labelled with a different form of the isobaric tagging reagent prior to pooling all the labelled peptides from each of the biological samples. After labelling, fractionation is performed, usually in two dimensions based on the physico-chemical properties of the peptides in order to reduce sample complexity. In the first dimension strong cation-exchange (SCX), hydrophilic interaction chromatography (HILIC),[37,38] mixed mode[39] or high pH reverse-phase separation[40] is usually performed prior to the second dimension separation. LC separation techniques are most commonly used;[9] however, non-LC-based methods, such as OFFGEL electrophoresis, can also be used to separate peptides on the basis of their isoelectric point.[41–43] Independent of the first-dimension separation, low pH reverse-phase is typically used either off-line or on-line to the MS analysis.

Typically, instrumentation applied in the iTRAQ/TMT workflow includes quadrupole time of flight (QToF) and hybrid Orbitrap™ instruments. Collision-induced dissociation (CID) is used as the main activation technique. Maximising the reporter ion signal (for quantification) without losing other product ions used for determining sequence is crucial and the collision energy has to be increased by 10–20% compared to what is typically recommended for underivatised peptides. Data-dependent acquisition (DDA) mass spectrometric methods are employed: a serial process of MS survey scans is followed by selection of a precursor *m*/*z* value for CID and MS/MS analysis,

Figure 3.2 From peptides to biological pathways. A representative workflow employing 8-plex iTRAQ (6-plex TMT follows exactly the same steps). Proteins are extracted from the biological samples (including replicates) and are separately digested (typically with trypsin). In total eight or six samples can be included in the workflow by iTRAQ or TMT, respectively. After proteolysis, each sample is labelled with one iTRAQ reagent (3 h incubation time) and mixed together. Combined samples are two-dimensionally fractionated by liquid chromatographic techniques: off-line separation by strong cation exchange (SCX) or hydrophilic interaction liquid chromatography (HILIC) followed by reverse-phase HPLC either off-line and prior to, or on-line and during MS analysed by MS. At each chromatographic elution time an MS survey scan is generated. Only ions that obey a pre-set charge state and intensity threshold are selected for tandem MS, typically using collision-induced dissociation (CID). Upon CID fragmentation, iTRAQ-labelled peptides generate reporter ions, which are used for quantification purposes (*m*/*z* 113–121), while the other product ions are used for peptide identification. Typically, Mascot software is used for identification of the labelled peptides and algorithms are used to extract quantitative information. Statistical tests (*e.g. t*-tests, ANOVA) are applied for differential expression analysis. Global protein function and interaction can be obtained, for example, from gene ontology and the Kyoto Encyclopedia of Gene and Genomes (KEGG).

defined for a specific period of time or until a specific ion current is achieved. The relevant *m*/*z* and retention time information for both precursor and product ions is then extracted from the raw data and processed for identification of peptides and proteins *via* sequence database searching, typically using a search engine such as Mascot.[44] Target-decoy approaches allow for an estimation of the false discovery rate (FDR), providing an indication of reliability of the peptide and protein identifications.[45–47]

Experimental replicates can be used to assess the consistency of the quantifications,[48] data which is reliant on the experimental, technical and biological variation due to instrumental variations, preparation processes, *etc*. It is important to establish the significance of the protein expression by applying statistical tests, such as analysis of variance (ANOVA), Student's *t*-test and/or a local pooled error (LPE)-test.[49] Finally, additional biochemical tests should be performed in order to validate the data and to have more confidence in the final result. Typical orthogonal methods include either biochemical assays, such as antibody-based Western blots or enzyme-linked immunoassays, or MS-based selected/multiple reaction monitoring (SRM/MRM) methods targeted to specific proteins.[50,51]

3.2.3 iTRAQ and TMT Variants

Amine-specific TMT exists as TMT^0, TMT 2-plex and TMT 6-plex formats. TMT^0-126 does not incorporate any heavy isotopes and is useful for protocol optimisation. Absolute quantification can be achieved using a TMT^0-126 labelled internal standard (of defined concentration), which is combined with any of TMT^6-126–131. A mass shift of 5 *m*/*z* units per labelled *N*-terminus or lysine residue is observed in the MS survey scan, resulting in a non-isobaric quantification.

A non-isobaric variant of iTRAQ, mTRAQ, enables absolute quantification. mTRAQ reagents were initially developed to overcome some of the problems associated with iTRAQ,[52] particularly quantitative inaccuracies.[53] mTRAQ provides accurate precursor ion-based quantification.[52,54,55] Quantification is achieved at both the MS and MS/MS levels due to different precursor *m*/*z* and unique mTRAQ reporter ions, respectively. mTRAQ is an amine specific reagent, which is supplied as a triplex ($\Delta 0$, $\Delta 4$ and $\Delta 8$ reagents), and is based on the same structure as the iTRAQ 4-plex system. Internal standards are easy to create and, typically, the internal standard is labelled with $\Delta 8$ and test samples are labelled with $\Delta 0$ and $\Delta 4$. mTRAQ is attractive since it eliminates the need for, and the high cost of, synthetic peptides.[52,56] mTRAQ has reduced sensitivity compared with iTRAQ (because samples are pooled and all contribute to the same product ions). Although less accurate (with regards to the closeness to the 'true' abundances) than mTRAQ, iTRAQ has been shown to be more precise (regarding the range of the measurements), mainly due to ratio compression effects.[52]

iTRAQ and TMT were initially designed to generate intense reporter ions under typical CID MS/MS conditions separated by 1 *m*/*z* unit in the low *m*/*z* region of the mass spectrum. Complementary to CID, electron transfer dissociation (ETD), either alone or in combination with CID (ETcaD), has successfully been applied with iTRAQ and is of particular value for multiply charged peptides.[57] ETD fragmentation mechanisms promote cleavage at the N–Cα bond (instead of the peptide bond); consequently, the multiplexing capability of iTRAQ and TMT is lower using ETD.[58] This should be factored into experimental workflows utilising ETD for isobaric tagging experiments. TMT^6-126 and TMT^6-127, as well as the TMT^6-128 and TMT^6-129, labelling reagents generate identical reporter ions at *m*/*z* 114 and 116, respectively; therefore, only four reporter ions are generated under ETD conditions (Figure 3.3).[59]

Similarly, $iTRAQ^8$-113 and $iTRAQ^8$-114 generate a reporter ion at *m*/*z* 101, $iTRAQ^8$-116 and $iTRAQ^8$-117 generate a reporter ion at *m*/*z* 104 and $iTRAQ^8$-118 and $iTRAQ^8$-119 generate a reporter ion at *m*/*z* 106 (Figure 3.3) so that only five reporter ions are formed.[57]

3.3 Technical Limitations of iTRAQ/TMT

iTRAQ and TMT-based quantification have been shown to yield data with good precision due to the equal distribution of errors across the low mass quantification region, even on medium-resolution Q-ToF platforms. However, the accuracy of iTRAQ and TMT quantitative data is subject to underestimation. This seems to be independent of the MS platform (Orbitrap, Q-TOF or TOF-TOF) used.[8,11] The detection of regulated proteins (up and down) is usually not compromised.[52,60,61] Interestingly, in spite of the underestimation associated with iTRAQ, compared to mTRAQ or other precursor-based quantification strategies (*e.g.*, SILAC), which exhibit little or no underestimation, the list of significantly regulated proteins obtained can be virtually identical regardless of the method adopted.[52]

Accuracy is reduced by the existence of isotopic contamination. However, the effect of isotopic impurities in the iTRAQ/TMT reagents can be corrected for during analysis when the manufacturer provides correction coefficients (this is usually the case for iTRAQ 4-plex and TMT 6-plex) as percentages of each reporter ion that have mass differences by −2, −1, +1, +2 *m*/*z* units. Non-reporter ions, such as immonium ions, can also be a source of interference with the reporter ions in the low mass region. For example, interference of the M + 1 peak of the phenylalanine (Phe) immonium ion (*m*/*z* 120.03) interferes with the *m*/*z* 121.1 iTRAQ reporter ion. The specific interference of the Phe immonium can be eliminated using correction coefficients.[53] The arginine-derived immonium ion at *m*/*z* 115.05 can interfere with the iTRAQ reporter ions at *m*/*z* 115.11.[62] Usually, ion contamination is weaker than the reporter ions but this can still have a dramatic effect on reporter ion intensities affecting the accuracy of the ratio measurements and quantification. Note that, if high resolving power MS platforms are used, it often becomes possible to confidently differentiate between these interfering ions and the reporter ions.[8]

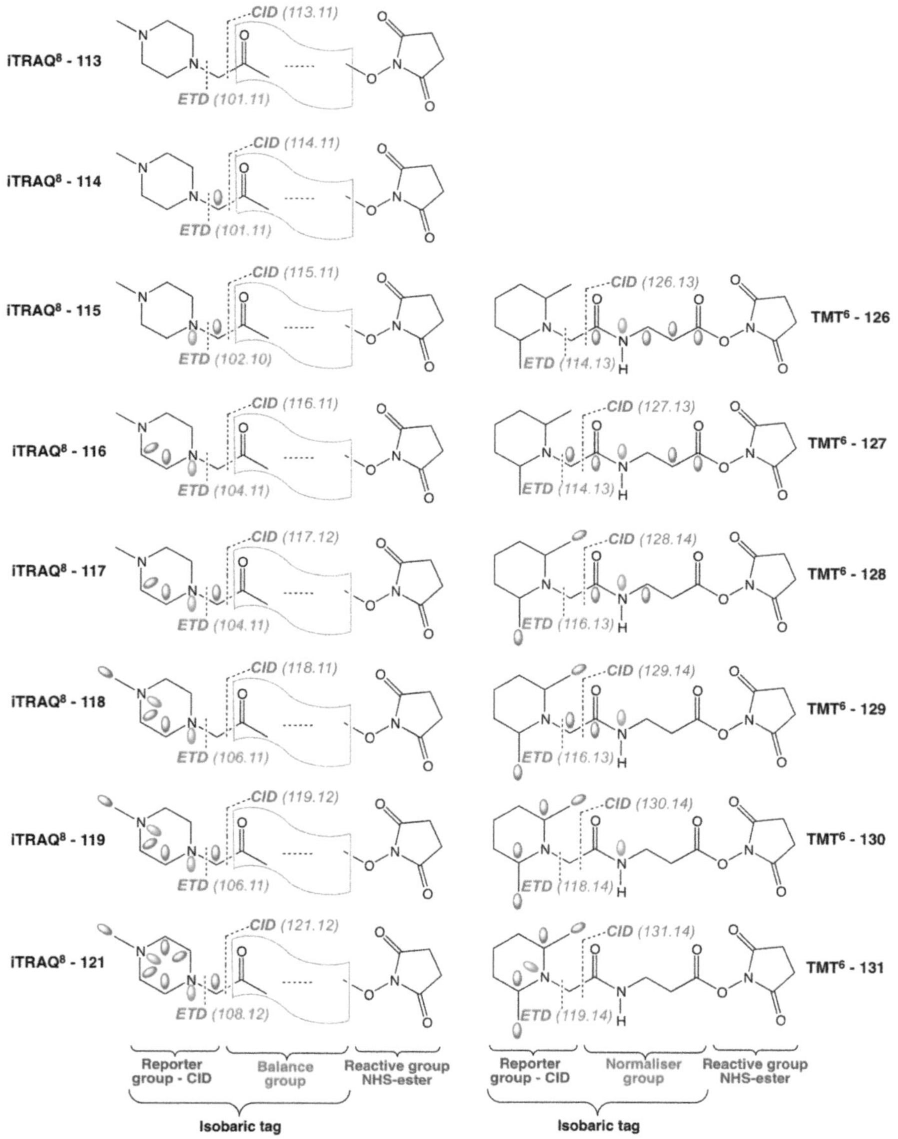

Experimental and data analysis strategies aimed at decreasing or compensating for experimental errors that affect the accuracy of iTRAQ and TMT measurements have been devised.[32,63–65] Mixed product ion spectra, arising from co-isolation and, therefore, the fragmentation of multiple precursor ions, are a direct result of sample complexity. Peptides are more effectively separated by HILIC than SCX and less underestimation is observed.[66,67] Increasing the gradient times and the number of fractions collected during the first dimensional separation also helps. Note that improved fractionation does not totally eliminate underestimation. At the MS level, narrowing the isolation window for precursor isolation would tend to reduce underestimation effects, but the effect is counterbalanced by a compromised sensitivity due to reduced MS signal.[68] Several MS-based strategies applied on Orbitrap instruments have been developed, including ion fragmentation methods and CID, to minimise underestimation. Gas-phase isolation steps, including proton transfer ion–ion reactions (PTR), and higher order MS3 scans, which function to minimise interference due to mixed MS/MS arising from co-isolation of precursors of the same *m*/*z* value, have shown particular utility.[32,65] Applying PTR prior to fragmentation reduces the charge state of the precursor ion and therefore removes the interfering ions with different charge states, which are co-isolated.[65] TMT-labelled tryptic peptides from yeast and human cells were analysed using this approach and the accuracy and precision was drastically improved; although this came at the expense of the number of identifications, mainly due to decreased sensitivity.[65] Using MS3 scans can also reduce the interference effect of contaminating ions.[32] MS2 spectra are used for identification, while the most abundant fragment ion is selected for higher collision dissociation (HCD) providing accurate quantification. The weaknesses of this approach are the lower sensitivity at the MS3 level, a reduced acquisition speed (12% fewer identifications), the requirement for specific instrumentation and the need for specialised software.[32]

An interference-free MS2 strategy has been recently reported; instead of using the reporter ions in the low *m*/*z* region, quantification can be performed using the complementary TMT fragment ion cluster (TMTc). The TMTc results from partial loss of the TMT tag, carries most of the TMT mass balance and is precursor ion specific.[64] The combination of the charge

---

Figure 3.3 Chemical structures of 8-plex iTRAQ (left) and TMT (right) isobaric tagging reagents. Positions where ^{15}N or ^{13}C are incorporated are labelled in black and light grey symbols, respectively. Typical cleavage sites (dashed lines) following collision-induced dissociation (CID) or electron transfer dissociation (ETD) are shown.

iTRAQ and TMT share the same reactive group, *N*-hydroxysuccinimide ester (NHS-ester), which has high specificity for primary amines. Peptides are chemically tagged *via* the *N*-terminus (unless blocked, *e.g.*, by acetylation) and *via* the epsilon side chain of lysine amino acid residues. The tags also contain a reporter group (*N*-methylpiperazine for iTRAQ and 1,5-dimethylpiperidine for TMT) and the mass balance/normaliser groups, which allow the tags to be isobaric (same overall mass).

Due to the isobaric nature of the tags, peptides derived from different samples, and therefore labelled with different forms of the isobaric tag, have the same mass and are thus indistinguishable in the MS spectrum. Upon CID, unique *m*/*z* reporter ions are generated, which are used for relative quantification. Fragmentation *via* ETD results in cleavage at the N–Cα bond; for this reason, iTRAQ only generates five reporter ions at *m*/*z* 101, 102, 104, 106 and 108, and TMT only generates four reporter ions at *m*/*z* 114, 116, 118, 119, thus reducing the multiplexing capability relative to CID.

state-specific TMTc ions, with a high mass accuracy measurement, allows for more accurate quantifications. Although the TMTc-130 and TMTc-129 are indistinguishable, this can be overcome by re-designing the label distribution within the TMT reagents. All information is extracted from the MS2 spectrum; therefore, high sensitivity is obtained in comparison with gas-phase manipulation strategies. However, the MS2 approach only allows for the quantification of similar numbers of peptides to the MS3 approaches due to insufficient formation of TMTc ions during MS2. In fact, the mass difference of the TMTc ions paves the way for parallel quantification of co-eluted peptides, as well as for data-independent acquisition methodologies.[64]

Software-based methods for dealing with underestimation of fold change are detailed in section 3.5. Mass spectrometers with improved acquisition speed and high resolving power capabilities increase the dynamic range of the quantified proteome in addition to improving the precision and accuracy of the quantification.[69,70]

3.4 Mass Spectrometry Platforms

As previously mentioned, iTRAQ and TMT reagents were initially designed to obtain simultaneous protein identification and quantification at the MS/MS level under CID conditions. Isobaric tags were designed in order to include reproducible fragmentation channels suitable for relative quantification.[30,71]

In contrast to beam-type collision activation typical of Q-ToFs and triple quadrupole mass spectrometers, CID in ion trap mass spectrometers is performed under resonance excitation. Ion trap instruments were not originally thought to be suitable to run iTRAQ experiments as the reporter ions cannot be observed due to the 'one-third rule'.[72–74] This shortcoming has encouraged the development of alternative activation methods, such as pulse Q dissociation (PQD) by Thermo Fisher Scientific, which is implemented in a linear ion trap,[75] and high-amplitude short-time excitation (HASTA).[76,77] The resonant excitation process was enhanced in order to trap low-mass ions: a high-pulse amplitude followed by a rapid decrease in the trapping voltage immediately after resonance excitation retains the low *m*/*z* ions.[76] iTRAQ-labelled peptides were analysed by PQD and similar spectra to those obtained with a triple quadrupole were obtained; however, the lower efficiency of conversion of the precursor to fragments compromises the quantification.[78]

In order to maximise reporter ions from iTRAQ- or TMT-labelled peptides, a high energy CID regime can be applied alone or in combination with low energy CID. A high collision dissociation (HCD) collision cell connected to a C-trap allows the linear trap quadrupole (LTQ)–Orbitrap instruments to perform HCD and/or CID.[34,79,80] In the C-trap no loss of low *m*/*z* ions as a function of the ejection q value is observed; therefore, low-mass diagnostic ions are fully detected.[34,81,82] HCD leads to a decrease in peptide identifications relative to CID due to poorer peptide sequence ion recovery.[83] Low *m*/*z* ions are usually the most intense peaks in the HCD spectrum, which is advantageous for quantification. A combined approach using HCD and CID can be used to perform and improve the quality of quantification and maximise protein identification.[84–86] As a result of the increased acquisition time, fewer peptides are analysed between each survey scan. Orbitrap mass spectrometers combining a quadrupole with an Orbitrap (Q-Exactive family) can successfully perform both HCD and CID activation techniques.[80]

CID and HCD activation methods, performed in a linear ion trap–Orbitrap hybrid instrument, allow for precise peptide quantification (from the HCD MS/MS spectrum) without compromising peptide identification with high sensitivity in the linear ion trap analyser.[87–89]

Whilst the hybrid mode of PQD and CID, as well as HCD and CID, appears to be advantageous for identification and quantification of unmodified peptides, post-translationally modified (PTM) peptides (*e.g.*, phosphopeptides and glycopeptides) are labile and the modification is easily lost during application of collisional activation methods.[57,90–93] The modification site is difficult to define and quantification is also compromised. ETD has proven to be an attractive complementary fragmentation technique for identification of such covalently modified peptides. ETD can be used for quantification with iTRAQ-labelled peptides even though the iTRAQ (4-plex) only generates three reporter ions at m/z 101, 102 and 104.[90] In addition, ETD and CID modes can be used to take full advantage of the eight-channel iTRAQ reporters.[94,95] Another hybrid combination taking advantage of ETD for PTM analysis and HCD for accurate peptide quantification was developed and ETD and HCD spectra of the same precursor ion can be acquired in parallel.

Infrared multiphoton dissociation (IRMPD) is an activation technique most often used on Fourier transform ion cyclotron resonance mass spectrometers. An infrared laser is used to fragment precursor ions. A modified dual cell linear ion trap was successfully used to fragment and quantitatively analyse iTRAQ-labelled peptides.[96]

Better performance of the mass spectrometers has also contributed to large improvements in protein identification and quantification. LTQ–FT and LTQ–Orbitrap instruments are equipped with dual detection (in contrast with single detectors for Q-ToFs or Q-FT-ICR); this feature allows for new data acquisition schemes and higher numbers of data-dependent acquisition MS/MS spectra can be achieved.[61,79,81,97] High resolving power mass spectrometers can also add additional information by distinguishing between reporter ions containing one extra neutron incorporated in either carbon or nitrogen (50 000 resolution).[98,99]

3.5 Downstream Analysis of Proteomic Data

In recent years there have been many developments in the analysis of quantitative proteomics data. It has been progressively realised that the idiosyncratic features of proteomic data had to be properly taken into account in order to be useful. Statistical methods were devised to deal with the hierarchical nature of the proteomics data produced: the top level is occupied by the various conditions that are under investigation, then biological replicates, then technical replicates, all the way down to individual MS/MS scans.[100,101]

Another key aspect of proteomics data is the fact that the signal-to-noise ratio decreases at low intensities.[102] One pragmatic answer to that issue has been to correct for it using variance-stabilising transforms.[103] Other authors have also attempted to deal with missing values, when for unknown experimental or biological reasons a reporter ion fails to materialise in a spectrum.[104,105] Another prevalent problem is the existence of 'shared peptides'. These are peptides whose sequence can be found in several proteins in a database and, therefore, cannot be used as conclusive evidence of the presence of a unique protein; they do, however, contain quantitative information that can be extracted.[106,107] ProteinPilot (ABSciex) attempts to deal with this using 'background compression', although there has been no rigorous and independent appraisal of its effectiveness to the best of our knowledge.

Interestingly, even though the issue of underestimation can be (partially) addressed experimentally (*e.g.*, through more effective separation) or technically (*e.g.*, by using different data-dependent selection of precursors),[108] modelling alone can compensate.[109] In the face of so many exciting developments in the numerical treatment of quantitative proteomics data (many *R* modules are now made available alongside a publication), a fair assessment of the merits of each one of the published methods is wanting. The other strategic question is whether iTRAQ will ever routinely provide the 'true value'. At this point, the answer to that remains unclear; it is, however, reasonable to expect methods whose quantifications are based on MS1 quantification (and less susceptible to compression), either label-free or metabolic labelling approaches, to gain ground in the meantime.

3.6 iTRAQ and TMT Relative Quantification Applied to the Analysis of Post-translational Modifications (PTMs)

Investigation of post-translational modification (PTM) events in the cell is an area of immense interest due to their ubiquitous role in cellular mechanisms regulating cell function. Moreover, it has been noted that a range of PTMs are also involved in cross-talk;[110–112] for example, lysine acetylation and phosphorylation events can both be involved in regulation of the same protein and the absence of one can affect the extent of the other.[111] To analyse PTMs, quantitative techniques in MS have generated the ability to systematically monitor the regulation of levels of various proteins.[59,113,114] To date, several recognised methods, including SILAC, $H_2{}^{18}O$ labelling, ICAT, iTRAQ and TMT, AQUA, and QconCAT, have been used to quantify PTM levels.[115–119] Over 300 PTM types have been reported[117] and we describe here how iTRAQ/TMT isobaric tagging-based MS workflows, as applied to protein phosphorylation and glycosylation as paradigm studies of PTM analysis, can be quantified by this approach.

3.6.1 Phosphoproteomic Analysis Using Isobaric Tagging

The analysis of protein phosphorylation by MS has largely benefited from phosphopeptide-enrichment strategies, which can easily be combined with iTRAQ/TMT workflows.[93,120,121] The (often) low abundance and low stoichiometry of phosphorylation requires an enrichment process before analysis to be able to detect the phosphorylation sites. Moreover, the mass spectrometric signals of phosphopeptides are usually affected (lower signals) by the presence of non-phosphorylated forms.[122] During the iTRAQ/TMT workflow, phosphopeptide enrichment is typically performed before the labelling step by strategies such as immunoprecipitation using generic phosphotyrosine antibodies, immobilised metal affinity chromatography (IMAC)[123,124] or metal oxide affinity chromatography.[125–128]

For example, a large scale phosphoproteomics work employed an Fe-IMAC phosphopeptide enrichment strategy followed by 4-plex iTRAQ labelling to identify (3766 phosphopeptides and 8309 phosphosites from 3402 phosphorylated proteins) and relatively quantify (133 phosphopeptides differentially expressed) phosphopeptides from low- and high-risk human breast cancer tissues.[129] As previously mentioned, iTRAQ and TMT only give relative values and further validation using alternative approaches are necessary. SRM/MRM was used as a complementary approach to the iTRAQ data in this case and, using this follow-on analysis, 15 phosphopeptides were successfully quantified and validated as potential biomarkers for breast cancer.[129]

The application of iTRAQ or TMT to the targeted analysis of specific proteins has been less popular than for global profiling; nevertheless, their potential has been demonstrated. A modified histidine biotin tag was used to purify the highly phosphorylated actin cytoskeleton regulatory complex protein (Pan1) from yeast. Following tryptic digestion, phosphopeptides were enriched on a TiO_2 column and similar quantitative information was obtained using SILAC and 4-plex iTRAQ analyses. For example, salt stress-induced changes in Pan1 were quantified as ratios (stress/unstressed): 2.73 *versus* 3.51 for Ser^{1003} phosphosite, 23.10 *versus* 16.56 for Thr^{1225} phosphosite, using SILAC and iTRAQ, respectively.[130]

In a different study 4-plex iTRAQ labelling was used to estimate phosphorylation stoichiometry from a Ccl1-Kin28-Tfb3 complex and an insulin receptor substrate homologue Chico complex from phosphatase-treated protein complexes from yeast cells and fruit flies, respectively.[131] Instead of enzymatic dephosphorylation (phosphatase treatment), an efficient chemical dephosphorylation process, using cerium oxide nanoparticles, in combination with TMT 2-plex was applied to obtain absolute stoichiometry of the phosphorylation S183 (70%) of the eukaryotic initiation factor 3H from HeLa cells using MALDI ToF/ToF and LTQ–Orbitrap XL platforms.[121]

The examples presented above show the iTRAQ/TMT potential in phosphoproteomics, including large (phosphoproteome), intermediate (interactome) or targeted studies.

3.6.2 Glycoproteomics Analysis Using Isobaric Tagging

Protein glycosylation analysis by MS is challenging mainly because of the plethora of different types and structures of glycans (sugars). Although different bio-analytical techniques have been applied in glycoproteomics analysis, MS-based methods are now widely applied in characterising and quantifying glycosylation patterns.[144] The application of iTRAQ in glycoproteomics has revealed novel biological insights. The generation of reliable and quantitative data here is crucial as a change in the concentration of glycoproteins can be responsible for a phenotypic variation, as can a change in glycan acceptor site occupancy and glycan structure.

Similar to phosphorylation, where phosphopeptides are usually affinity enriched before labelling and MS analysis, glycoproteins are also usually enriched before downstream analysis. Using hydrazide-functionalised resins,[132,133] *N*-linked glycoproteins were enriched from the extracellular fluid of epithelial cells covering the surface of the eyes of patients with climatic droplet keratopathy. By covalently linking oxidised glycans with hydrazine functional groups in the resin, *N*-linked glycopeptides could be enriched by selectively cleaving them from the resin using peptide-*N*-glycosidase F (PNGase F), an enzyme that specifically deglycosylates *N*-linked glycoproteins and not *O*-linked glycoproteins.[150] Forty three unique *N*-glycoproteins were identified, 19 of which had not previously been identified in tear fluid.

By combining the structure-specific enrichment technology IGEL (isotopic glycosidase elution and labelling on lectin-column chromatography) with iTRAQ labelling (8-plex), 107 glycopeptides were identified and quantified from serum samples.[134] IGEL employs sugar-specific lectins to enrich for *N*-linked glycopeptides that are then eluted using *N*-glycosidase whilst being labelled by $H_2{}^{18}O$ water. The complexity of glycopeptides in a sample means many different workflows can optimise the identification ability and hence the depth of glycoproteomics coverage. Using multiple digestion proteases combined with parallel enrichment (hydrazine capture, titanium dioxide and ionic and non-ionic HILIC), 1556 non-redundant *N*-linked glycosylation sites were identified. This represented 972 protein groups. iTRAQ added the important quantitative element to the study and 80 out of 437 glycopeptides were found to be significantly regulated.[152] In 2012, a new technique enabled quantitative analysis of glycopeptides and non-glycopeptides from the same samples.[153] This strategy employed a combination of PNGase-catalysed labelling of *N*-glycosylation sites, $H_2{}^{16}O$ or $H_2{}^{18}O$ isotopic labelling of different phenotypes and iTRAQ labelling for quantitative analysis.[153] Glycosite ratios were shown to change significantly in patients with hepatocellular carcinoma compared to patients suffering from liver cirrhosis and the hepatitis B virus.[153]

TMT has also been used for quantifying both *O*-linked and *N*-linked glycopeptides using the reporter intensities of the 6-plex TMT tags for quantification of the exemplar protein-modified bovine crystallin. This was the first report to assess the utility of combining TMT labelling and ETD for relative quantification of labile PTMs.[53]

Carbonyl-reactive TMTs for labelling *N*-glycopeptides have been proposed.[154] Although two types of reagent were initially considered based on hydrazide and amino-oxy-functionalised chemistry, the latter was found to be the most suitable to characterise *N*-glycosylation patterns in human colon carcinoma cell lines. They revealed significant differences in the quantities of high-abundance high-mannose glycans in the metastatic and non-metastatic lines.[154] iTRAQ/TMT isobaric labelling strategies have emerged as the key drivers of technological developments in glycosylation analysis.

3.7 Bio-engineering and Biomedical Applications of Isobaric Tagging Technology

3.7.1 Application to Biological Engineering and Systems Research

Quantification of the proteome and an understanding of how metabolic and regulatory pathways work are extremely important for biological engineering for biotechnological applications. Cells in this context can be viewed as factories that can be genetically manipulated using metabolic engineering and/or synthetic biology approaches to generate high titres of products of technological interest, such as therapeutic proteins. In this section, we show several examples where quantitative proteomics has been or is being used as a tool to aid in forward engineering of cells for 'useful purposes'.

The iTRAQ methodology has been used, for example, in a targeted approach to quantify recombinant glycoprotein production for metabolic engineering purposes.[135] An 8-plex iTRAQ experiment was performed in the initial discovery stage of a four-part iterative engineering cycle to look for protein abundance changes in *Escherichia coli* cells overexpressing *Campylobacter* glycosylation molecular machinery. However, an aliquot (20%) of iTRAQ labels was used in the final stage of this four-part iterative metabolic engineering strategy to quantify recombinant glycoprotein production using targeting *pseudo* SRM.[135] A *ca*. 300% increase in glycosylation efficiency was observed using these iTRAQ labels, which was supported by Western blot analysis and the determination of protein amounts.

Discovery proteomics provides a means to elucidate protein pathways for metabolic engineering to optimise production. A paradigm is biofuel production by algal organisms, which produce lipids, starch and hydrogen: compounds with high potential application in biofuels. A combination of iTRAQ and label-free approaches compared protein abundance in two mutants of *Chlamydomonas reinhardtii* (sta6, a starch-less mutant of cw15, and cw15, a cell wall-deficient strain).[136,137] Both quantification strategies showed similar regulation trends; for example, proteins involved in glycolysis, the Krebs cycle, fatty acid metabolism, intracellular signalling (calcium-dependent protein kinase) and thiamine metabolism were up-regulated in the sta6 strain. On the other hand, enzymes involved in chlorophyll biosynthetic processes, glycogen and starch metabolism, and the Calvin cycle were down-regulated in the sta6 strain. An 8-plex iTRAQ analysis assessed biofuel production in *C. reinhardtii* under conditions of nitrogen starvation and demonstrated up-regulation of proteins involved in carbohydrate and lipid metabolism, leading to the conclusion that lipid is accumulated at the expense of normal cell function, providing insights into future strategies for the optimisation of biofuel production.[136,137]

Quantitative studies with *Aspergillus niger* revealed a high level of hydrolytic activity in their secretome, including cellulases and hemicellulases with potential for biomass degradation and biofuel applications. The cellulolytic enzyme machinery in a thermophilic bacterium (*Thermobifida fusca*)[138] and the secretome of *A. niger*[139] were studied using an iTRAQ-based quantitative proteomic approach that yielded the identification of novel hydrolytic enzymes. Another example focussed on *Arcobacter butzleri*, which generates electron fluxes with a potential application in bioelectrochemical cells (as an energy source) and biosensors (for wastewater treatment). In this case iTRAQ was used to profile the metabolic pathways involved.[140] A key finding of the study was the increased abundance of a flagellin protein under anaerobic growth conditions on an insoluble electrode, with data suggesting that this protein mediates communication with the anode surface and electrogenesis rather than affecting cell motility.[140] This information can be used as a lead for future engineering to improve performance.

In terms of bioremediation, *cis* 1,2-dichloroethylene (DCE) is known to be a persistent pollutant in groundwater systems. This pollutant can be degraded by metabolically engineering *Escherichia coli* expressing an evolved toluene *ortho*-monooxygenase along with either (i) glutathione *S*-transferase and altered gamma-glutamylcysteinesynthetase or (ii) a rationally engineered epoxide hydrolase. The effect on the proteome and transcriptome were analysed by combined iTRAQ and DNA microarrays analyses and the study identified the stress proteins YhcN and YchH as important for the degradation of *cis*-DCE.[141,142] The application of iTRAQ technology is thus providing useful insights in the area of chemical and biological engineering and is generating a map of metabolism that can aid in cellular engineering. Bioengeering applications are not particularly widespread at present, but these examples show how utilising this information has proven effective[135] and other cases have determined useful targets.[136]

3.7.2 Applications to Medical Research

iTRAQ and TMT workflows have been applied to the analysis of a range of sample types, including cell line models, tissues from animal disease models and clinical samples. Animal models enable system-level analysis and mammalian models are widely employed, typically through the use of transgenic mouse models and gene targeting by homologous recombination. The zebrafish (*Danio rerio*) provides a more genetically tractable and less costly vertebrate model system, which is amenable to iTRAQ analysis at the level of proteins[143] and phosphoproteins[144] in drug discovery applications.

In order to illustrate the types of analyses that iTRAQ and TMT can be directed to, their application to research into disease are outlined in this section with specific focus on cancer research. The identification of biomarkers for diagnostic and prognostic applications and chemopreventive and therapeutic target proteins are key drivers in cancer research. The utility of cell line models lies in the relative ease of generating large amounts of material, typically in mg quantities, for analysis. The use of panels of isogenic cell lines (with the same genetic background) can provide both mechanistic information and potential biomarker candidates, for example, in prostate cancer,[145] including eukaryotic translation elongation factor 1 alpha 1 (eEF1A1), which was subsequently confirmed in a serum biomarker study.[146] Cell line studies can also be used to identify modes of action of drugs, for example, the combinatorial effects of gossypol and valproic acid for increased efficacy in the treatment of prostate cancer.[147]

Isobaric tagging of tissue samples can also be very informative, for example, the comparison of prostate cancer biopsy material with non-malignant benign prostate hyperplasia (control) samples has identified prostate cancer-associated proteins.[148,149] Similar approaches have identified alterations in protein expression associated with altered prostate cancer risk *via* chemopreventive agents[150] and ageing.[151] It should be noted that clinical samples (tissue, biological fluids) are inherently more variable due to their differing genetic background and the potential confounding factors of the age and sex of the subjects and individualised disease progression. The use of age- and sex-matched material from control subjects[146,152] or (same) patient paired adjacent normal tissue[153] can control for these factors. Phosphoproteome and glycosylation analysis of cancer samples, which may be particularly interesting due to the known changes in these PTMs in cancer, are discussed in sections 3.6.1 and 3.6.2 of this chapter.

Biological fluids present a particular challenge to proteomics analysis due to the large dynamic range and high number of proteins present. For example, plasma and serum (derived from blood) contain over 10 000 protein species whose concentrations cover a dynamic range over ten orders of magnitude.[154] Albumin predominates since it is present at >99%, with 14 proteins comprising 95% of the total protein mass. Strategies for removal of these proteins are required, including antibody-based affinity depletion of serum and other abundant proteins.[155] Biomarker analysis of serum using peptide labelling by 4-plex and 8-plex iTRAQ has been compared using the typical workflow, *i.e.*, to two-dimensional LC-based protein separation prior to MS analysis. In this analysis 8-plex iTRAQ was more effective than 4-plex in terms of more consistent ratiometric measurements.[156] A detailed analysis of the optimal study design concluded that six samples per experimental group provided sufficient statistical power for most proteins with >2-fold changes for analysis of plasma biomarkers for pancreatic cancer.[157] An alternative biomarker analysis strategy employing protein level labelling combined with 1D gel separation and MS analysis of gel slices has been applied to samples that pre-date diagnosis of pancreatic cancer. The study was based on the assumption that cancer-specific biomarkers are present during early-stage tumour development, avoiding the confounding effects of inflammation associated with late-stage tumours, which result in non-specific alterations in acute-phase reactant proteins.[146,158]

Data from the techniques described above for biomedical applications are derived from isobaric tagging at the peptide level. Protein labelling can also be performed and a powerful example of the utility of this approach is analysis of the cancer 'degradome' (cancer-specific proteolysis products with (novel) 'neo' *N*-termini and altered protein structure function). iTRAQ TAILS employs terminal amino isotope labelling of substrates. TAILS labels all primary amines, including protein *N*-termini and lysine side chains. By this means, free *N*-termini, including neo *N*-termini, are labelled and this is combined with negative selection to enrich for labelled proteins. iTRAQ labelling blocks the lysine residue so that subsequent tryptic cleavage occurs at arginine only, effectively elongating the proteolytically truncated peptides, which has the benefit of improved MS/MS analysis and peptide identification.[159] iTRAQ and TMT thus provide valuable tools in biomedical research.

3.8 Conclusions

MS-based quantitative proteomics approaches have largely benefited from improvements in instrumental platforms (better separation strategies, speed of acquisition, mass accuracy) and data analysis tools to provide depth of protein coverage and information content for biological samples. Whilst different quantitative methods have been developed, none to date completely satisfy all the requirements for universal application. Of those methods, iTRAQ and TMT have benefited from easy implementation in the majority of labs due to their robust chemistry and simple one-step labelling. This, together with their multiplexing ability (8-plex iTRAQ and 6-plex TMT) enabling parallel comparison of samples in a single experiment, has led to widespread implementation.

TMT and iTRAQ are expensive due to their design, which is based on the multistep incorporation of expensive isotopes. Naturally, alternatives have been sought. For example, deuterium has been used as a cheaper alternative to carbon and nitrogen stable isotopes.[160] iTRAQ and TMT were specifically designed to label free amine groups. However, other isobaric tags have been developed to label other reactive groups,[161] including cysteine reactive 6-plex TMT reagents.[162–164] iTRAQ and TMT show good precision, but the accuracy is strongly affected by contamination from different sources and errors result from the compression of the ratios used for quantification proposes, mainly because of the cross-contamination of different peptide ions during the isolation process. Several experimental and technical approaches have been developed to minimise or overcome underestimation issues, including MS-based approaches (PTR and MS3) or software algorithms incorporating statistical tests to take into account underestimation.

TMTc-based relative quantification has emerged as a novel, low mass reporter ion-independent strategy for quantification.[64] Isobaric peptide termini labelling (IPTL) also avoids low mass reporter ion-based quantification and instead uses all of the product ions in the MS/MS spectrum.[165–167] A triplex version of this reagent has been reported recently.[168] The advantage of this approach is that the sequence-determining ions are usually free from contamination, but the number of samples that can be compared in a single experiment (multiplexing) is limited by the number of different reporter ions that can be generated by differential stable isotope incorporation.

In terms of increasing multiplex capability, whilst TMT reagents can be used for between eight to ten samples,[99] further increases pose significant technical challenges. For example, tag size is an important factor affecting the CID behaviour of labelled peptides,[60,156] probably reflecting the increased mass of the balance group and increased charging during ESI.[169] If new isobaric tags with increasing multiplexing rates are to be designed, they must have suitable sizes for efficient fragmentation.

In summary, iTRAQ and TMT remain valuable methodologies in the proteomics toolkit with promise for the future.

Acknowledgements

The authors would like to acknowledge financial support from Engineering and Physical Sciences Research Council, EPSRC (EP/E036252/1) and EPSRC ROADBLOCK (EP/I031812/1). Wen Qiu also acknowledges The UK–China Scholarships for Excellence Programme for financial support of her PhD. The authors also express their gratitude to Esther Karunakaran (PhD) for her help during the preparation of this manuscript.

References

1. N. Couto, J. Barber and S. J. Gaskell, *J. Mass Spectrom.*, 2011, **46**, 1233.
2. C. E. Eyers, C. Lawless, D. C. Wedge, K. W. Lau, S. J. Gaskell and S. J. Hubbard, *Mol. Cell. Proteomics*, 2011, **10**, M110.003384.
3. P. Rawlins, *Drug Discov. World*, 2010, **10**, 17.
4. S. A. Gerber, J. Rush, O. Stemman, M. W. Kirschner and S. P. Gygi, *Proc. Natl. Acad. Sci.*, 2003, **100**, 6940.
5. R. J. Beynon, M. K. Doherty, J. M. Pratt and S. J. Gaskell, *Nat. Methods*, 2005, **2**, 587.

6. M. Zeiler, W. L. Straube, E. Lundberg, M. Uhlen and M. Mann, *Mol. Cell. Proteomics*, 2012, **11**, O111.009613.
7. G. Picard, D. Lebert, M. Louwagie, A. Adrait, C. Huillet, F. Vandenesch, C. Bruley, J. Garin, M. Jaquinod and V. Brun, *J. Mass Spectrom.*, 2012, **47**, 1353.
8. C. Evans, J. Noirel, S. Y. Ow, M. Salim, A. G. Pereira-Medrano, N. Couto, J. Pandhal, D. Smith, T. K. Pham, E. Karunakaran, X. Zou, C. A. Biggs and P. C. Wright, *Anal. Bioanal. Chem.*, 2012, **404**, 1011.
9. J. Noirel, C. Evans, M. Salim, J. Mukherjee, S. Y. Ow, J. Pandhal, T. K. Pham, C. A. Biggs and P. C. Wright, *Curr. Proteomics*, 2011, **8**, 17.
10. A. Treumann and B. Thiede, *Expert Rev. Proteomics*, 2010, **7**, 647.
11. A. L. Christoforou and K. S. Lilley, *Anal. Bioanal. Chem.*, 2012, **404**, 1029.
12. G. L. Finney, A. R. Blackler, M. R. Hoopmann, J. D. Canterbury, C. C. Wu and M. J. MacCoss, *Anal. Chem.*, 2008, **80**, 961.
13. A. S. Benk and C. Roesli, *Anal. Bioanal. Chem.*, 2012, **404**, 1039.
14. G. Chen and B. N. Pramanik, *Drug Discovery Today*, 2009, **14**, 465.
15. P. A. Everley, J. Krijgsveld, B. R. Zetter and S. P. Gygi, *Mol. Cell. Proteomics*, 2004, **3**, 729.
16. S. E. Ong, B. Blagoev, I. Kratchmarova, D. B. Kristensen, H. Steen, A. Pandey and M. Mann, *Mol. Cell. Proteomics*, 2002, **1**, 376.
17. X. Yao, A. Freas, J. Ramirez, P. A. Demirev and C. Fenselau, *Anal. Chem.*, 2001, **73**, 2836.
18. K. J. Reynolds, X. Yao and C. Fenselau, *J. Proteome Res.*, 2002, **1**, 27.
19. K. L. Johnson and D. C. Muddiman, *J. Am. Chem. Soc.*, 2004, **15**, 437.
20. F. Y. Che, R. Biswas and L. D. Fricker, *J. Mass Spectrom.*, 2005, **40**, 227.
21. J. L. Hsu, S. Y. Huang, N. H. Chow and S. H. Chen, *Anal. Chem.*, 2003, **75**, 6843.
22. P. J. Boersema, R. Raijmakers, S. Lemeer, S. Mohammed and A. J. R. Heck, *Nat. Protoc.*, 2009, **4**, 484.
23. M. Geng, J. Ji and F. E. Regnier, *J. Chrom. A*, 2000, **870**, 295.
24. J. Ji, A. Chakraborty, M. Geng, X. Zhang, A. Amini, M. Bina and F. Regnier, *J. Chrom. B: Biomed. Sci. Appl.*, 2000, **745**, 197.
25. M. Münchbach, M. Quadroni, G. Miotto and P. James, *Anal. Chem.*, 2000, **72**, 4047.
26. S. P. Gygi, B. Rist, S. A. Gerber, F. Turecek, M. H. Gelb and R. Aebersold, *Nat. Biotechnol.*, 1999, **17**, 994.
27. R. Zhang, C. S. Sioma, R. A. Thompson, L. Xiong and F. E. Regnier, *Anal. Chem.*, 2002, **74**, 3662.
28. H. Zhou, J. A. Ranish, J. D. Watts and R. Aebersold, *Nat. Biotechnol.*, 2002, **20**, 512.
29. K. C. Hansen, G. Schmitt-Ulms, R. J. Chalkley, J. Hirsch, M. A. Baldwin and A. Burlingame, *Mol. Cell. Proteomics*, 2003, **2**, 299.
30. P. L. Ross, Y. N. Huang, J. N. Marchese, B. Williamson, K. Parker, S. Hattan, N. Khainovski, S. Pillai, S. Dey and S. Daniels, *Mol. Cell. Proteomics*, 2004, **3**, 1154.

31. L. Choe, M. D'Ascenzo, N. R. Relkin, D. Pappin, P. Ross, B. Williamson, S. Guertin, P. Pribil and K. H. Lee, *Proteomics*, 2007, 7, 3651.
32. L. Ting, R. Rad, S. P. Gygi and W. Haas, *Nat. Methods*, 2011, **8**, 937.
33. C. J. Koehler, M. Ø. Arntzen, M. Strozynski, A. Treumann and B. Thiede, *Anal. Chem.*, 2011, **83**, 4775.
34. G. C. McAlister, D. H. Phanstiel, J. Brumbaugh, M. S. Westphall and J. J. Coon, *Mol. Cell. Proteomics*, 2011, **10**, O111.009456.
35. D. Baeumlisberger, T. N. Arrey, B. Rietschel, M. Rohmer, D. G. Papasotiriou, B. Mueller, T. Beckhaus and M. Karas, *Proteomics*, 2010, **10**, 3905.
36. M. Rohmer, D. Baeumlisberger and M. Karas, *Int. J. Mass Spectrom.*, 2013, **345**, 37.
37. P. J. Boersema, S. Mohammed and A. J. R. Heck, *Anal. Bioanal. Chem.*, 2008, **391**, 151.
38. S. Di Palma, P. J. Boersema, A. J. R. Heck and S. Mohammed, *Anal. Chem.*, 2011, **83**, 3440.
39. H. L. Phillips, J. C. Williamson, K. A. van Elburg, A. P. Snijders, P. C. Wright and M. J. Dickman, *Proteomics*, 2010, **10**, 2950.
40. C. Song, M. Ye, G. Han, X. Jiang, F. Wang, Z. Yu, R. Chen and H. Zou, *Anal. Chem.*, 2009, **82**, 53.
41. C. M. Warren, D. L. Geenen, D. L. Helseth, H. Xu and R. J. Solaro, *J. Proteomics*, 2010, **73**, 1551.
42. N. C. Hubner, S. Ren and M. Mann, *Proteomics*, 2008, **8**, 4862.
43. Y. Yang, X. Qiang, K. Owsiany, S. Zhang, T. W. Thannhauser and L. Li, *J. Proteome Res.*, 2011, **10**, 4647.
44. D. N. Perkins, D. J. C. Pappin, D. M. Creasy and J. S. Cottrell, *Electrophoresis*, 1999, **20**, 3551.
45. H. Choi and A. I. Nesvizhskii, *J. Proteome Res.*, 2007, **7**, 47.
46. A. I. Nesvizhskii, O. Vitek and R. Aebersold, *Nat. Methods*, 2007, **4**, 787.
47. J. Listgarten and A. Emili, *Mol. Cell. Proteomics*, 2005, **4**, 419.
48. C. S. Gan, P. K. Chong, T. K. Pham and P. C. Wright, *J. Proteome Res.*, 2007, **6**, 821.
49. C. Murie and R. Nadon, *Bioinformatics*, 2008, **24**, 1735.
50. S. Pan, R. Chen, R. E. Brand, S. Hawley, Y. Tamura, P. R. Gafken, B. P. Milless, D. R. Goodlett, J. Rush and T. A. Brentnall, *J. Proteome Res.*, 2012, **11**, 1937.
51. S. Muraoka, H. Kume, S. Watanabe, J. Adachi, M. Kuwano, M. Sato, N. Kawasaki, Y. Kodera, M. Ishitobi and H. Inaji, *J. Proteome Res.*, 2012, **11**, 4201.
52. P. Mertins, N. D. Udeshi, K. R. Clauser, D. Mani, J. Patel, S. Ong, J. D. Jaffe and S. A. Carr, *Mol. Cell. Proteomics*, 2012, **11**, M111.014423.
53. S. Y. Ow, M. Salim, J. Noirel, C. Evans, I. Rehman and P. C. Wright, *J. Proteome Res.*, 2009, **8**, 5347.
54. J. Y. Yoon, J. Yeom, H. Lee, K. Kim, S. Na, K. Park, E. Paek and C. Lee, *BMC Bioinformatics*, 2011, **12**, S46.

55. J. Holzmann, P. Pichler, M. Madalinski, R. Kurzbauer and K. Mechtler, *Anal. Chem.*, 2009, **81**, 10254.
56. F. Pailleux and F. Beaudry, *Biomed. Chromatogr.*, 2012, **26**, 881.
57. D. Phanstiel, R. Unwin, G. C. McAlister and J. J. Coon, *Anal. Chem.*, 2009, **81**, 1693.
58. J. E. P. Syka, J. J. Coon, M. J. Schroeder, J. Shabanowitz and D. F. Hunt, *Proc. Natl. Acad. Sci. U. S. A.*, 2004, **101**, 9528.
59. R. I. Viner, T. Zhang, T. Second and V. Zabrouskov, *J. Proteomics*, 2009, **72**, 874.
60. P. Pichler, T. Köcher, J. Holzmann, M. Mazanek, T. Taus, G. Ammerer and K. Mechtler, *Anal. Chem.*, 2010, **82**, 6549.
61. Z. Li, R. M. Adams, K. Chourey, G. B. Hurst, R. L. Hettich and C. Pan, *J. Proteome Res.*, 2012, **11**, 1582.
62. P. M. Gehrig, P. E. Hunziker, S. Zahariev and S. Pongor, *J. Am. Soc. Mass Spectrom.*, 2004, **15**, 142.
63. L. Dayon, B. Sonderegger and M. Kussmann, *J. Proteome Res.*, 2012, **11**, 5081.
64. M. Wühr, W. Haas, G. C. McAlister, L. Peshkin, R. Rad, M. W. Kirschner and S. P. Gygi, *Anal. Chem.*, 2012, **84**, 9214.
65. C. D. Wenger, M. V. Lee, A. S. Hebert, G. C. McAlister, D. H. Phanstiel, M. S. Westphall and J. J. Coon, *Nat. Methods*, 2011, **8**, 933.
66. P. Hao, J. Qian, Y. Ren and S. K. Sze, *J. Proteome Res.*, 2011, **10**, 5568.
67. S. Y. Ow, M. Salim, J. Noirel, C. Evans and P. Wright, *Proteomics*, 2011, **11**, 2341.
68. M. M. Savitski, G. Sweetman, M. Askenazi, J. A. Marto, M. Lang, N. Zinn and M. Bantscheff, *Anal. Chem.*, 2011, **83**, 8959.
69. M. Bantscheff, M. Boesche, D. Eberhard, T. Matthieson, G. Sweetman and B. Kuster, *Mol. Cell. Proteomics*, 2008, **7**, 1702.
70. S. Y. Ow, J. Noirel, M. Salim, C. Evans, R. Watson and P. C. Wright, *Proteomics*, 2010, **10**, 2205.
71. A. Thompson, J. Schäfer, K. Kuhn, S. Kienle, J. Schwarz, G. Schmidt, T. Neumann and C. Hamon, *Anal. Chem.*, 2003, **75**, 1895.
72. R. E. March, *J. Mass Spectrom.*, 1997, **32**, 351.
73. R. E. March and J. F. Todd, *Quadrupole ion trap mass spectrometry*, Wiley-Interscience, Hoboken, New Jersey, 2005.
74. J. Wilson and R. W. Vachet, *Anal. Chem.*, 2004, **76**, 7346.
75. T. Guo, C. S. Gan, H. Zhang, Y. Zhu, O. L. Kon and S. K. Sze, *J. Proteome Res.*, 2008, **7**, 4831.
76. C. Cunningham, G. L. Glish and D. J. Burinsky, *J. Am. Soc. Mass Spectrom.*, 2006, **17**, 81.
77. Ü. A. Laskay and G. P. Jackson, *Rapid Commun. Mass Spectrom.*, 2008, **22**, 2342.
78. W. W. Wu, G. Wang, P. A. Insel, C. T. Hsiao, S. Zou, B. Martin, S. Maudsley and R. F. Shen, *J. Proteomics*, 2012, **75**, 2480.
79. P. Pichler, T. Köcher, J. Holzmann, T. Möhring, G. Ammerer and K. Mechtler, *Anal. Chem.*, 2011, **83**, 1469.

80. A. Michalski, E. Damoc, J. P. Hauschild, O. Lange, A. Wieghaus, A. Makarov, N. Nagaraj, J. Cox, M. Mann and S. Horning, *Mol. Cell. Proteomics*, 2011, **10**, O111.013698.
81. T. Köcher, P. Pichler, M. Schutzbier, C. Stingl, A. Kaul, N. Teucher, G. Hasenfuss, J. M. Penninger and K. Mechtler, *J. Proteome Res.*, 2009, **8**, 4743.
82. M. P. Jedrychowski, E. L. Huttlin, W. Haas, M. E. Sowa, R. Rad and S. P. Gygi, *Mol. Cell. Proteomics*, 2011, **10**, M111.009910.
83. L. Dayon, C. Pasquarello, C. Hoogland, J. C. Sanchez and A. Scherl, *J. Proteomics*, 2009, **73**, 769.
84. J. V. Olsen, B. Macek, O. Lange, A. Makarov, S. Horning and M. Mann, *Nat. Methods*, 2007, **4**, 709.
85. N. Nagaraj, R. C. J. D'Souza, J. Cox, J. V. Olsen and M. Mann, *J. Proteome Res.*, 2010, **9**, 6786.
86. M. Rohmer, D. Baeumlisberger, B. Stahl, U. Bahr and M. Karas, *Int. J. Mass Spectrom.*, 2011, **305**, 199.
87. A. Michalski, E. Damoc, O. Lange, E. Denisov, D. Nolting, M. Müller, R. Viner, J. Schwartz, P. Remes and M. Belford, *Mol. Cell. Proteomics*, 2012, **11**, O111.013698.
88. C. D. Kelstrup, C. Young, R. Lavallee, M. L. Nielsen and J. V. Olsen, *J. Proteome Res.*, 2012, **11**, 3487.
89. E. Denisov, E. Damoc, O. Lange and A. Makarov, *Int. J. Mass Spectrom.*, 2012, **325–327**, 80.
90. H. Han, D. J. Pappin, P. L. Ross and S. A. McLuckey, *J. Proteome Res.*, 2008, 7, 3643.
91. D. Phanstiel, Y. Zhang, J. A. Marto and J. J. Coon, *J. Am. Soc. Mass Spectrom.*, 2008, **19**, 1255.
92. M. L. Hennrich, P. J. Boersema, H. van den Toorn, N. Mischerikow, A. J. R. Heck and S. Mohammed, *Anal. Chem. Columbus*, 2009, **81**, 7814.
93. F. Yang, S. Wu, D. L. Stenoien, R. Zhao, M. E. Monroe, M. A. Gritsenko, S. O. Purvine, A. D. Polpitiya, N. Tolic and Q. Zhang, *Anal. Chem.*, 2009, **81**, 4137.
94. N. Mischerikow, P. van Nierop, K. W. Li, H. G. Bernstein, A. B. Smit, A. J. Heck and A. F. Altelaar, *Analyst*, 2010, **135**, 2643.
95. D. L. Swaney, G. C. McAlister and J. J. Coon, *Nat. Methods*, 2008, **5**, 959.
96. A. R. Ledvina, M. V. Lee, G. C. McAlister, M. S. Westphall and J. J. Coon, *Anal. Chem.*, 2012, **84**, 4513.
97. M. S. Kim, K. Kandasamy, R. Chaerkady and A. Pandey, *J. Am. Soc. Mass Spectrom.*, 2010, **21**, 1606.
98. G. C. McAlister, E. L. Huttlin, W. Haas, L. Ting, M. P. Jedrychowski, J. C. Rogers, K. Kuhn, I. Pike, R. A. Grothe and J. D. Blethrow, *Anal. Chem.*, 2012, **84**, 7469.
99. T. Werner, I. Becher, G. Sweetman, C. Doce, M. M. Savitski and M. Bantscheff, *Anal. Chem.*, 2012, **84**, 7188.

100. E. G. Hill, J. H. Schwacke, S. Comte-Walters, E. H. Slate, A. L. Oberg, J. E. Eckel-Passow, T. M. Therneau and K. L. Schey, *J. Proteome Res.*, 2008, 7, 3091.
101. J. H. Schwacke, E. G. Hill, E. L. Krug, S. Comte-Walters and K. L. Schey, *BMC Bioinformatics*, 2009, **10**, 342.
102. C. Hundertmark, R. Fischer, T. Reinl, S. May, F. Klawonn and L. Jänsch, *Bioinformatics*, 2009, **25**, 1004.
103. N. A. Karp, W. Huber, P. G. Sadowski, P. D. Charles, S. V. Hester and K. S. Lilley, *Mol. Cell. Proteomics*, 2010, **9**, 1885.
104. J. Noirel, G. Sanguinetti and P. C. Wright, *Bioinformatics*, 2008, **24**, 2792.
105. Y. V. Karpievitch, T. Taverner, J. N. Adkins, S. J. Callister, G. A. Anderson, R. D. Smith and A. R. Dabney, *Bioinformatics*, 2009, **25**, 2573.
106. B. Dost, N. Bandeira, X. Li, Z. Shen, S. P. Briggs and V. Bafna, *J. Comput. Biol.*, 2012, **19**, 337.
107. M. Blein-Nicolas, H. Xu, D. de Vienne, C. Giraud, S. Huet and M. Zivy, *Proteomics*, 2012, **12**, 2797.
108. M. M. Savitski, F. Fischer, T. Mathieson, G. Sweetman, M. Lang and M. Bantscheff, *J. Am. Soc. Mass Spectrom.*, 2010, **21**, 1668.
109. F. P. Breitwieser, A. Müller, L. Dayon, T. Köcher, A. Hainard, P. Pichler, U. Schmidt-Erfurth, G. Superti-Furga, J. C. Sanchez and K. Mechtler, *J. Proteome Res.*, 2011, **10**, 2758.
110. J. A. Latham and S. Y. R. Dent, *Nat. Struct. Mol. Biol.*, 2007, **14**, 1017.
111. X. J. Yang and E. Seto, *Mol. Cell*, 2008, **31**, 449.
112. J. S. Lee, E. Smith and A. Shilatifard, *Cell*, 2010, **142**, 682.
113. H. C. Beck, E. C. Nielsen, R. Matthiesen, L. H. Jensen, M. Sehested, P. Finn, M. Grauslund, A. M. Hansen and O. N. Jensen, *Mol. Cell. Proteomics*, 2006, **5**, 1314.
114. F. M. White, *Curr. Opin. Biotechnol.*, 2008, **19**, 404.
115. S. E. Ong, G. Mittler and M. Mann, *Nat. Methods*, 2004, **1**, 119.
116. C. Peng, S. Tang, C. Pi, J. Liu, F. Wang, L. Wang, W. Zhou and A. Xu, *Peptides*, 2006, **27**, 2174.
117. Y. Zhao and O. N. Jensen, *Proteomics*, 2009, **9**, 4632.
118. M. Mann, S. E. Ong, M. Gronborg, H. Steen, O. N. Jensen and A. Pandey, *Trends Biotechnol.*, 2002, **20**, 261.
119. H. Zhang, W. Yan and R. Aebersold, *Curr. Opin. Chem. Biol.*, 2004, **8**, 66.
120. J. Wu, P. Warren, Q. Shakey, E. Sousa, A. Hill, T. E. Ryan and T. He, *Proteomics*, 2010, **10**, 2224.
121. W. Jia, A. Andaya and J. A. Leary, *Anal. Chem.*, 2012, **84**, 2466.
122. T. E. Thingholm, O. N. Jensen and M. R. Larsen, *Proteomics*, 2009, **9**, 1451.
123. J. Lee, Y. Xu, Y. Chen, R. Sprung, S. C. Kim, S. Xie and Y. Zhao, *Mol. Cell. Proteomics*, 2007, **6**, 669.
124. K. Moser and F. M. White, *J. Proteome Res.*, 2006, **5**, 98.

125. A. Leitner, *Trends Anal. Chem.*, 2010, **29**, 177.
126. S. Feng, M. Ye, H. Zhou, X. Jiang, X. Jiang, H. Zou and B. Gong, *Mol. Cell. Proteomics*, 2007, **6**, 1656.
127. M. R. Larsen, T. E. Thingholm, O. N. Jensen, P. Roepstorff and T. J. D. Jørgensen, *Mol. Cell. Proteomics*, 2005, **4**, 873.
128. J. Rush, A. Moritz, K. A. Lee, A. Guo, V. L. Goss, E. J. Spek, H. Zhang, X. M. Zha, R. D. Polakiewicz and M. J. Comb, *Nat. Biotechnol.*, 2004, **23**, 94.
129. R. Narumi, T. Murakami, T. Kuga, J. Adachi, T. Shiromizu, S. Muraoka, H. Kume, Y. Kodera, M. Matsumoto and K. Nakayama, *J. Proteome Res.*, 2012, **11**, 5311.
130. W. Reiter, D. Anrather, I. Dohnal, P. Pichler, J. Veis, M. Grøtli, F. Posas and G. Ammerer, *Proteomics*, 2012, **12**, 3030.
131. D. Pflieger, M. A. Jünger, M. Müller, O. Rinner, H. Lee, P. M. Gehrig, M. Gstaiger and R. Aebersold, *Mol. Cell. Proteomics*, 2008, **7**, 326.
132. L. Zhou, R. W. Beuerman, A. P. Chew, S. K. Koh, T. A. Cafaro, E. A. Urrets-Zavalia, J. A. Urrets-Zavalia, S. F. Y. Li and H. M. Serra, *J. Proteome Res.*, 2009, **8**, 1992.
133. H. Zhang, X. Li, D. B. Martin and R. Aebersold, *Nat. Biotechnol.*, 2003, **21**, 660.
134. K. Ueda, S. Takami, N. Saichi, Y. Daigo, N. Ishikawa, N. Kohno, M. Katsumata, A. Yamane, M. Ota and T. A. Sato, *Mol. Cell. Proteomics*, 2010, **9**, 1819.
135. J. Pandhal, S. Y. Ow, J. Noirel and P. C. Wright, *Biotechnol. Bioeng.*, 2011, **108**, 902.
136. H. Wang, S. Alvarez and L. M. Hicks, *J. Proteome Res.*, 2012, **11**, 487.
137. J. Longworth, J. Noirel, J. Pandhal, P. C. Wright and S. Vaidyanathan, *J. Proteome Res.*, 2012, **11**, 5959.
138. S. S. Adav, C. S. Ng and S. K. Sze, *J. Proteomics*, 2011, **74**, 2112.
139. S. S. Adav, A. A. Li, A. Manavalan, P. Punt and S. K. Sze, *J. Proteome Res.*, 2010, **9**, 3932.
140. A. G. Pereira-Medrano, M. Knighton, G. J. S. Fowler, Z. Y. Ler, T. K. Pham, S. Y. Ow, A. Free, B. Ward and P. C. Wright, *J. Proteomics*, 2012, **78**, 197.
141. J. Lee, L. Cao, S. Y. Ow, M. E. Barrios-Llerena, W. Chen, T. K. Wood and P. C. Wright, *J. Proteomics Res.*, 2006, **5**, 1388.
142. J. Lee, S. R. Hiibel, K. F. Reardon and T. K. Wood, *J. Appl. Microbiol.*, 2010, **108**, 2088.
143. S. Cuello, P. Ximénez-Embún, I. Ruppen, H. B. Schonthaler, K. Ashman, Y. Madrid, J. L. Luque-Garcia and C. Cámara, *Analyst*, 2012, **137**, 5302.
144. S. Lemeer, C. Jopling, J. Gouw, S. Mohammed, A. J. Heck, M. Slijper and J. den Hertog, *Mol. Cell. Proteomics*, 2008, **7**, 2176.
145. A. Glen, C. A. Evans, C. S. Gan, S. S. Cross, F. C. Hamdy, J. Gibbins, J. Lippitt, C. L. Eaton, J. Noirel, P. C. Wright and I. Rehman, *Prostate*, 2010, **70**, 1313.

146. I. Rehman, C. A. Evans, A. Glen, S. S. Cross, C. L. Eaton, J. Down, G. Pesce, J. T. Phillips, O. S. Yen, G. N. Thalmann, P. C. Wright and F. C. Hamdy, *PLoS One*, 2012, 7, e30885.
147. D. Y. Ouyang, Y. H. Ji, M. Saltis, L. H. Xu, Y. T. Zhang, Q. B. Zha, J. Y. Cai and X. H. He, *J. Proteomics*, 2011, **74**, 2180.
148. S. D. Garbis, S. I. Tyritzis, T. Roumeliotis, P. Zerefos, E. G. Giannopoulou, A. Vlahou, S. Kossida, J. Diaz, S. Vourekas, C. Tamvakopoulos, K. Pavlakis, D. Sanoudou and C. A. Constantinides, *J. Proteome Res.*, 2008, 7, 3146.
149. C. Sun, C. Song, Z. Ma, K. Xu, Y. Zhang, H. Jin, S. Tong, W. Ding, G. Xia and Q. Ding, *Proteome Sci.*, 2011, **9**, 22.
150. J. Zhang, K. Nkhata, A. A. Shaik, L. Wang, L. Li, Y. Zhang, L. A. Higgins, K. H. Kim, J. D. Liao, C. Xing, S. H. Kim and J. Lu, *Curr. Cancer Drug Targets*, 2011, **11**, 787.
151. A. Das, J. J. D. Bortner, C. A. Aliaga, A. Baker, A. Stanley, B. A. Stanley, M. Kaag, J. J. P. Richie and K. El-Bayoumy, *Prostate*, 2012, **73**, 363.
152. J. Sinclair and J. F. Timms, *Methods*, 2011, **54**, 361.
153. S. Zhang, X. Liu X, X. Kang, C. Sun, H. Lu, P. Yang and Y. Liu, *Talanta*, 2012, **91**, 122.
154. N. L. Anderson and N. G. Anderson, *Mol. Cell. Proteomics*, 2002, **1**, 845.
155. J. E. Bandow, *Proteomics*, 2010, **10**, 1416.
156. G. Pottiez, J. Wiederin, H. S. Fox and P. Ciborowski, *J. Proteome Res.*, 2012, **11**, 3774.
157. C. Zhou, K. L. Simpson, L. J. Lancashire, M. J. Walker, M. J. Dawson, R. D. Unwin, A. Rembielak, P. Price, C. West, C. Dive and A. D. Whetton, *J. Proteome Res.*, 2012, **11**, 2103.
158. L. Yan, S. Tonack, R. Smith, S. Dodd, R. E. Jenkins, N. Kitteringham, W. Greenhalf, P. Ghaneh, J. P. Neoptolemos and E. Costello, *J. Proteome Res.*, 2009, **8**, 142.
159. A. Prudova, U. auf dem Keller, G. S. Butler and C. M. Overall, *Mol. Cell. Proteomics*, 2010, **9**, 894.
160. W. Yuan, J. Zhang, S. Li and J. L. Edwards, *J. Proteome Res.*, 2011, **10**, 5242.
161. S. Li and D. Zeng, *Chem. Commun.*, 2007, 2181.
162. H. S. Chung, S. B. Wang, V. Venkatraman, C. I. Murray and J. E. Van Eyk, *Circ. Res.*, 2013, **112**, 382.
163. P. T. Doulias, M. Tenopoulou, J. L. Greene, K. Raju and H. Ischiropoulos, *Sci. Signaling*, 2013, **6**, rs1.
164. C. I. Murray, H. Uhrigshardt, R. N. O'Meally, R. N. Cole and J. E. Van Eyk, *Mol. Cell. Proteomics*, 2012, **11**, M111.013441.
165. C. J. Koehler, M. Strozynski, F. Kozielski, A. Treumann and B. Thiede, *J. Proteome Res.*, 2009, **8**, 4333.
166. M. Arntzen, C. J. Koehler, A. Treumann and B. Thiede, *Methods Mol. Biol.*, 2011, **753**, 65.

167. C. J. Koehler, M. Ø. Arntzen, A. Treumann and B. Thiede, *Anal. Bioanal. Chem.*, 2012, **404**, 1103.
168. C. J. Koehler, M. Ø. Arntzen, G. A. de Souza and B. Thiede, *Anal. Chem.*, 2013, **85**, 2487.
169. T. E. Thingholm, G. Palmisano, F. Kjeldsen and M. R. Larsen, *J. Proteome Res.*, 2010, **9**, 4045.

CHAPTER 4

Getting Absolute: Determining Absolute Protein Quantities via Selected Reaction Monitoring Mass Spectrometry

CHRISTINA LUDWIG*[a] AND RUEDI AEBERSOLD[a,b]

[a] Institute of Molecular Systems Biology, Department of Biology, ETH Zurich, 8093 Zurich, Switzerland; [b] Faculty of Science, University of Zurich, Switzerland
*Email: ludwig@imsb.biol.ethz.ch

4.1 Introduction

4.1.1 Why Do We Need Absolute Protein Quantification?

Accurate quantification of proteins is important for a wide range of questions in molecular and cell biology, systems biology, and clinical research. The development of quantitative proteomics techniques has been a major advance in the field over the last decade. Depending on the specific research question asked, either relative protein changes, where the relative abundance of a protein is compared between samples, or absolute quantification, where the concentration of specific proteins is determined in a sample, are most informative.

The result of relative quantification is a dimensionless protein ratio describing an abundance change of the same protein between a usually limited

New Developments in Mass Spectrometry No. 1
Quantitative Proteomics
Edited by Claire Eyers and Simon J Gaskell

Published by the Royal Society of Chemistry, www.rsc.org

number of samples. The vast majority of quantitative protein data generated to date uses relative quantification for extensive comparisons of protein profiles between, for example, cellular states,[1] organisms[2] and disease states.[3] However, for samples of different compositions and sample amounts, statistical constraints complicate meaningful relative quantification[4] and, therefore, relative quantification is most frequently limited to relatively few samples that are prepared as uniformly as possible within the scope of a specific study.

Absolute protein abundances are typically expressed in units like "protein copies per cell", "picomoles of protein per volume of body fluid" or "protein weight per tissue or cell extract weight" and a wide range of scientific questions require information about the absolute quantity of a protein in a given sample (Figure 4.1). For example, studies on cellular protein complexes and their subunit stoichiometry require absolute concentrations of all complex-forming protein components[5] and the stoichiometry of a complex can, therefore, not be extracted from relative quantitative data. Further, mathematical modeling to simulate and predict the behaviour of biological processes, an essential component of systems biology studies,[6] is dependent on the absolute protein concentrations to determine, for example, rate constants or kinetic fluxes. As in every relative proteomics study, protein abundances are relative to an arbitrary "reference sample", comparisons

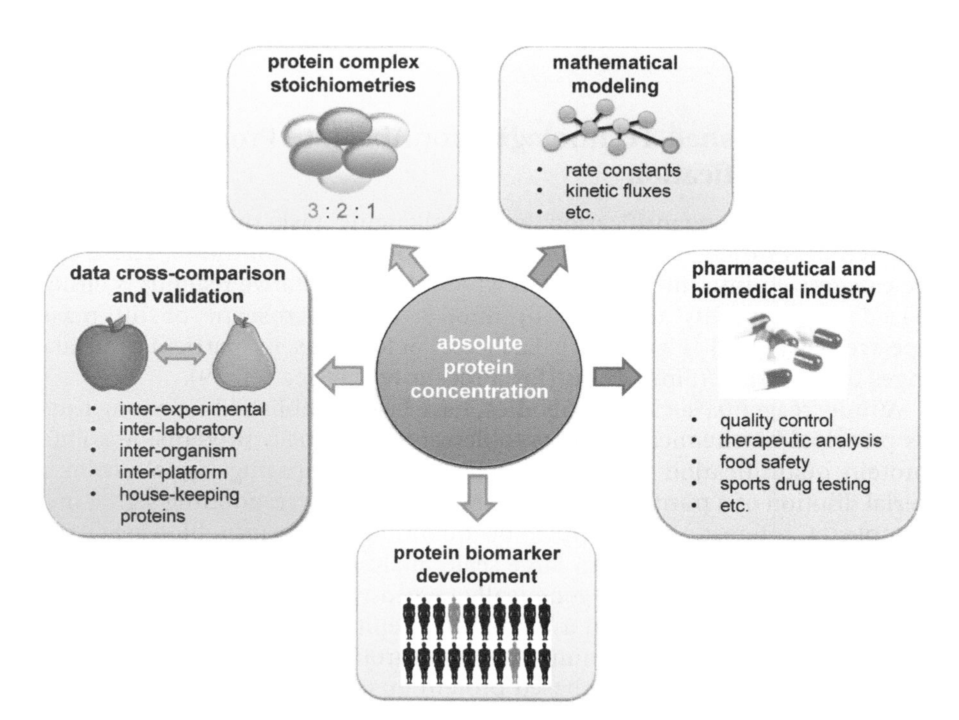

Figure 4.1 Fields of application for absolute quantitative protein data.

between datasets generated with different "reference samples" are impossible. Therefore, a third important advantage of absolute data is its suitability for comprehensive comparisons across datasets. As illustrated by several recent studies, information about absolute protein quantities offers the unique possibility not only to compare proteomes at different experimental states,[7] but to also correlate protein abundances with other data types, such as absolute mRNA abundances and protein turn-over rates.[8–10] Furthermore, inter-organism, inter-laboratory or inter-proteomic platform comparisons become feasible.[11] Therefore, absolute protein quantification is well suited for data validation, standardization and the identification of stable and invariant "benchmarking" or "housekeeping" proteins for normalization.[12,13] In addition, discovery and development of new therapeutic and diagnostic protein biomarkers benefit from absolute protein quantification.[14] Knowledge about the absolute concentration of a protein biomarker in a patient's blood or urine is important for defining thresholds and cut-offs for clinical biomarker tests and for diagnostic standardization. Finally, precise and accurate absolute protein quantification is of interest to the biomedical and pharmaceutical industry, for instance, in areas such as therapeutic proteins,[15] food safety,[16] food allergies[17] or sports drug testing.[18]

In summary, a wide range of important applications and benefits arise from the absolute measurements of protein abundances compared to relative quantitative values, justifying the usually more time-, cost- and labor-intensive workflows required.

4.1.2 Established Technologies for Absolute Protein Quantification

Absolute protein quantification is a challenging task that has been addressed by a variety of experimental technologies and workflows. These can be categorized into three groups, in which the quantitative readout is either based on i) affinity reagents, ii) fluorescence microscopy or iii) mass spectrometry (MS) (Figure 4.2). Each technology has its particular advantages and technical limitations (for a recent review see ref. 19).

Affinity reagents, such as antibodies, have been established for a long time as principal biochemical reagents to detect and quantify proteins. Absolute protein quantification is typically achieved by comparing signals from a serial dilution of a purified protein standard to the corresponding signal in a sample of interest, for example, by quantitative Western blotting.[20] An adaption of this technique to large-scale analyses has been carried out by Ghaemmaghami *et al.*, who genetically tagged each open reading frame in *Saccharomyces cerevisiae* with a high-affinity epitope and absolutely quantified protein amounts by immunodetection through this common tag.[21] The major advantage of antibody-based protein quantification is the potentially high sensitivity and accuracy if an optimally performing antibody is available. Hence, protein quantification with affinity reagents is limited by the

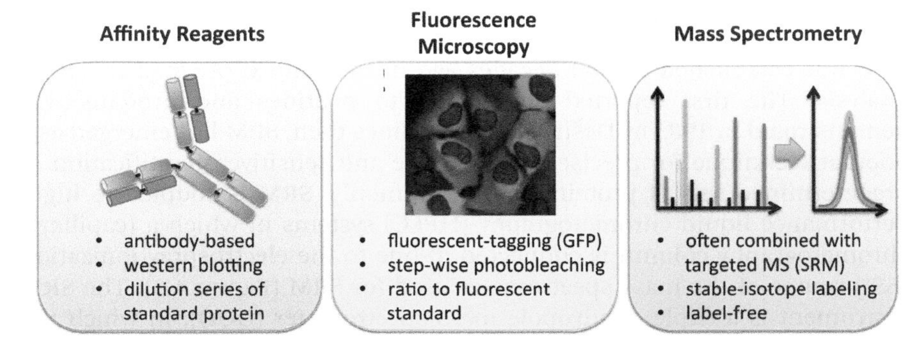

Figure 4.2 Established technologies for absolute protein quantification. (GFP: green fluorescent protein; MS: mass spectrometry; SRM: selected reaction monitoring).

availability and quality of specific antibodies. Furthermore, quantitative biases due to signal saturation must be controlled carefully by performing quantitative Western blot analyses with optimal sample amounts.

In recent years, absolute quantification of cellular proteins using fluorescence microscopy has become feasible. Two major methods are being used wherein quantification is either based on stepwise photobleaching or on ratio comparisons to a known fluorescent protein standard (for a recent review see ref. 22). Fluorescence microscopy measurements offer the unique possibility to study molecular processes at the single-cell or even single-molecule level. Drawbacks are the arduous creation of genetically modified strains in which the target gene needs to be fused to a fluorescent tag, limiting this technology to organisms that allow genetic manipulation. Furthermore, the fluorescent tag can potentially impact protein function, quantity, stability and localization and thereby lead to experimental artefacts.

The most widely used technology for absolute protein quantification is based on tandem mass spectrometry coupled to liquid chromatography (LC-MS/MS) (for reviews see refs. 23–26) and this will be discussed in more detail in the following sections.

4.2 Targeted Mass Spectrometry

The large and diverse field of MS includes a multitude of different instrument types and operation modes, such as discovery-driven (shotgun) and directed MS, data-independent acquisition methods, and targeted MS. Each technology and platform has its advantages and limitations, as reviewed by Domon and Aebersold.[27] The targeted proteomic method selected reaction monitoring (SRM), also referred to as multiple reaction monitoring (MRM), has proven most suitable for absolute protein quantification and has been applied most frequently where quantification precision, accuracy, sensitivity, repeatability and reproducibility are important.[19,28,29]

4.2.1 Principles of SRM

SRM was established several decades ago in the context of small molecule analysis. The first reported application to peptides and proteins was demonstrated in 1983 by Desiderio *et al.*[30] Since then, SRM has emerged as a popular technique for precise, reproducible and sensitive quantification of predetermined sets of proteins. Most commonly, SRM is coupled to high-performance liquid chromatography (HPLC) systems in which a (capillary) chromatography column is connected in-line to the electrospray ionization (ESI) source of the mass spectrometer used for SRM (Figure 4.3). The SRM instrument is a triple quadrupole mass spectrometer (QQQ), in which the first and third quadrupole are used as mass filters with selection windows typically in the range of 0.7–1 Da. The second quadrupole serves as the collision cell. For an SRM experiment, the QQQ is pre-programmed such that the first quadrupole, at a given time, selectively transmits precursor ions of a given *m*/*z* value to the collision cell. After collision-induced dissociation (CID) in Q2, one or more specific product ions are selectively transmitted through the third quadrupole. During the selections in Q1 and Q3 all ions outside the specified values are discarded. Typically, the specific windows for a given precursor–product ion pair termed "transition" are selected for a time period termed "dwell time" in the millisecond scale, after which the instrument switches to the acquisition of the next ion. A detector finally counts the number of product ions over time, resulting in an MS peak that is associated with a specific chromatographic retention time and an intensity value. All ions that are generated but are not on the target list will not be detected. The process of precursor ion selection, fragmentation, product ion selection and signal recording is then repeated, either over the whole chromatographic range or a segment thereof (scheduled SRM) with a

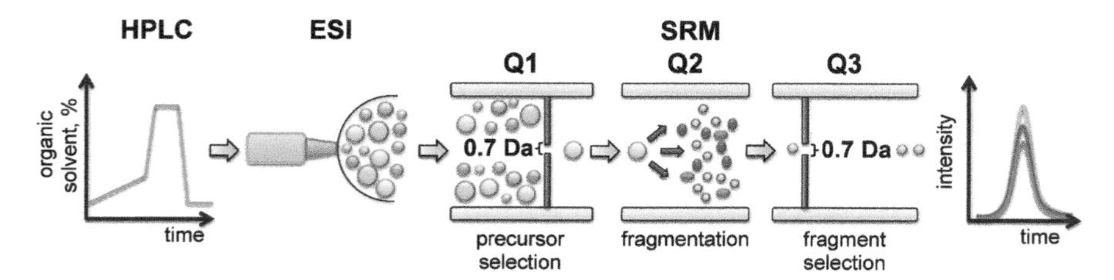

Figure 4.3 Principle of SRM using a triple quadrupole mass spectrometer. In a typical SRM experiment, a complex peptide sample is chromatographically separated on an HPLC column that is connected in-line to an ESI source. Through a narrow *m*/*z* filter (typically in the range of 0.7 Da) in the first quadrupole (Q1), fragmentation (Q2) and a second *m*/*z* filter (0.7 Da) in the third quadrupole (Q3), a single transition gets recorded by a detector over time. Commonly, several transitions are monitored per peptide and several peptides per protein. (HPLC: high-performance liquid chromatography; ESI: electrospray ionization. Figure modified from ref. 19.)

periodicity termed "cycle time" that allows the reconstruction of a peak for each product ion along the chromatographic time axis (Figure 4.3).

In a typical proteomic SRM experiment several transitions are recorded for a given peptide (typically three to six) and several peptides per protein are monitored (typically two to five). The number of recordable peptides and transitions within a single SRM run is limited by two parameters. First, the dwell time; the signal-to-noise ratio of the detected peaks is, to some extent, dependent on the period of time spent acquiring a specific transition signal. To achieve a good signal-to-noise ratio, dwell times in the range of 10 to 100 ms are typically applied. Second, the cycle time; to achieve accurate reconstruction of the intensity profile across the time axis, and thus to support accurate quantification, at least eight to ten data points across the chromatographic peak are required to reconstitute the peak shape.[29] That means that with a typical chromatographic peak width for a peptide of *ca.* 30 s the same precursor ion window needs to be selected approximately every 3 s. However, it should be noted that a peptide's peak width depends on the chromatographic system used, the applied gradient settings and the concentration of the peptide. Conclusively, the maximal number of recordable transitions within a single SRM run can be increased either by decreasing the dwell time (reduced limit of detection) or by increasing the cycle time (reduced quantitative accuracy). Therefore, in every SRM experiment an optimum between the number of monitored transitions, cycle and dwell time must be found. An approach used to significantly increase the number of targetable proteins, peptides and transitions in a single run is scheduled SRM.[31] In scheduled SRM transitions for a specific peptide are exclusively monitored in a time window around its expected chromatographic elution time, which allows routine measurements in the order of 1000 transitions per run with good analytical performance.[19]

Of final note, SRM represents an exclusively hypothesis-driven method, meaning that, for each protein of interest, *a priori* knowledge (SRM assay) needs to be available in order to define the transitions and program the QQQ instrument and to subsequently analyze the data. No new proteins or peptides are being discovered using such an SRM approach.

4.2.2 The SRM Assay

To perform an SRM experiment, *a priori* information about the proteins of interest is strictly required. This information includes the sequence of peptides uniquely representing the targeted protein, peptide fragmentation behaviour, suitable transitions, chromatographic retention time and optimal QQQ instrument settings.

The entity of such information is called an "SRM assay". Design, optimization and validation of peptide-specific SRM assays upfront of any SRM measurement are often the most time-, cost- and labor-consuming steps of a project. The selection of several representative peptides for a given protein is of critical importance as those peptides have to reflect the true protein abundance and provide optimal sensitivity and precision for the SRM measurement (for a detailed discussion, see section 4.4.3). After peptide selection, an SRM assay needs to be generated for each peptide individually. The assay consists of: i) precursor–product ion (transition) coordinates, ii) collision energy, iii) chromatographic retention time information and iv) relative intensities of the selected transitions per peptide. While transition coordinates and collision energies are strictly required, retention time information is only needed for scheduled SRM acquisition.[31] Furthermore, retention time and relative transition intensities are valuable parameters for unambiguous peptide identification during data analysis.

How are SRM assays generated in practice? As mentioned above, SRM assay generation necessitates selections on three levels, including i) the protein target set, ii) optimal peptides representing each target protein and iii) suitable transitions for each peptide. While the selection of proteins is exclusively project-dependent and might be carried out based on previous experiments or literature knowledge, several general strategies for peptide and transition selection at high throughput and reasonable costs have been reported.

Optimal peptides representing a protein of interest can be extracted, for example, from large proteomic repositories, such as "PeptideAtlas"[32] or "PRIDE".[33] However, the varying data quality within such databases, as well as technological differences between various instrument platforms and measurement modes, leading to differing precursor charge state distributions and fragmentation patterns, might be problematic. To alleviate some of these variables, the PeptideAtlas database re-computes all submitted datasets in a consistent manner using the Trans-Proteomic Pipeline software suite.[34] Alternatively, optimal peptides can be obtained from *in vitro* synthesized complete proteins.[35] In addition to peptide selection based on such empirical strategies, bioinformatics tools can also be used as a starting point for optimal peptide selection,[36–38] which are especially useful if no empirical data is available. Such algorithms are typically trained on large-scale datasets from which optimal physicochemical properties for peptide detection are experimentally determined and employ machine-learning approaches to predict the most suitable peptides for any given protein solely based on the protein sequence.

Selection of the most intense transitions per peptide is typically also carried out based on experimental data, exploiting, for example, the same repositories or experiments as used for the peptide selection. Another high-throughput strategy uses crude synthetic peptide libraries that are analyzed directly on QQQ instruments permitting high-quality SRM assays to be designed in a reasonable timeframe.[39] Recently, with this strategy a compendium of reference spectra for >97% of the predicted *Saccharomyces cerevisiae* open reading frames was published, demonstrating the possibility of generating assay libraries for whole proteomes.[40] Furthermore, where no experimental peptide fragmentation information is available, bioinformatics tools have been established for transition selection using data-driven machine-learning approaches,[41,42] as well as for the determination of transition uniqueness within a given sample background.[43] Importantly, the success rate of detecting peptides and transitions based on prediction tools is typically lower than when it is based on empirical data. However, in cases where no previous experimental data is available, prediction algorithms remain valuable and time-saving.

Another important SRM assay parameter is the peptide-specific chromatographic elution time, which is a strictly required coordinate to carry out scheduled SRM. Peptide elution times can either be predicted[44] or determined experimentally, the latter leading to more accurate retention time coordinates. By reporting peptide elution times relative to a set of standard peptides it becomes possible to reproduce and exchange experimental values between experiments, platforms and laboratories.[45]

Once an SRM assay has been established, optimized and validated it can be used universally. To avoid replicating assay generation, huge efforts are currently underway to build up publicly accessible databases for the storage of validated SRM assays (www.srmatlas.org),[46] as well as for the collection and representation of SRM data (www.peptideatlas.org/passel[47] and https://panoramaweb.org).

4.2.3 Why is SRM Suited for Absolute Protein Quantification?

Compared to other MS techniques, SRM is particularly well suited for precise and accurate absolute quantification of predetermined proteins of interest. With its two levels of mass filtering (Q1 and Q3), SRM achieves high selectivity and sensitivity, excellent signal-to-noise levels, and a wide dynamic quantification range spanning four to five orders of magnitude down to concentrations in the low attomole range.[48] Importantly, compared to non-targeted MS technologies, these favourable performance characteristics are only minimally affected by sample complexity. Further, in an SRM measurement the complete measurement time (duty cycle) is focused exclusively on the pre-selected peptide and product ions of interest. Hence, SRM measurements provide data of unprecedented reproducibility. In a comprehensive multi-laboratory study the excellent reproducibility and precision of SRM measurements have been demonstrated across laboratories and instrument platforms.[11]

4.2.4 Next-generation Targeted Mass Spectrometry

SRM is the most mature technology for targeted quantitative proteomics. Recently, new strategies have been emerging in the field, which focus on the usage of high-resolution and high-mass-accuracy MS instruments for targeted protein quantification. A developed technology by Gillet *et al.* termed "SWATH-MS" uses high resolution and high mass accuracy in combination with data-independent acquisition on a hybrid quadrupole time-of-flight instrument for proteome-wide targeted analysis.[49] In another approach called parallel reaction monitoring (PRM), targeted measurements are carried out using an Orbitrap mass analyzer equipped with a quadrupole at the front-end for precursor ion selection.[50,133]

With both of these technologies the acquired data consists of transition intensities recorded across the chromatographic time axis. These can be used for absolute protein quantification in the same way as SRM data. The major advantage of SWATH-MS comes from its unlimited number of targetable proteins in a single run, allowing truly proteome-wide analyses, as peptide ions are not pre-selected for monitoring. Also, with PRM, a large number of proteins can be targeted and, due to the higher mass accuracy and resolution of the instrumentation used, better specificity of measured transitions can be achieved, leading to fewer interferences and false-positive identifications.

4.3 Absolute Protein Quantification Combined with SRM

MS-based absolute protein quantification techniques can be divided into two categories depending on whether they quantify proteins based on stable isotope labeling or operate label-free (Figure 4.4). In the following section, we discuss the two strategies.

4.3.1 Stable Isotope Labeling

Stable isotope labeling in combination with SRM is widely accepted as the "gold standard" for absolute protein quantification. Its underlying principle is the addition of a heavy stable isotope-labeled reference molecule (standard) at a known amount to the sample of interest. The standard must resemble as closely as possible the targeted analyte in its physico-chemical properties. This is best achieved if the standard is a chemically identical but isotopically labeled form of the analyte (typically using ^{13}C and/or ^{15}N stable isotopes), making the analyte and standard distinguishable by their molecular weights provided that the mass difference is sufficiently large and can be resolved by the applied mass spectrometer without interference.

Absolute quantification is based on the ratio between the SRM signals of the analyte and standard (area under the curve). As the concentration of the standard added to the sample is known, the absolute concentration of the analyte can be directly calculated through linear correlation. Hereby it is important to consider that the required linear dependency between the SRM signal intensity and absolute concentration only holds true within a certain dynamic range of the mass spectrometer. Therefore, for each analyte of interest, the linear dynamic quantification range should be determined *a priori* by serial dilution experiments. Furthermore, it is recommendable

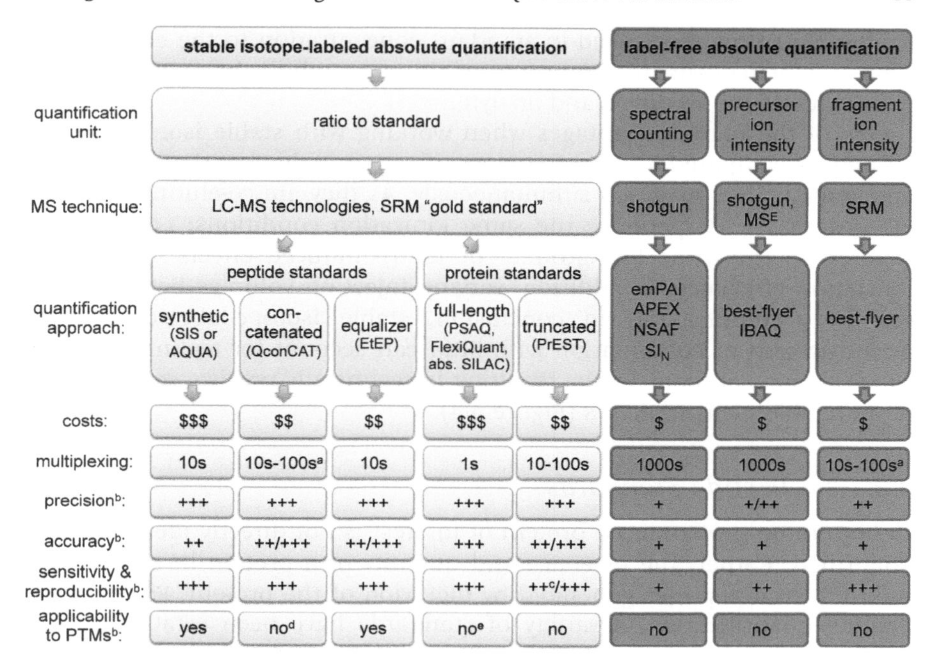

Figure 4.4 **MS-based absolute quantification approaches.** Absolute protein quantification can be achieved based on stable isotope labeling, for which known amounts of either peptide or protein standards are spiked into the analyte sample. In combination with SRM, stable isotope labeling represents the accepted "gold standard" for absolute protein quantification. Label-free absolute protein quantification methods have been established based on various quantification units depending on the applied MS technique. (SIS: stable isotope-labeled standard peptide, AQUA: absolute quantification peptide; QconCAT: quantitative concatamer; EtEP: equimolarity through equalizer peptide; PSAQ: protein standard absolute quantification; FLEXIQuant: full-length expressed stable isotope-labeled proteins for quantification; abs. SILAC: absolute stable isotope labeling through amino acids in cell culture; PrEST: protein epitope signature tags; emPAI: exponentially modified protein abundance index; APEX: absolute protein abundance index; NSAF: normalized spectral abundance factor; SI_N: normalized spectral index; best-flyer: combined intensities from the three typically most intense peptides per protein; IBAQ: intensity-based absolute quantification; PTM: post-translational modification)

a) In large-scale studies these approaches have been applied to target 1000s of proteins.[87,112]
b) For all stable isotope-labeled approaches, numbers are based on the assumption that the SRM technology is applied.
c) Sensitivity might be negatively affected in case the selected epitope region does not include well-suited peptides for MS-detection.
d) QconCATs are applicable to indirect PTM quantification.[65,66]
e) Full-length protein standards are typically not suited for direct PTM studies and are only applicable if an *in vitro* method for the specific and quantitative modification of the standard protein is available.[74]

to roughly adjust the spiked-in standard concentration to the endogenous protein level to ensure the highest possible comparability between the signal intensities of the standard and analyte.

One of the major advantages when working with stable isotope-labeled references compared to label-free quantification is the fact that the analyte and standard are analyzed simultaneously. As they are co-eluting from the LC system, they experience the same ionization conditions; *i.e.*, they are exposed to the same background- or matrix-induced effects that govern ionization efficiency. In addition, varying injection volumes between runs can be accounted for. In conclusion, stable isotope-labeled absolute quantification in combination with SRM can account for technical run-to-run variations at many levels, resulting in technically precise, sensitive and reproducible measurements (Figure 4.4).

4.3.1.1 Peptide Standards

In bottom-up proteomics a protein of interest is not directly identified and quantified. Rather, the presence and quantity of a protein in a sample is inferred from peptides generated by digestion of the protein. For absolute peptide quantification, a variety of standards have been established, including synthetic peptides, concatenated peptides and peptides carrying an equalizer sequence (Figure 4.4).

4.3.1.1.1 Synthetic Peptides. Stable isotope-labeled standard (SIS) peptides,[30] also termed absolute quantification (AQUA) peptides,[51] are chemically synthesized peptides usually containing a single isotopically labeled amino acid. Most commonly, the *C*-terminal lysine or arginine residue is labeled with ^{13}C and ^{15}N isotopes, resulting in a mass shift of 8 Da and 10 Da, respectively. Each SIS peptide must be purified to a very high degree and the absolute concentration determined accurately, which is typically carried out by amino acid analysis (AAA). Custom-defined SIS peptides are commercially available from various companies. However, SIS peptides come at substantial financial costs, limiting their application to, typically, a few protein targets per study. On the other hand, a single SIS peptide synthesis batch is sufficient for thousands of MS injections. A major advantage of SIS peptides is their applicability to the direct study of post-translational modifications (PTMs), for example, phosphorylation,[52–54] acetylation[55] or methylation.[56] Such PTMs can be chemically introduced into the SIS peptide at the required positions to study a particular biologically relevant modification.

While the analytical precision of SRM measurements with SIS peptides is generally high (in the range of 5–15% error),[57] the accuracy of a determined absolute protein concentration can be limited by several factors. First, SIS peptides can neither account for variations arising from incomplete or unspecific proteolysis of the endogenous protein, nor for the occurrence of PTMs, unless they are specifically targeted. Second, as SIS peptides are spiked into the sample either before or after proteolysis, they cannot correct for protein or peptide losses that occur in upstream sample preparation steps, for example, during cell lysis, protein extraction or protein fractionation procedures. Third, any type of chemical modification (such as oxidation, deamidation, *etc.*) that does not occur to the same extent on SIS and analyte peptides will impede accurate quantification. Fourth, the stability and solubility of the SIS peptides themselves can also distort quantification. Effects, such as hydrophobic interactions with surfaces or peptide instability, can affect the concentration of a SIS peptide dramatically and thereby result in erroneous quantification.

To minimize such errors, certain precautions should be taken: i) after the accurate determination of the SIS peptide concentration, the peptide should be kept in solution to minimize problems associated with re-solubilization (ideally solubilized rather than lyophilized peptides should be purchased); ii) freezing–thawing cycles should be kept minimal by appropriate aliquoting of the initial SIS peptide stock solution; iii) depending on the primary sequence, peptides may stick to vessel walls and other hydrophobic surfaces, an issue which becomes more severe the lower the handled SIS peptide concentration. Utilization of low-binding plasticware and glass vessels can help to keep such peptide losses minimal, as well as usage of a "pre-saturation reagent", for example, BSA,[58] or a complex sample background.

In conclusion, SIS peptides are currently the most commonly applied standard molecules for absolute protein quantification due to their easy commercial availability, as well as their straightforward mode of application, despite their apparent limitations in terms of accuracy and costs.

4.3.1.1.2 Concatenated Peptides. Concatenated stable isotope-labeled reference peptides (QconCATs) represent an alternative approach to the use of synthetic peptides as standards.[59] QconCATs (quantification concatamers) are artificial isotope labeled proteins produced by heterologous expression of a synthetic gene consisting of several (up to ~50) concatenated peptide sequences in cellular or cell-free systems.[60,61] QconCATs are spiked into the sample before proteolysis and are then co-digested with the analyte proteins. This simultaneously leads to the generation of a large number of equimolar isotope labeled standard peptides.

One major advantage of the QconCAT approach is its multiplexing capability, *i.e.*, with this technology, in the range of 50 peptides can be generated from a single QconCAT protein omitting the cost-intensive accurate quantification of every single standard peptide based on the assumption that the peptides derived from a single QconCAT are equimolar. Currently, the costs of QconCATs are still relatively high, although they are decreasing rapidly and ready-to-use QconCAT kits for specific systems, both commercial and from academic groups, are also available. Other advantages of QconCATs over SIS peptides are reduced problems such as inaccessibility of certain peptide sequences to chemical synthesis[62] and loss of peptides due to surface attachment, solubility and stability. However, the issue of incomplete proteolysis remains as the complete digestion of the analyte protein and also of the QconCAT itself is crucial for accurate quantification and necessitates careful optimization of the peptides selected as quantification surrogates and the digestion procedure.[62–64] Another limitation of QconCATs is the imposed equimolarity of each peptide, leaving no possibility to specifically adjust the added standard concentration to the individual endogenous peptide level. Furthermore, QconCATs cannot be used to directly investigate biological PTMs as heterologous expression in *E. coli* or cell-free systems does not allow specific and quantitative incorporation of PTMs. Nevertheless, studies have been carried out in which PTMs were investigated indirectly (Figure 4.4).[65,66]

4.3.1.1.3 Equalizer Peptides. An alternative strategy to QconCATs in producing cost-effective equimolar isotope labeled peptides is "equimolarity through equalizer peptide" (EtEP).[67] In this approach isotope labeled standard peptides are chemically synthesized with each one including the same additional *N*-terminal sequence, the equalizer peptide. Upon digestion, the equalizer peptide is cleaved and quantified and its quantity is used to determine the concentration of the released standard peptide. Once the standard peptide is quantified it can be used for absolute quantification in the same way as a SIS peptide. A further decrease in synthesis costs can be achieved through chemical isotopic labeling of the EtEP peptides, for example, using mTRAQ reagents[67] or dimethyl labeling.[68] Apart from the cost improvements, the EtEP approach performs similarly to the SIS peptide approach, including the already discussed advantages and limitations (Figure 4.4).

4.3.1.2 Protein Standards

In bottom-up MS analyses, protein quantification is always carried out at the peptide-level, even when a protein standard is used. However, the main advantage of protein standards is that they can be spiked into the analyte sample at an early step of the sample preparation workflow, for example, directly after cell lysis. Early combination of the standard with the analyte accounts for various experimental biases originating from the sample preparation procedure. In recent years various absolute quantification approaches based on protein standards have been developed that operate with labeled full-length proteins, as well as truncated protein versions (Figure 4.4).

4.3.1.2.1 Full-length Protein Standards. The generation of isotope labeled full-length protein standards, typically carrying ^{13}C- and/or ^{15}N-labeled arginine and lysine residues, has been described by various groups that applied slightly different protocols in terms of protein expression and absolute quantification of the standard protein (for a recent review see ref. 69). In an approach called "protein standard absolute quantification"

(PSAQ) Brun *et al.* heterologously produced isotope labeled full-length proteins using a cell-free *in vitro* system.[70] In an alternative approach termed "absolute SILAC" Hanke *et al.* adapted the "stable isotope labeling by amino acids in cell culture" (SILAC) technology to generate full-length labeled proteins *in vivo* using an auxotrophic *E. coli* strain.[58] In a third approach, termed "full-length expressed stable isotope-labeled proteins for quantification" (FLEXIQuant), protein standards have been produced in wheat germ extracts.[71]

The general benefits of full-length protein standards lie in their capability to: i) account for differential digestion efficiencies between the standard and analyte; ii) account for protein losses during sample preparation and protein fractionation, the latter of which is frequently applied for complex biological samples and iii) monitor a maximal number of peptides per protein. These advantages generally increase precision, accuracy and statistical reliability, thus making absolute quantification with full-length protein standards the most accurate strategy (Figure 4.4).

However, to exploit these advantages to their fullest extent, a full-length protein standard must resemble the analyte protein in structure and modification status, otherwise equal performance in terms of protein fractionation and digestion is not guaranteed.[72,73] Hence, accurate absolute quantification of every full-length protein standard is an important requirement and characteristics, such as labeling status, three-dimensional structure, biological activity and stability of the standard under the applied storage conditions, should be checked carefully.[69] A further limitation of full-length proteins is that every standard has to be generated, quantified and quality-controlled separately, which is highly cost- and labor-intensive and limits the multiplexing capability. Typically, full-length protein standards are not used for the direct study of PTMs as they are not expressed under native conditions and hence do not carry biologically relevant PTMs. However, if an *in vitro* assay for the quantitative modification of a purified protein standard is available, direct absolute quantitative analysis of PTMs becomes feasible, although this is dependent on first ascertaining the stoichiometry of the induced *in vitro* modification.[74]

4.3.1.2.2 **Truncated Protein Standards.** The principle of truncated protein standards is based on the use of a large human library of protein epitope signature tags (PrESTs), which has been developed in a high-throughput manner in the course of the Human Protein Atlas Project.[75] Every PrEST represents a heterologously expressed fusion of a unique region from a protein of interest (between 50 and 150 amino acids in size) with a purification and solubilization tag.[76]

Advantages of the PrEST approach are the rather large size of the protein epitope tag, which allows several peptides per protein to be monitored, resulting in improved statistical reliability, as well as the accessibility of the human epitope library at high throughput within the frame of the Human Protein Atlas Project. On the other hand, it is a limitation that the epitope region of the target protein in the PrEST library was selected for use as an antigen rather than for optimal MS detectability, which potentially results in a suboptimal MS detection sensitivity of the generated peptides. Furthermore, the benefit of full-length protein standards to account for digestion, protein fractionation and modification effects does not hold true for PrESTs as a truncated version of a protein does not necessarily fold and interact like the native protein (Figure 4.4).

4.3.2 Absolute Label-free Protein Quantity Estimation

The state-of-the-art approach for accurate quantification of absolute protein abundances relies on the use of stable isotope labeling, which provides optimal performance in terms of measurement precision and accuracy. However, this benefit comes at the price of producing cost-intensive labeled standards, which hampers large-scale applications. Consequently, in recent years, label-free quantification approaches have become more and more popular. While the vast majority of label-free quantitative analyses published to date are based on relative quantification, a few methods for label-free absolute protein abundance estimation have been developed (Figure 4.4), which all necessitate a translation of the label-free determined MS intensity value into an absolute protein quantity unit (for details see section 4.4.4).

The general benefits of label-free methods are their applicability to any sample type at high throughput and with minimized costs, as well as their capability to multiplex quantification of large numbers of proteins. On the other hand, label-free workflows cannot account for discrepancies during the preparation of different samples or technical run-to-run variations, which inevitably decrease quantification precision and accuracy. To keep technical variations in label-free quantification as small as possible, thoroughly optimized sample preparation workflows and reproducible MS measurements, as well as robust data analysis procedures, including alignment and normalization steps, are critical requirements (for reviews see refs. 23 and 77).

4.3.2.1 Label-free Quantification Units

Absolute protein quantification based on label-free proteomic data exploit different MS-derived indicators of the protein quantity, such as "spectral counts", "precursor ion intensities" or "product ion intensities" (Figure 4.4).

Ishihama *et al.* were the first to demonstrate a linear correlation between the absolute abundance of a protein in a sample and the number of peptides identified from an LC-MS/MS analysis in shotgun mode normalized by the number of theoretically observable peptides for a given protein ("exponentially modified protein abundance index", emPAI).[78] In a more sophisticated approach, Lu *et al.* used the number of identified MS/MS spectra for a given protein and accounted for differences in MS detectability and ionization efficiency by applying a predictive algorithm ("absolute protein abundance index", APEX).[79] Other developments in the field of spectral counting additionally consider the length of the protein ("normalized spectral abundance factor", NSAF)[80] or the product ion intensity of each MS/MS spectra ("normalized spectral index", SI_N).[81] These strategies are relatively easy to implement, useable under high throughput and can be applied to datasets even retrospectively. However, spectral counting methods remain controversial as no direct physical property of the quantified peptides is measured. Furthermore, spectral counting datasets can be biased by undersampling and saturation effects, as well as by the applied dynamic exclusion setting (for a detailed review see ref. 82).

Next to spectral counts, chromatographic precursor ion intensities obtained from MS analyses in data-dependent,[83] as well as data-independent, acquisition modes[84] have successfully been explored as a root for absolute quantification. It is well known that precursor ion intensities vary significantly among peptides present at an equimolar level depending on their specific MS response factor. Silva *et al.* were the first who aimed to reduce this effect of variance by focusing exclusively on the most intense (best "flyer") peptides per protein. The authors reported a linear relationship between the averaged extracted precursor ion intensity of the three best flyer peptides per protein to the total protein abundances.[84] In an alternative approach Schwanhäusser *et al.* demonstrated optimal protein abundance estimation from combined precursor ion intensities of all detectable peptides per protein normalized by the number of theoretically observable peptides ("intensity-based absolute quantification", IBAQ).[10] In recent years several other studies have applied these approaches or variations thereof and estimated absolute protein abundances on proteome-wide scales, for example, in *Saccharomyces cerevisiae,*[1] *Leptospira interrogans,*[83] *Mycoplasma pneumonia,*[8] *Schizosaccharomyces pombe*[9] and human tissue culture cell lines.[85]

Recently, absolute label-free protein quantification has also been established based on product ion intensities measured by SRM.[86] In this work Ludwig *et al.* demonstrated a good linear correlation between the absolute abundance of a protein and its SRM signals by summing up most intense transitions of best flyer peptides. The authors reported an average two-fold error and a measureable linear abundance range over three orders of magnitude. In a follow up study this targeted approach was applied to estimate absolute protein quantities in *Mycobacterium tuberculosis* on a proteome-wide scale.[87]

In various comparative studies spectral counting, as well as ion intensity-based approaches, have been shown to provide good protein abundance estimates[85,88–90] and average errors in the range of two-fold have repeatedly been reported.[7,78,85,86] However, label-free quantification cannot provide the same quantitative precision and accuracy as stable isotope labeling (Figure 4.4). Hence, depending on the project-specific question of interest and the requirements for measurement precision and accuracy, the most suitable approach should be selected accordingly.

4.4 Challenges of Absolute Protein Quantification

Absolute protein quantification represents a more challenging task than relative quantification because a variety of difficult-to-control factors and parameters can directly affect absolute quantitative results. While workflow-specific issues have already been addressed above, this section will focus on discussing four prominent and more general challenges in the context of absolute quantification, including the challenge of complete protein extraction, of specific and complete protein digestion, of optimal peptide selection and of translating label-free MS intensities into absolute protein quantities.

4.4.1 The Challenge of Complete Protein Extraction

For the determination of "protein copies per cell" it is critical to know the efficiency of all sample preparation steps, including cell lysis and protein extraction. If the cells are not lysed completely, or if a significant amount of protein precipitates or remains bound to the insoluble cell membrane fraction, the ratio between the extracted protein amount and the processed cell number gets distorted and protein copy numbers will be globally underestimated. Even if quantification is based on labeled full-length protein standards, this remains a problem because spiking of a standard protein can, at the earliest, happen after cell lysis. Hence, a thorough and optimized protein extraction protocol is a very important requirement for every accurate absolute quantitative analysis.

Depending on the organism, cell type or sample preparation workflow, optimal protocols may vary substantially. For organisms with a rigid cell wall, such as yeast, complete cell breakage is particularly difficult to achieve and has been addressed, for example, by Carroll *et al.* through multiple rounds of successive cell disruption cycles with glass beads.[63] Another emerging approach in this context is the ultra-high pressure cycling technology, also known as barocycling, for which improved protein extraction and digestion efficiencies have been reported.[91,92] Furthermore, considerable efforts have been undertaken to develop protein extraction protocols with improved protein solubilization efficiencies using surfactants (for example, RapiGest™ or sodium dodecyl sulfate), chaotropic reagents (such as guanidine hydrochloride or urea) or organic solvents (like methanol or acetonitrile). In particular, application of the acid-label surfactant RapiGest has been reported as beneficial for the solubilization of membrane proteins without interfering with the MS analysis.[93,94] However, these comparative studies have further demonstrated complementary results for the various tested protocols, highlighting that protein losses do not always occur uniformly for all proteins in a sample, but that some proteins are better solubilized with a given protocol than others. Conclusively, even if cell lysis and protein extraction workflows have been optimized carefully on a global scale, significant losses, and hence erroneous absolute quantification results of specific proteins, cannot be excluded, especially if the lysates are clarified at any point. This issue is further complicated by the fact that hardly any technique exists to reliably assess the completeness of protein extraction. One option might be the use of radioactive labeling, for example, with ^{35}S-methionine or sulphate,[95] which after protein extraction allows determination of radioactivity in the soluble, as well as in the insoluble, fraction. However, even using such an approach, only global protein losses can be corrected, while protein-specific losses cannot be accounted for.

In conclusion, as long as the above mentioned issues regarding complete protein extraction remain unresolved, a question mark will remain on how accurate our absolute analyses indeed are.

4.4.2 The Challenge of Specific and Complete Protein Digestion

Bottom-up proteomic analyses rely on the use of proteolytic enzymes to digest proteins into peptides. Trypsin is the most commonly used enzyme for this purpose. However, trypsin, as well as alternative proteases, such as Lys-C, Lys-N, Asp-N or chymotrypsin, does not always achieve complete and specific digestion of all proteins in a sample.[96] Cleavage sites may be inaccessible or cleaved at a significantly reduced rate due to a lower protease-to-substrate affinity or due to steric hindrances, for example, by insufficient denaturation of the protein, formation of secondary structures or the presence of bulky amino acid modifications close to the cleavage site. Various studies have shown that the local sequence context around the immediate cleavage site can strongly influence digestion efficiencies and kinetics.[97,98] In the case of trypsin, vicinal acidic residues, *i.e.*, aspartic and glutamic acid, negatively affect the protease-to-substrate affinity, resulting in missed-cleaved peptides.[64,99] A *C*-terminal proline residue may completely block tryptic cleavage. However, there are documented cases where trypsin does cleave such peptide bonds to a significant extent.[100,101] Furthermore, dibasic sequences are known to result in mixtures of fully- and missed-cleaved tryptic peptides.

While, for protein identification, incomplete proteolysis can be tolerated or can even improve proteome coverage, for absolute protein quantification, complete and specific proteolysis is crucial. Numerous comparative studies have aimed at identifying optimal proteolytic reaction conditions by assessing digestion yields and kinetics upon varying proteases, temperature, incubation time, buffers and digestion-enhancing additives, such as organic solvents, chaotropic agents or surfactants (exemplified by refs. 98,102–105). Among the multitude of different digestion protocols tested, the usage of the denaturing agents urea and RapiGest[106,107] and a combinatorial digest with Lys-C and trypsin[98,108,109] have repeatedly been shown to increase proteolytic efficiency for many applications in proteomic research. However, from the plethora of published studies, it is also obvious that, depending on the organism, sample type, applied protease and targeted proteins, different digestion protocols may lead to optimal results. Hence, a thorough optimization of the applied proteolytic procedure is highly recommended for each individual project as it can substantially affect the absolute quantitative outcome.

4.4.3 The Challenge of Optimal Peptide Selection

The average protein in any given proteome theoretically gives rise to tens or even hundreds of peptides when subjected to proteolysis; however, typically, only a small proportion of those peptides are consistently observed by ESI-MS.[110,111] Hence, one important necessity for a targeted MS analysis is the selection of a few (two to five) peptides optimally representing the underlying protein. In this context, the term "proteotypic peptide" was coined,[38,111] describing fully-cleaved peptides that fulfil two selection criteria. First, the peptides have to be unique within their proteome of interest. Typically, peptide uniqueness is defined based on the genome; however, splice isoforms, single nucleotide polymorphisms or isobaric peptides (different primary sequences but highly similar masses) can also be taken into account. Second, proteotypic peptides need a good MS detectability. This can be determined either from empirical data or with computational prediction tools (see also section 4.2.2). A good MS detectability can be expected for peptides with a length between 8 and 25 amino acids and which predominantly form doubly and/or triply charged precursor ions in a mass range accessible to the QQQ instrument under the applied ESI condition.

For accurate absolute quantification it is advisable to consider additional factors, such as complete excision of the peptide upon proteolysis and resistance to biological and chemical modifications (Table 4.1). Peptides fulfilling these tightened criteria have been termed "quantotypic".[112] In the case of trypsin, complete proteolytic cleavage can be improved by avoiding peptides with acidic residues in close proximity to the cleavage site, by omitting dibasic cleavage sites, and by excluding peptides with an arginine–proline or lysine–proline motif (see also section 4.4.2). The exclusion of peptides potentially carrying biological PTMs is challenging because, to date, a substantial number of PTMs still remain undiscovered or only occur under specific conditions. However, a reconciliation of potential quantotypic peptides with available PTM databases (for example, UniProt[113] and Phospho.ELM[114]) or PTM prediction tools[115] can be helpful. A rather straightforward example is the exclusion of peptides containing the glycosylation motive -N–X–S/T-, in which the asparagine residue is often glycosylated *in vivo*. Artifactual chemical modifications can be introduced into proteins and peptides during sample processing. Frequently occurring chemical modifications are the oxidation of methionine and tryptophan residues,[116] deamidation of glutamine and asparagine (mainly when followed by a small residue like glycine),[117] incomplete alkylation of cysteine, internal

Table 4.1 Overview of selection criteria for quantotypic peptides. (K: lysine; R: arginine; P: proline; N: asparagine; D: aspartic acid; G: glycine; Q: glutamine; E: glutamic acid; C: cysteine; M: methionine; W: tryptophan; S: serine; T: threonine; X: arbitrary amino acid; PTM: post-translational modification).

Selection criteria	*Comments*
fully-cleaved	The peptide should not contain a missed or a non-specific cleavage site according to the applied protease.
unique amino acid sequence in the proteome	Usually defined based on the genome, but also splice isoforms, single nucleotide polymorphisms or isobaric peptides might be considered.
MS detectability	Depends on specific physico-chemical properties, as well as on the chemical composition of the background (matrix effect). Information about the best peptide detectabilities per protein can be obtained either from empirical data or by using bioinformatic prediction tools.
peptide length	Ideally in the range of 8 to 25 amino acids.
optimal mass range	2+ and/or 3+ precursor *m/z* values should be in the accessible mass range of the MS instrument used.
complete proteolysis	For trypsin avoid: • dibasic cleavage sites: -K–K-, -R–R-, -R–K-, -K–R- • acidic residues (D and E) in close proximity to the cleavage site • *N*-terminal proline cleavage: -K–P-, -R–P-
no biological modifications (PTMs)	Avoid: • peptides carrying a PTM; search available databases or use prediction tools • glycosylation motive: -N–X–S/T- • *N*- and *C*-terminal peptides because these are more easily susceptible to protein proteolysis and degradation
no chemical modifications	Avoid: • oxidation: M and W • deamidation: -N–G-, -Q–G- • incomplete alkylation: C • internal peptide backbone fragmentation: -D–P- • *N*-terminal cyclization: *N*-terminal Q and E (pyroglutamate), *N*-terminal carbamidomethylated C

fragmentation of the aspartate–proline bond[118] and cyclization of *N*-terminal glutamine, glutamic acid or carbamidomethylated cysteine (Table 4.1).[119]

In conclusion, a quantotypic peptide, equipped exclusively with optimal properties for accurate absolute quantification, is defined by a long list of selection criteria. However, when applying these criteria onto proteins of interest, the number of suitable quantotypic peptides diminishes very rapidly,[112] often resulting in cases of problematic proteins for which not a single peptide meets the required criteria. In these cases, relaxation of specific selection filters might be necessary to find the best trade-off between sensitivity and quantitative accuracy.

4.4.4 The Challenge of Translating Label-free MS Intensities into Absolute Protein Quantities

All of the absolute label-free approaches illustrated in Figure 4.4 ultimately lead to tables containing protein-specific MS intensity values based on spectral counts or chromatographic peak intensities. These values can be used directly for relative comparative analyses between samples. However, if absolute quantification is the aim of the study, a further translation into absolute protein concentrations is necessary. This can be achieved either by applying an estimate of the total protein content per cell[79,120,134] or by linear regression based on a small number of anchor proteins (Figure 4.5).[83,85,86,134] The first approach estimates absolute protein abundances by distributing the total protein content per cell among all quantified proteins according to their label-free MS intensities (Figure 4.5A). Consequently it is restricted to proteome-wide MS studies and necessitates an accurate estimate of the total cellular protein content or the total number of protein molecules in a cell (for information see: http://bionumbers.hms.harvard.edu/). The alternative strategy based on linear regression necessitates accurate absolute abundance information for a small number of anchor proteins (~15 to 50) ideally spanning the whole protein abundance range of interest (Figure 4.5B). Typically, this is conducted by spike-in experiments with purified proteins (for example, the commercially available UPS2 standard)[10] or SIS peptides,[83] although any alternative absolute quantification approach can be applied. Through linear correlation of the log-transformed absolute abundances of the anchor proteins and their corresponding log-transformed label-free MS intensities, an experiment-specific linear calibration curve is generated and subsequently used to convert all measured MS intensities into absolute protein concentrations. Compared to the approach based on the total cellular content, this method necessitates usage of a few stable isotope-labeled standards and is therefore less straight-forwardly applicable and more cost-intensive; however, on the other hand, it can account for variances in the total cellular protein content and it does not necessitate a proteome-wide quantitative coverage.

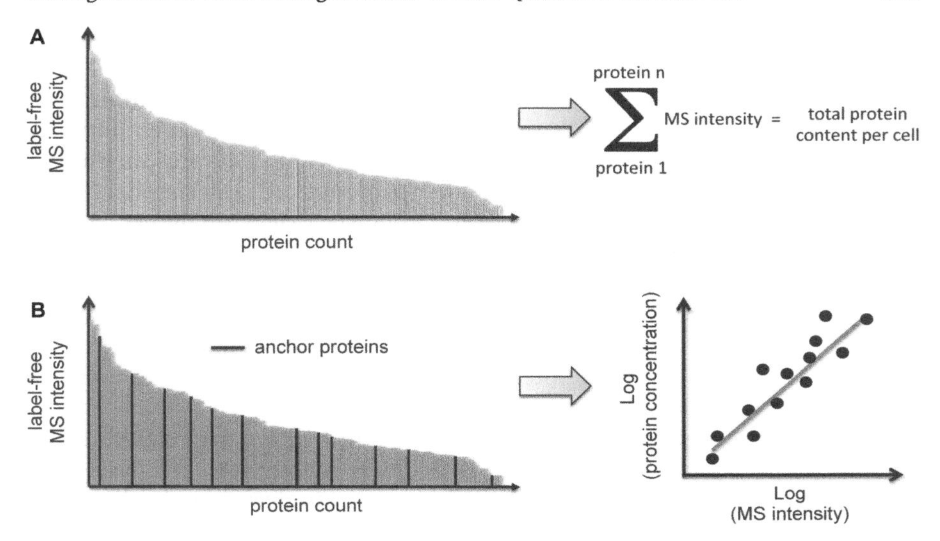

Figure 4.5 Translating label-free MS intensities into absolute protein concentrations. A) The first strategy necessitates an accurate estimate of the total protein content per cell. By equalizing the sum of all measured MS intensities to the total protein content per cell, the fractional contribution of every single protein can be calculated. B) The second strategy necessitates accurate knowledge about the absolute protein concentration of a few selected anchor proteins (black), ideally spanning the whole MS intensity range. Based on these anchor proteins, a linear calibration curve is generated and subsequently used to estimate absolute concentrations for all quantified proteins.

4.5 Conclusions and Further Perspectives

Within the variety of MS-based quantification methods, absolute protein quantification in combination with SRM is the method of choice for projects where absolute concentrations for pre-defined sets of proteins are required. In particular, fields like systems biology and clinical proteomics, which are rapidly evolving research areas, necessitate information about absolute protein quantities that are determined accurately and reproducibly across multiple samples (Figure 4.1). Hence, the demand for absolute protein quantification increases steadily and a multitude of workflows and technologies have been established to achieve it (Figure 4.4).

However, it is important to keep in mind that absolute protein quantification still remains a challenging task due to the technical and practical issues discussed above. Challenges related to the SRM technology include: i) selection of truly representative peptides per protein, ii) SRM assay generation and validation, and iii) reliable and automated SRM data analysis, as well as iv) the inherent limitation in the number of targetable proteins, peptides and transitions per SRM run. Over the past few years, these issues have been recognized and solutions and improvements are being developed.

For example, new experimental workflows[35,39] and bioinformatics tools[36,38] enable rapid proteotypic and/or quantotypic peptide selection and SRM assay development. Validated transitions and other SRM assay parameters can be stored in centralized databases and are accessible to the community (www.srmatlas.org). Software has been introduced that facilitates manual[121] or automated[122] analysis of SRM data, detection of interferences and unreliable transitions,[123] as well as reliable statistical analyses.[124] New concepts in optimal scheduling of SRM transitions[125] and reporting empirical peptide retention times[45] have helped to increase the number of measurable transitions per SRM run. Finally, new targeted MS technologies exploiting high-resolution and high-mass-accuracy instruments are emerging,[49,50,133] which have the potential to carry out proteome-wide analysis and to minimize upfront assay development.

Practical challenges for absolute protein quantification are related to the sample preparation procedure and include: i) incomplete cell lysis and protein extraction protocols, ii) unspecific or incomplete protein digestion procedures and iii) high costs associated with the use of stable isotope-labeled standards. While the extent of incomplete proteolysis can be assessed with full-length protein standards, high expenses for isotope labeled and purified standards remain. To reduce these costs, collaborative efforts within the proteomic community and/or industrial suppliers would be highly desirable to make synthesized standards storable and accessible to the community.[126] A step in this direction has been taken for QconCATs, which are becoming commercially available as ready-to-use kits for functionally related protein classes. Alternatively, label-free absolute quantification strategies significantly reduce the expenses associated to standards, albeit at the expense of precision and accuracy. As outlined in section 4.4.1, the issue of ensuring complete cell lysis and protein extraction procedures still remains to be solved.

The majority of published projects involving absolute protein quantification monitor total protein quantities, which is not necessarily the required piece of information to address all biological questions. Instead, information about a protein's absolute activity or relative abundance between conditions might be more informative. However, protein activity can be influenced by various factors, such as PTMs, interaction partners or sub-cellular localization. The SRM technology allows identification and absolute quantification of PTMs, but currently only a few cost-intensive approaches are compatible with the direct analysis of PTMs (Figure 4.4). Another complicating factor for targeted PTM analysis is the requirement of tedious PTM-specific enrichment steps,[127] which reduce quantitative precision, accuracy and sample throughput. Also the study of protein complexes[5] and sub-cellular protein localization[128] necessitate specialized sample preparation procedures. Despite these challenges, several pioneering studies have demonstrated the great potential of absolute SRM quantification to study phosphorylation site occupancies,[53,129] to characterize the inter-connectivity of ubiquitin,[130] or to accurately determine protein complex stoichiometries.[5,67] A limitation of absolute quantification with SRM is its impracticality to single cell analysis as the results from SRM always represent averages over a whole cell population. Recently, a promising targeted MS technology, termed mass cytometry, has been emerging, which allows quantitative analysis of proteins on the single cell level.[131,132] However, mass cytometry requires sensitive and specific antibodies and its applicability to absolute protein quantification still needs to be demonstrated.

Conclusively, in recent years substantial improvements in technology, workflows, software tools and available standards for SRM-based absolute protein quantification have been achieved. In the future these progresses will contribute to the routine and worldwide application of SRM also by non-specialized laboratories. While, so far, most efforts have been made to absolutely quantify total protein abundances averaged over a cell population, monitoring absolute protein activity as determined by its PTM status, interaction partners or subcellular localization is still far from being routinely applied and leaves ample room for further improvements and innovative technical ideas.

Acknowledgements

The authors thank Olga Schubert for critical reading of the manuscript and Ben Collins for helpful discussions. We also thank Lilli Amelie Ludwig for being patient. Funding by SystemsX.ch (YeastX project), the Swiss initiative for systems biology, is gratefully acknowledged.

References

1. L. M. de Godoy, J. V. Olsen, J. Cox, M. L. Nielsen, N. C. Hubner, F. Frohlich, T. C. Walther and M. Mann, *Nature*, 2008, **455**, 1251.
2. J. Krijgsveld, R. F. Ketting, T. Mahmoudi, J. Johansen, M. Artal-Sanz, C. P. Verrijzer, R. H. Plasterk and A. J. Heck, *Nat. Biotechnol.*, 2003, **21**, 927.
3. V. M. Faca, K. S. Song, H. Wang, Q. Zhang, A. L. Krasnoselsky, L. F. Newcomb, R. R. Plentz, S. Gurumurthy, M. S. Redston, S. J. Pitteri, S. R. Pereira-Faca, R. C. Ireton, H. Katayama, V. Glukhova, D. Phanstiel, D. E. Brenner, M. A. Anderson, D. Misek, N. Scholler, N. D. Urban, M. J. Barnett, C. Edelstein, G. E. Goodman, M. D. Thornquist, M. W. McIntosh, R. A. DePinho, N. Bardeesy and S. M. Hanash, *PLoS Med.*, 2008, **5**, e123.
4. D. Lovell, W. Müller, J. Taylor, A. Zwart, C. Helliwell, in *Proportions, Percentages, ppm: do the Molecular Biosciences Treat Compositional Data Right?*, eds. V. Pawlowsky-Glahn and A. Buccianti, John Wiley & Sons, New York, 2011, pp. 193–206.
5. C. Schmidt, C. Lenz, M. Grote, R. Luhrmann and H. Urlaub, *Anal. Chem.*, 2010, **82**, 2784.

6. A. Bensimon, A. J. Heck and R. Aebersold, *Annu. Rev. Biochem.*, 2012, **81**, 379.
7. A. Schmidt, M. Beck, J. Malmstrom, H. Lam, M. Claassen, D. Campbell and R. Aebersold, *Mol. Syst. Biol.*, 2011, 7, 510.
8. T. Maier, A. Schmidt, M. Guell, S. Kuhner, A. C. Gavin, R. Aebersold and L. Serrano, *Mol. Syst. Biol.*, 2011, 7, 511.
9. S. Marguerat, A. Schmidt, S. Codlin, W. Chen, R. Aebersold and J. Bahler, *Cell*, 2012, **151**, 671.
10. B. Schwanhausser, D. Busse, N. Li, G. Dittmar, J. Schuchhardt, J. Wolf, W. Chen and M. Selbach, *Nature*, 2011, **473**, 337.
11. T. A. Addona, S. E. Abbatiello, B. Schilling, S. J. Skates, D. R. Mani, D. M. Bunk, C. H. Spiegelman, L. J. Zimmerman, A. J. Ham, H. Keshishian, S. C. Hall, S. Allen, R. K. Blackman, C. H. Borchers, C. Buck, H. L. Cardasis, M. P. Cusack, N. G. Dodder, B. W. Gibson, J. M. Held, T. Hiltke, A. Jackson, E. B. Johansen, C. R. Kinsinger, J. Li, M. Mesri, T. A. Neubert, R. K. Niles, T. C. Pulsipher, D. Ransohoff, H. Rodriguez, P. A. Rudnick, D. Smith, D. L. Tabb, T. J. Tegeler, A. M. Variyath, L. J. Vega-Montoto, A. Wahlander, S. Waldemarson, M. Wang, J. R. Whiteaker, L. Zhao, N. L. Anderson, S. J. Fisher, D. C. Liebler, A. G. Paulovich, F. E. Regnier, P. Tempst and S. A. Carr, *Nat. Biotechnol.*, 2009, **27**, 633.
12. K. Bluemlein and M. Ralser, *Nat. Protoc.*, 2011, **6**, 859.
13. J. R. Whiteaker, C. Lin, J. Kennedy, L. Hou, M. Trute, I. Sokal, P. Yan, R. M. Schoenherr, L. Zhao, U. J. Voytovich, K. S. Kelly-Spratt, A. Krasnoselsky, P. R. Gafken, J. M. Hogan, L. A. Jones, P. Wang, L. Amon, L. A. Chodosh, P. S. Nelson, M. W. McIntosh, C. J. Kemp and A. G. Paulovich, *Nat. Biotechnol.*, 2011, **29**, 625.
14. N. Rifai, M. A. Gillette and S. A. Carr, *Nat. Biotechnol.*, 2006, **24**, 971.
15. O. Heudi, S. Barteau, D. Zimmer, J. Schmidt, K. Bill, N. Lehmann, C. Bauer and O. Kretz, *Anal. Chem.*, 2008, **80**, 4200.
16. A. Dupuis, J. A. Hennekinne, J. Garin and V. Brun, *Proteomics*, 2008, **8**, 4633.
17. S. E. Stevenson, N. L. Houston and J. J. Thelen, *Regul. Toxicol. Pharmacol.*, 2010, **58**, S36.
18. M. Bredehoft, W. Schanzer and M. Thevis, *Rapid Commun. Mass Spectrom.*, 2008, **22**, 477.
19. P. Picotti and R. Aebersold, *Nat. Methods*, 2012, **9**, 555.
20. M. J. Emanuele, M. L. McCleland, D. L. Satinover and P. T. Stukenberg, *Mol. Biol. Cell*, 2005, **16**, 4882.
21. S. Ghaemmaghami, W. K. Huh, K. Bower, R. W. Howson, A. Belle, N. Dephoure, E. K. O'Shea and J. S. Weissman, *Nature*, 2003, **425**, 737.
22. V. C. Coffman and J. Q. Wu, *Trends Biochem. Sci.*, 2012, **37**, 499.
23. M. Bantscheff, M. Schirle, G. Sweetman, J. Rick and B. Kuster, *Anal. Bioanal. Chem.*, 2007, **389**, 1017.
24. M. Bronstrup, *Expert Rev. Proteomics*, 2004, **1**, 503.

25. V. Brun, C. Masselon, J. Garin and A. Dupuis, *J. Proteomics*, 2009, **72**, 740.
26. K. Kito and T. Ito, *Curr. Genomics*, 2008, **9**, 263.
27. B. Domon and R. Aebersold, *Nat. Biotechnol.*, 2010, **28**, 710.
28. S. W. Holman, P. F. Sims and C. E. Eyers, *Bioanalysis*, 2012, **4**, 1763.
29. V. Lange, P. Picotti, B. Domon and R. Aebersold, *Mol. Syst. Biol.*, 2008, **4**, 222.
30. D. M. Desiderio and M. Kai, *Biomed. Mass Spectrom.*, 1983, **10**, 471.
31. J. Stahl-Zeng, V. Lange, R. Ossola, K. Eckhardt, W. Krek, R. Aebersold and B. Domon, *Mol. Cell Proteomics*, 2007, **6**, 1809.
32. E. W. Deutsch, H. Lam and R. Aebersold, *EMBO Rep.*, 2008, **9**, 429.
33. P. Jones, R. G. Cote, S. Y. Cho, S. Klie, L. Martens, A. F. Quinn, D. Thorneycroft and H. Hermjakob, *Nucleic Acids Res.*, 2008, **36**, D878.
34. P. G. Pedrioli, *Methods Mol. Biol.*, 2010, **604**, 213.
35. A. B. Stergachis, B. MacLean, K. Lee, J. A. Stamatoyannopoulos and M. J. MacCoss, *Nat. Methods*, 2011, **8**, 1041.
36. C. E. Eyers, C. Lawless, D. C. Wedge, K. W. Lau, S. J. Gaskell and S. J. Hubbard, *Mol. Cell Proteomics*, 2011, **10**, M110.003384.
37. V. A. Fusaro, D. R. Mani, J. P. Mesirov and S. A. Carr, *Nat. Biotechnol.*, 2009, **27**, 190.
38. P. Mallick, M. Schirle, S. S. Chen, M. R. Flory, H. Lee, D. Martin, J. Ranish, B. Raught, R. Schmitt, T. Werner, B. Kuster and R. Aebersold, *Nat. Biotechnol.*, 2007, **25**, 125.
39. P. Picotti, O. Rinner, R. Stallmach, F. Dautel, T. Farrah, B. Domon, H. Wenschuh and R. Aebersold, *Nat. Methods*, 2010, **7**, 43.
40. P. Picotti, M. Clement-Ziza, H. Lam, D. S. Campbell, A. Schmidt, E. W. Deutsch, H. Rost, Z. Sun, O. Rinner, L. Reiter, Q. Shen, J. J. Michaelson, A. Frei, S. Alberti, U. Kusebauch, B. Wollscheid, R. L. Moritz, A. Beyer and R. Aebersold, *Nature*, 2013, **494**, 266.
41. A. Bertsch, S. Jung, A. Zerck, N. Pfeifer, S. Nahnsen, C. Henneges, A. Nordheim and O. Kohlbacher, *J. Proteome Res.*, 2010, **9**, 2696.
42. S. Li, R. J. Arnold, H. Tang and P. Radivojac, *Anal. Chem.*, 2011, **83**, 790.
43. H. Rost, L. Malmstrom and R. Aebersold, *Mol. Cell Proteomics*, 2012, **11**, 540.
44. O. V. Krokhin, R. Craig, V. Spicer, W. Ens, K. G. Standing, R. C. Beavis and J. A. Wilkins, *Mol. Cell Proteomics*, 2004, **3**, 908.
45. C. Escher, L. Reiter, B. Maclean, R. Ossola, F. Herzog, J. Chilton, M. J. Maccoss and O. Rinner, *Proteomics*, 2012, **12**, 1111.
46. P. Picotti, H. Lam, D. Campbell, E. W. Deutsch, H. Mirzaei, J. Ranish, B. Domon and R. Aebersold, *Nat. Methods*, 2008, **5**, 913.
47. T. Farrah, E. W. Deutsch, R. Kreisberg, Z. Sun, D. S. Campbell, L. Mendoza, U. Kusebauch, M. Y. Brusniak, R. Huttenhain, R. Schiess, N. Selevsek, R. Aebersold and R. L. Moritz, *Proteomics*, 2012, **12**, 1170.
48. P. Picotti, B. Bodenmiller, L. N. Mueller, B. Domon and R. Aebersold, *Cell*, 2009, **138**, 795.

49. L. C. Gillet, P. Navarro, S. Tate, H. Rost, N. Selevsek, L. Reiter, R. Bonner and R. Aebersold, *Mol. Cell Proteomics*, 2012, **11**, O111.016717.
50. S. Gallien, E. Duriez, C. Crone, M. Kellmann, T. Moehring and B. Domon, *Mol. Cell Proteomics*, 2012, **11**, 1709.
51. S. A. Gerber, J. Rush, O. Stemman, M. W. Kirschner and S. P. Gygi, *Proc. Natl. Acad. Sci. USA*, 2003, **100**, 6940.
52. V. Mayya, K. Rezual, L. Wu, M. B. Fong and D. K. Han, *Mol. Cell Proteomics*, 2006, **5**, 1146.
53. A. P. Oliveira, C. Ludwig, P. Picotti, M. Kogadeeva, R. Aebersold and U. Sauer, *Mol. Syst. Biol.*, 2012, **8**, 623.
54. A. Wolf-Yadlin, S. Hautaniemi, D. A. Lauffenburger and F. M. White, *Proc. Natl. Acad. Sci. USA*, 2007, **104**, 5860.
55. K. Zhang, M. Schrag, A. Crofton, R. Trivedi, H. Vinters and W. Kirsch, *Proteomics*, 2012, **12**, 1261.
56. A. Darwanto, M. P. Curtis, M. Schrag, W. Kirsch, P. Liu, G. Xu, J. W. Neidigh and K. Zhang, *J. Biol. Chem.*, 2010, **285**, 21868.
57. H. Keshishian, T. Addona, M. Burgess, E. Kuhn and S. A. Carr, *Mol. Cell Proteomics*, 2007, **6**, 2212.
58. S. Hanke, H. Besir, D. Oesterhelt and M. Mann, *J. Proteome Res.*, 2008, **7**, 1118.
59. R. J. Beynon, M. K. Doherty, J. M. Pratt and S. J. Gaskell, *Nat. Methods*, 2005, **2**, 587.
60. J. M. Pratt, D. M. Simpson, M. K. Doherty, J. Rivers, S. J. Gaskell and R. J. Beynon, *Nat. Protoc.*, 2006, **1**, 1029.
61. D. M. Simpson and R. J. Beynon, *Anal. Bioanal. Chem.*, 2012, **404**, 977.
62. H. Mirzaei, J. K. McBee, J. Watts and R. Aebersold, *Mol. Cell Proteomics*, 2008, **7**, 813.
63. K. M. Carroll, D. M. Simpson, C. E. Eyers, C. G. Knight, P. Brownridge, W. B. Dunn, C. L. Winder, K. Lanthaler, P. Pir, N. Malys, D. B. Kell, S. G. Oliver, S. J. Gaskell and R. J. Beynon, *Mol. Cell Proteomics*, 2011, **10**, M111.007633.
64. C. Lawless and S. J. Hubbard, *OMICS*, 2012, **16**, 449.
65. H. Johnson, C. E. Eyers, P. A. Eyers, R. J. Beynon and S. J. Gaskell, *J. Am. Soc. Mass Spectrom.*, 2009, **20**, 2211.
66. C. Ding, Y. Li, B. J. Kim, A. Malovannaya, S. Y. Jung, Y. Wang and J. Qin, *J. Proteome Res.*, 2011, **10**, 3652.
67. J. Holzmann, P. Pichler, M. Madalinski, R. Kurzbauer and K. Mechtler, *Anal. Chem.*, 2009, **81**, 10254.
68. D. Kovanich, S. Cappadona, R. Raijmakers, S. Mohammed, A. Scholten and A. J. Heck, *Anal. Bioanal. Chem.*, 2012, **404**, 991.
69. G. Picard, D. Lebert, M. Louwagie, A. Adrait, C. Huillet, F. Vandenesch, C. Bruley, J. Garin, M. Jaquinod and V. Brun, *J. Mass Spectrom.*, 2012, **47**, 1353.
70. V. Brun, A. Dupuis, A. Adrait, M. Marcellin, D. Thomas, M. Court, F. Vandenesch and J. Garin, *Mol. Cell Proteomics*, 2007, **6**, 2139.

71. S. Singh, M. Springer, J. Steen, M. W. Kirschner and H. Steen, *J. Proteome Res.*, 2009, **8**, 2201.
72. C. Huillet, A. Adrait, D. Lebert, G. Picard, M. Trauchessec, M. Louwagie, A. Dupuis, L. Hittinger, B. Ghaleh, P. Le Corvoisier, M. Jaquinod, J. Garin, C. Bruley and V. Brun, *Mol. Cell Proteomics*, 2012, **11**, M111.008235.
73. C. Pritchard, M. Quaglia, A. E. Ashcroft and G. O'Connor, *Bioanalysis*, 2011, **3**, 2797.
74. E. Ciccimaro, S. K. Hanks, K. H. Yu and I. A. Blair, *Anal. Chem.*, 2009, **81**, 3304.
75. L. Berglund, E. Bjorling, P. Oksvold, L. Fagerberg, A. Asplund, C. A. Szigyarto, A. Persson, J. Ottosson, H. Wernerus, P. Nilsson, E. Lundberg, A. Sivertsson, S. Navani, K. Wester, C. Kampf, S. Hober, F. Ponten and M. Uhlen, *Mol. Cell Proteomics*, 2008, **7**, 2019.
76. M. Zeiler, W. L. Straube, E. Lundberg, M. Uhlen and M. Mann, *Mol. Cell Proteomics*, 2012, **11**, O111.009613.
77. K. A. Neilson, N. A. Ali, S. Muralidharan, M. Mirzaei, M. Mariani, G. Assadourian, A. Lee, S. C. van Sluyter and P. A. Haynes, *Proteomics*, 2011, **11**, 535.
78. Y. Ishihama, Y. Oda, T. Tabata, T. Sato, T. Nagasu, J. Rappsilber and M. Mann, *Mol. Cell Proteomics*, 2005, **4**, 1265.
79. P. Lu, C. Vogel, R. Wang, X. Yao and E. M. Marcotte, *Nat. Biotechnol.*, 2007, **25**, 117.
80. B. Zybailov, A. L. Mosley, M. E. Sardiu, M. K. Coleman, L. Florens and M. P. Washburn, *J. Proteome Res.*, 2006, **5**, 2339.
81. N. M. Griffin, J. Yu, F. Long, P. Oh, S. Shore, Y. Li, J. A. Koziol and J. E. Schnitzer, *Nat. Biotechnol.*, 2010, **28**, 83.
82. D. H. Lundgren, S. I. Hwang, L. Wu and D. K. Han, *Expert Rev. Proteomics*, 2010, 7, 39.
83. J. Malmstrom, M. Beck, A. Schmidt, V. Lange, E. W. Deutsch and R. Aebersold, *Nature*, 2009, **460**, 762.
84. J. C. Silva, M. V. Gorenstein, G. Z. Li, J. P. Vissers and S. J. Geromanos, *Mol. Cell Proteomics*, 2006, **5**, 144.
85. M. Beck, A. Schmidt, J. Malmstroem, M. Claassen, A. Ori, A. Szymborska, F. Herzog, O. Rinner, J. Ellenberg and R. Aebersold, *Mol. Syst. Biol.*, 2011, 7, 549.
86. C. Ludwig, M. Claassen, A. Schmidt and R. Aebersold, *Mol. Cell Proteomics*, 2012, **11**, M111.013987.
87. O. T. Schubert, J. Mouritsen, C. Ludwig, H. L. Rost, G. Rosenberger, P. K. Arthur, M. Claassen, D. S. Campbell, Z. Sun, T. Farrah, M. Gengenbacher, A. Maiolica, S. H. Kaufmann, R. L. Moritz and R. Aebersold, *Cell Host Microbe*, 2013, **13**, 602.
88. J. Grossmann, B. Roschitzki, C. Panse, C. Fortes, S. Barkow-Oesterreicher, D. Rutishauser and R. Schlapbach, *J. Proteomics*, 2010, **73**, 1740.
89. K. Ning, D. Fermin and A. I. Nesvizhskii, *J. Proteome Res.*, 2012, **11**, 2261.

90. S. Ryu, B. Gallis, Y. A. Goo, S. A. Shaffer, D. Radulovic and D. R. Goodlett, *Cancer Inform.*, 2008, **6**, 243.
91. E. Freeman and A. R. Ivanov, *J. Proteome Res.*, 2011, **10**, 5536.
92. B. S. Powell, A. V. Lazarev, G. Carlson, A. R. Ivanov and D. A. Rozak, *Methods Mol. Biol.*, 2012, **881**, 27.
93. F. Mbeunkui and M. B. Goshe, *Proteomics*, 2011, **11**, 898.
94. F. Wu, D. Sun, N. Wang, Y. Gong and L. Li, *Anal. Chim. Acta*, 2011, **698**, 36.
95. P. Baudouin-Cornu, G. Lagniel, S. Chedin and J. Labarre, *Proteomics*, 2009, **9**, 4606.
96. P. Picotti, R. Aebersold and B. Domon, *Mol. Cell Proteomics*, 2007, **6**, 1589.
97. P. Brownridge and R. J. Beynon, *Methods*, 2011, **54**, 351.
98. T. Glatter, C. Ludwig, E. Ahrne, R. Aebersold, A. J. Heck and A. Schmidt, *J. Proteome Res.*, 2012, **11**, 5145.
99. J. A. Siepen, E. J. Keevil, D. Knight and S. J. Hubbard, *J. Proteome Res.*, 2007, **6**, 399.
100. R. Raijmakers, P. Neerincx, S. Mohammed and A. J. Heck, *Chem. Commun.*, 2010, **46**, 8827.
101. J. Rodriguez, N. Gupta, R. D. Smith and P. A. Pevzner, *J. Proteome Res.*, 2008, 7, 300.
102. G. Choudhary, S. L. Wu, P. Shieh and W. S. Hancock, *J. Proteome Res.*, 2003, **2**, 59.
103. W. J. t. Hervey, M. B. Strader and G. B. Hurst, *J. Proteome Res.*, 2007, **6**, 3054.
104. M. Peng, N. Taouatas, S. Cappadona, B. van Breukelen, S. Mohammed, A. Scholten and A. J. Heck, *Nat. Methods*, 2012, **9**, 524.
105. J. L. Proc, M. A. Kuzyk, D. B. Hardie, J. Yang, D. S. Smith, A. M. Jackson, C. E. Parker and C. H. Borchers, *J. Proteome Res.*, 2010, **9**, 5422.
106. E. I. Chen, D. Cociorva, J. L. Norris and J. R. r. Yates, *J. Proteome Res.*, 2007, **6**, 2529.
107. K. R. Rebecchi, E. P. Go, L. Xu, C. L. Woodin, M. Mure and H. Desaire, *Anal. Chem.*, 2011, **83**, 8484.
108. A. A. Klammer and M. J. MacCoss, *J. Proteome Res.*, 2006, **5**, 695.
109. M. P. Washburn, D. Wolters and J. R. r. Yates, *Nat. Biotechnol.*, 2001, **19**, 242.
110. R. Aebersold and M. Mann, *Nature*, 2003, **422**, 198.
111. B. Kuster, M. Schirle, P. Mallick and R. Aebersold, *Nat. Rev. Mol. Cell Biol.*, 2005, **6**, 577.
112. P. Brownridge, S. W. Holman, S. J. Gaskell, C. M. Grant, V. M. Harman, S. J. Hubbard, K. Lanthaler, C. Lawless, R. O'Cualain, P. Sims, R. Watkins and R. J. Beynon, *Proteomics*, 2011, **11**, 2957.
113. A. Bairoch, R. Apweiler, C. H. Wu, W. C. Barker, B. Boeckmann, S. Ferro, E. Gasteiger, H. Huang, R. Lopez, M. Magrane, M. J. Martin, D. A. Natale, C. O'Donovan, N. Redaschi and L. S. Yeh, *Nucleic Acids Res.*, 2005, **33**, D154.

114. H. Dinkel, C. Chica, A. Via, C. M. Gould, L. J. Jensen, T. J. Gibson and F. Diella, *Nucleic Acids Res.*, 2011, **39**, D261.
115. B. Eisenhaber and F. Eisenhaber, *Methods Mol. Biol.*, 2010, **609**, 365.
116. J. M. Froelich and G. E. Reid, *Proteomics*, 2008, **8**, 1334.
117. R. Tyler-Cross and V. Schirch, *J. Biol. Chem.*, 1991, **266**, 22549.
118. N. Li, F. Fort, K. Kessler and W. Wang, *J. Pharm. Biomed. Anal.*, 2009, **50**, 73.
119. J. Reimer, D. Shamshurin, M. Harder, A. Yamchuk, V. Spicer and O. V. Krokhin, *J. Chromatogr. A*, 2011, **1218**, 5101.
120. J. R. Wisniewski, P. Ostasiewicz, K. Dus, D. F. Zielinska, F. Gnad and M. Mann, *Mol. Syst. Biol.*, 2012, **8**, 611.
121. B. MacLean, D. M. Tomazela, N. Shulman, M. Chambers, G. L. Finney, B. Frewen, R. Kern, D. L. Tabb, D. C. Liebler and M. J. MacCoss, *Bioinformatics*, 2010, **26**, 966.
122. L. Reiter, O. Rinner, P. Picotti, R. Huttenhain, M. Beck, M. Y. Brusniak, M. O. Hengartner and R. Aebersold, *Nat. Methods*, 2011, **8**, 430.
123. S. E. Abbatiello, D. R. Mani, H. Keshishian and S. A. Carr, *Clin. Chem.*, 2010, **56**, 291.
124. C. Y. Chang, P. Picotti, R. Huttenhain, V. Heinzelmann-Schwarz, M. Jovanovic, R. Aebersold and O. Vitek, *Mol. Cell Proteomics*, 2012, **11**, M111.014662.
125. R. Kiyonami, A. Schoen, A. Prakash, S. Peterman, V. Zabrouskov, P. Picotti, R. Aebersold, A. Huhmer and B. Domon, *Mol. Cell Proteomics*, 2011, **10**, M110.002931.
126. R. Aebersold, *Nature*, 2003, **422**, 115.
127. Y. Zhao and O. N. Jensen, *Proteomics*, 2009, **9**, 4632.
128. M. Trost, G. Bridon, M. Desjardins and P. Thibault, *Mass Spectrom. Rev.*, 2010, **29**, 962.
129. D. Domanski, L. C. Murphy and C. H. Borchers, *Anal. Chem.*, 2010, **82**, 5610.
130. H. Mirzaei, R. S. Rogers, B. Grimes, J. Eng, A. Aderem and R. Aebersold, *Mol. Biosyst.*, 2010, **6**, 2004.
131. S. C. Bendall, E. F. Simonds, P. Qiu, e.-A. D. Amir, P. O. Krutzik, R. Finck, R. V. Bruggner, R. Melamed, A. Trejo, O. I. Ornatsky, R. S. Balderas, S. K. Plevritis, K. Sachs, D. Pe'er, S. D. Tanner and G. P. Nolan, *Science*, 2011, **332**, 687.
132. B. Bodenmiller, E. R. Zunder, R. Finck, T. J. Chen, E. S. Savig, R. V. Bruggner, E. F. Simonds, S. C. Bendall, K. Sachs, P. O. Krutzik and G. P. Nolan, *Nat. Biotechnol.*, 2012, **30**, 858.
133. A. C. Peterson, J. D. Russell, D. J. Bailey, M. S. Westphall and J. J. Coon, *Mol. Cell Proteomics*, 2012, **11**, 1475.
134. E. Ahrne, L. Molzahn, T. Glatter and A. Schmidt, *Proteomics*, 2013, **13**, 2567.

CHAPTER 5

Proteomics Standards with Controllable Trueness—Absolute Quantification of Peptides, Phosphopeptides and Proteins Using ICP- and ESI-MS

ANNA KONOPKA, CHRISTINA WILD, MARTIN E. BOEHM AND WOLF D. LEHMANN*

Molecular Structure Analysis, German Cancer Research Center (DKFZ), Heidelberg, Germany
*Email: wolf.lehmann@dkfz.de

5.1 Introduction

Quantitative data are of outstanding importance in the life sciences,[1] in particular for testing scientific hypotheses.[2] Strong focus has been put on quantitative proteomics based on mass spectrometry (MS) over the last decade.[3–7] Due to the initial lack of standards, quantitative proteomics was initially nearly synonymous with relative quantification. Relative measurements can be performed without standards since they are only quantitative comparisons between two (or more) samples. A common scenario in relative quantification is to compare two (or more) cell populations subjected to

New Developments in Mass Spectrometry No. 1
Quantitative Proteomics
Edited by Claire Eyers and Simon J Gaskell

Published by the Royal Society of Chemistry, www.rsc.org

different biochemical perturbation. Protein signal intensity differences observed can be interpreted as protein concentration differences so that up- and down-regulated proteins can be recognized. In this way fold changes in protein concentrations can be measured without knowledge of their actual concentrations. When performed on the peptide level (the bottom-up proteomics strategy) with differently labeled samples, statistical analyses implied that significant up- and down-regulation events are connected with ratio changes in excess of a factor of 1.5 to 2.[8,9] This, in turn, complies with an standard deviation (S.D.) value of around 30% for individual peptide intensity ratios,[10] so that a data range of ± 3 S.D. roughly corresponds to a factor of 2.

In contrast, absolute quantification requires peptide or protein standards (stable isotope-labeled peptides or proteins with known concentrations) to be added. By this approach concentrations are measured, from which concentration ratios between different samples can then be derived. To summarize, relative data only represent unitless concentration ratios, whereas absolute data additionally provide insight into the actual concentrations, *e.g.*, in the chemical unit (mol/volume) or in the common biological unit (copy number/cell). Compared to relative data, absolute quantification allows for better data comparability between cell types, organisms, and laboratories and, in addition, is a much better basis for model calculations.

Development, validation, and interconnection of quantitative proteomics methods strongly depend on the availability of standards. Figure 5.1 summarizes currently available peptide and protein standards and the points of their addition within a bottom-up proteomic workflow.

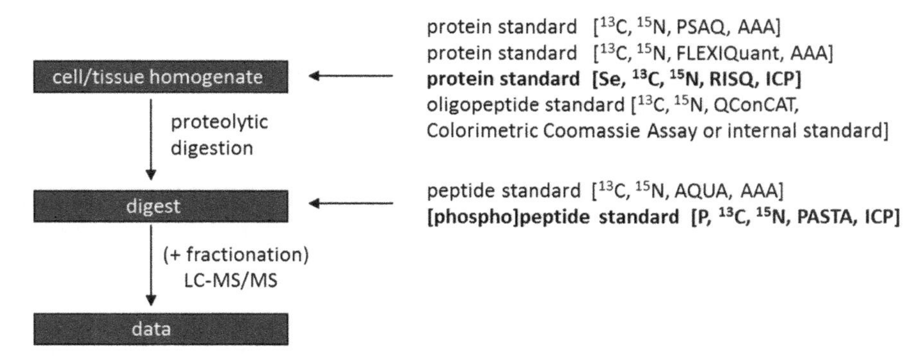

Figure 5.1 Workflow of quantitative bottom-up proteomics with annotation of the typical levels of addition of the internal standard. PSAQ, protein standards for absolute quantification;[17,18] FLEXIQuant, full-length expressed stable-isotope-labeled proteins for quantification;[10] RISQ, recombinant isotope-labeled and selenium quantified;[19] AQUA, peptides for absolute quantification;[11,12] PASTA, phosphorus-based absolutely quantified standards;[13] QConCAT, quantification concatamer.[14–16] Standards quantified by use of ICP-MS and ESI-MS are printed in bold.

The standards listed in Figure 5.1 can be described, briefly, as follows: AQUA peptides are stable isotope-labeled peptides quantified by amino acid analysis (AAA).[11,12] PASTA (phosphorus-based absolutely quantified standards) peptides are stable isotope-labeled phosphopeptides quantified by inductively coupled plasma mass spectrometry (ICP-MS) with phosphorus detection.[13] QconCAT (quantification concatamer) proteins are recombinantly expressed constructs of concatenated tryptic peptides from different target proteins quantified by a colorimetric assay[14,15] or using an internal peptide standard and stable isotope dilution.[16] PSAQ (protein standards for absolute quantification) are recombinantly expressed stable isotope-labeled full-length protein standards, which are quantified by AAA.[17,18] FLEXIQuant (full-length expressed stable isotope-labeled proteins for quantification) describes recombinant proteins with an *N*-terminal short peptide extension as a quantification tag.[10] After enzymatic digestion this extension is quantified relative to a stable isotope-labeled analogue named heavy FLEX-peptide quantified by AAA. RISQ (recombinant isotope-labeled and selenium quantified) protein standards are generated by cell-free synthesis in the presence of L-SeMet replacing L-Met and selected stable isotope-labeled amino acids.[19] Quantification of RISQ standards is achieved by ICP-MS and selenium detection.

Standards can be added at different levels of the workflow, as indicated in Figure 5.1. Peptide standards are added together with protease(s) or are added to the mixture of proteolytic peptides. QconCAT standards and protein standards are usually added directly to the cell homogenate prior to addition of the protease(s). In the bottom-up proteomic workflow protein standards are superior compared to peptide standards since they are added at the earliest point of sample processing. Irrespective of this detail, accurate standard quantification is the basic premise to achieve results with a high level of trueness. This point is a major focus of this chapter.

The majority of standards are quantified either by colorimetric methods or by quantitative AAA. In the latter technique amino acids are quantified subsequent to complete acid hydrolysis, which is optionally followed by a derivatization step.[20,21] The final quantification can be achieved either by fluorescence or UV detection. Both AAA and colorimetric methods quantify all peptides or proteins present in the sample so that the standard samples need to be pure to obtain accurate data. Preparation of pure synthetic peptides can be challenging due to the occurrence of deletion sequences and other artefacts during peptide synthesis and the concomitant difficulty to completely separate side products from the peptide of interest. Production of pure protein samples is even more difficult due to the moderate separation power of protein purification methods, which often is insufficient in relation to the typical complexity of protein samples. Therefore, the use of quantification methods with rateable trueness, which are not compromised by standard purity, is recommended. These methods are based on ICP-MS and electrospray ionization (ESI)-MS and are described in detail in the following sections.

5.2 μLC-ICP-MS as an Absolute Quantification Method for Peptides

Since its invention in the mid 1970s, ICP-MS has developed into a powerful and versatile element-specific quantification technique for metals and semi-metals.[22] The rough ionization conditions in the ICP source completely break down analyte molecules to atoms and atomic ions. Therefore, the response of the method is almost independent of the chemical nature of the analyte and is only weakly matrix-dependent. Due to this feature of ICP-MS, internal standards can be selected mainly on the basis of their optimal chemical purity, stability or suitable behavior during liquid chromatography (LC) and are independent of their chemical similarity to the analyte. Moreover, for multi-isotopic elements, ICP-MS can be combined with stable-isotope dilution to obtain data with optimal precision of around 1%. Direct coupling of ICP-MS to various separation methods,[23,24] including nanoUPLC (nano ultra performance liquid chromatography),[25] has created the basis for its broader application in the field of bioinorganic analysis.[26–30] In the following, the use of LC-ICP-MS for peptide quantification is described in detail.

5.2.1 Phosphorus-quantified Phosphopeptide Standards

So far, two approaches have been reported for peptide quantification using LC-ICP-MS. In peptides containing tyrosine the tyrosine residue was quantitatively converted into diiodotyrosine and, subsequently, LC-ICP-MS with iodine detection was used for quantification relative to an internal standard, such as 2-iodobenzoic acid.[31] In the second strategy synthetic phosphopetides containing phosphoserine, phosphothreonine or phosphotyrosine (pSer, pThr, or pTyr) were quantified directly by ^{31}P detection relative to an internal standard, such as bis(4-nitrophenyl)phosphate (BNPP).[13,32]

Phosphopeptides can be synthesized in a straightforward manner since pSer, pThr, and pTyr residues are available for chemical peptide synthesis in protected forms. The basic sensitivity of ICP-MS for detection of P is not as high as for metals, but still excellent for the detection of phosphopeptides and phospholipids.[33,34] In combination with solvents with low phosphate content (<1–10 ppb) a limit of detection in the femtomole range (50 fmol) can be achieved by μLC-ICP-MS. The available amounts of synthetic phosphopeptide standards typically exceed this limit by several orders of magnitude. Phosphorus is a monoisotopic element (^{31}P) precluding isotope dilution analysis, in which optimal quantification accuracy around 1% can be achieved. However, the use of internal standardization with a phosphorus-containing compound of high purity (*e.g.*, BNPP) provides data with accuracy between 1–6%.[32]

Regarding the positive features of LC-ICP-MS for the quantification of peptides containing iodine or phosphorus mentioned above, the corresponding peptide LC-peak purity remains a major challenge. Information on the standard purity can be obtained by analysis of the same sample by LC-ESI-MS. In the majority of investigations reporting both LC-ESI and LC-ICP data for the same sample, results of separate runs performed with identical LC conditions were aligned according to retention times.[31–33,35–37] However, exact alignment of two LC runs over a broader retention time window is a complex problem due to non-linear retention time variations that are intrinsically present in all gradient LC runs, which tend to increase with decreasing LC flow rate. The comparison between two LC runs connected with two different detection systems, as in the use of LC-ICP-MS and LC-ESI-MS, is even more challenging since the recognition of similar peak patterns (features) may be impossible over wide retention time regions. This problem is best solved by installing a y-split at the end of the LC column, which directs the eluent to both an ICP-MS and an ESI-MS system so that both elemental and molecular information is obtained within a single LC run.[13,38,39] The set up and typical results of such a y-split μLC-[ICP/ESI]-MS system obtained for analysis of an equimolar mixture of four phosphopeptides is shown in Figure 5.2.

The ^{31}P trace in Figure 5.2 shows the equimolar composition since the peak area of the doubly phosphorylated peptide is about twice as large compared to those of the other three singly phosphorylated species. Quantification of the phosphopeptides can be performed relative to the internal standard BNPP, which elutes after the phosphopeptides as a separate peak. Due to the low ionization efficiency of BNPP in positive ion ESI-MS, this

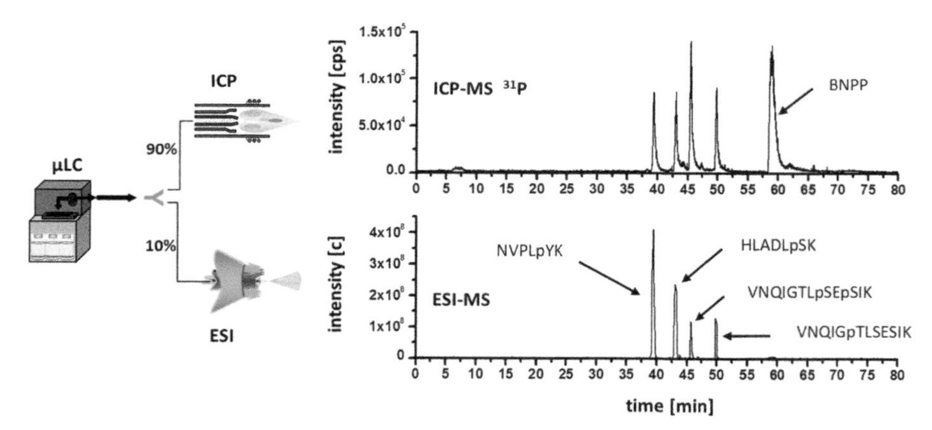

Figure 5.2 Set up of a combined μLC-[ICP/ESI]-MS system and typical data.[13] Shown is the analysis of a commercially available standard (MassPREP™ Phosphopeptide Standard Enolase, Waters) containing equimolar amounts of the four annotated phosphopeptides. A capLC system (Waters, Milford, MA) was used for HPLC separations coupled online to a sector field ICP-MS type Element 2 (Thermo, Bremen, Germany) through a low-flow microconcentric nebulizer CEI-100 (CETAC, Omaha, USA) and to an ion trap LCQ Deca XP Plus (Thermo, Bremen, Germany) at a flow rate of 5 μL min^{-1} (reproduced from ref. 13).

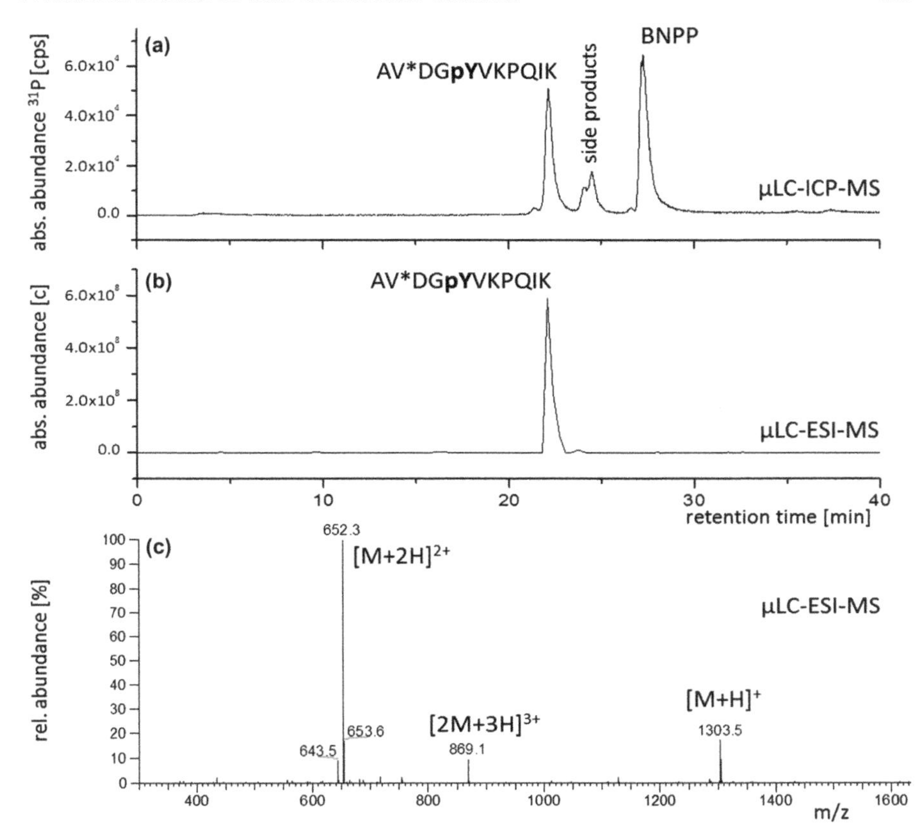

Figure 5.3 Specific phosphopeptide quantification *via* y-split μLC-[ICP/ESI]-MS. a) ICP-MS trace with ^{31}P detection; b) ESI-MS selected ion current for *m*/*z* 652.3 ($[M + 2H]^{2+}$ ion of AV*DGpYVKPQIK); c) ESI mass spectrum at a retention time of 22.1 min, corresponding to the peak in panel b) (V*, L-[$^{13}C_5$,^{15}N]-Val). pY in bold indicates the phosphorylated tyrosine residue.

compound does not show up in the LC-ESI-MS chromatogram. Results obtained with this y-split μLC-[ICP/ESI]-MS system for the quantification of a phosphopeptide are shown in Figure 5.3. This phosphopeptide is a stable-isotope-labeled analogue of the tryptic peptide AVDGpYVKPQIK of the activated signaling protein STAT5 that contains its activation motif. In activated STAT5 the tyrosine residue in this peptide is phosphorylated (designated as pY).

Figure 5.3a gives the ^{31}P trace showing the elution of the standard phosphopeptide, phosphorus-containing impurities (from the synthesis or degradation), and the added internal standard BNPP, as indicated. The phosphopeptide standard can be quantified by comparing its peak area to that of the internal standard. A correction factor has to be included in the quantitative evaluation, which corrects the moderate increase of ionization efficiency with retention time. This effect is caused by the increasing acetonitrile content during gradient elution, which causes a higher plasma temperature, which in turn leads to a more effective $[^{31}P]^{+}$ formation.[33] The phosphopeptide standard represents the first-eluting compound, as indicated by the selected ion chromatogram for the *m/z* value of its $[M+2H]^{2+}$ ion displayed in Figure 5.3b. The corresponding mass spectrum recorded at the top of this LC-peak is shown in Figure 5.3c. At a relative abundance level >3%, this mass spectrum only shows ion signals of the standard compound of interest. Therefore, we conclude that the phosphorus signal obtained by μLC-ICP-MS (Figure 5.3a) provides the correct quantitative value for the standard phosphopeptide of interest. Quantification by LC-MS may be biased by incomplete recovery of the phosphopeptide from the LC column. In this context we have shown that virtually complete recovery of small to medium size phosphopeptides is observed when sample amounts >5 pmol are injected onto a capillary column.[13] To obtain a high recovery of multi-phosphorylated peptides, addition of a metal ion complexing agent, such as citrate, to the sample is recommended to avoid metal ion-mediated IMAC-like adsorption of phosphopeptides on the reversed-phase column.[40,41] In summary, we conclude that μLC-[ICP/ESI]-MS with a chelator spike-in is a reliable method to prepare absolutely quantified phosphopeptide standards.

5.2.2 Phosphorus-quantified but Phosphorus-free Peptide Standards

The phosphoryl group of pSer, pThr and pTyr phosphopeptides can be selectively and quantitatively removed by incubation with a phosphatase. This step opens the possibility for the generation of peptide standards *via* quantitative dephosphorylation of phosphopeptide standards. Obviously, this concept requires the dephosphorylation step to be complete. Enzymatic dephosphorylation of phosphopeptides is less problematic than phosphoproteins due to their lower ability to form secondary structures that may sterically interfere with the dephosphorylation step. Nevertheless, control of the completeness of phosphoryl group removal should be performed as an essential quality check of a peptide standard generated in this way. Figure 5.4 shows data of such a control experiment. NanoESI mass spectra of a phosphopeptide standard before (Figure 5.4a) and after (Figure 5.4b) enzymatic dephosphorylation are shown. The nanoESI mass spectrum in Figure 5.4a exclusively shows phosphopeptide signals and no peptide signals, whereas in Figure 5.4b the reverse situation is observed. When such a result is obtained, it can be concluded that the concentration of the dephosphorylated peptide equals exactly that of the phosphopeptide. Thus, the phosphopeptide standard is converted quantitatively into the corresponding peptide standard.

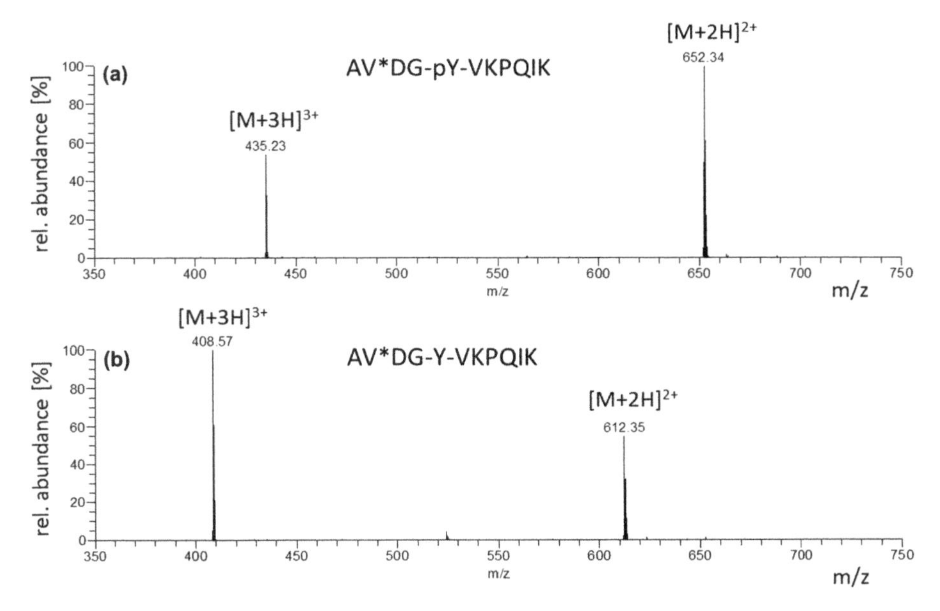

Figure 5.4 Generation of a peptide standard *via* complete dephosphorylation of a phosphopeptide standard. The peptide AV*DG–Y–VKPQIK contains the activation motif of the signaling protein STAT5. a) ESI-MS spectrum of the phosphopeptide standard before dephosphorylation; b) ESI-MS spectrum after dephosphorylation. (V*, L–[$^{13}C_5$,^{15}N]–Val).

In analogy to the discussion above, it may be objected that the result shown in Figure 5.4b is not sufficient to prove complete dephosphorylation because of the different ionization efficiencies between peptides and the corresponding phosphopeptides. While this point was of great concern in the early days of phosphopeptide analysis, it has been shown in the meantime that preferred ionization can occur for both species and that the differences are smaller than originally anticipated. Typically, the ionization efficiency ratios for peptide/phosphopeptide pairs lie in the range from 0.5 to 1.5.[40,42,43] Therefore, the spectra in Figure 5.4a and 5.4b are highly useful for the purity control of peptide and phosphopeptide samples before and after dephosphorylation. A minor limitation of preparing peptide standards *via* dephosphorylation of the phospho-forms is that it requires the presence of a Ser, Thr or Tyr residue in the peptide sequence. However, this confinement can be overcome by adding an enzyme-cleavable extension, which contains a phosphoamino acid residue. Tryptic peptides can be synthesized, *e.g.*, with a *C*-terminal sequence [K/R]XXpY. Following phosphopeptide standard quantification, the phosphoamino acid is then removed by tryptic digestion. The latter strategy resembles the concept of "equalizer peptides"[44] or "tandem peptides",[45] where also equimolar mixtures of two peptides are produced by enzymatic peptide cleavage.

5.2.3 Peptide/Phosphopeptide Ratio Standards

Reversible protein phosphorylation is a widespread regulation principle connected with a large variety of cellular processes, such as mitosis, metabolic regulation or receptor-mediated signal transduction.[46] Signaling proteins are frequently activated/deactivated by phosphorylation/dephosphorylation of Ser, Thr or Tyr residues positioned in an activation motif. In numerous cases the activation status of a signaling protein is represented by the degree of phosphorylation of its activation motif. Therefore, site-specific phosphorylation status measurements are of high importance for signaling dynamics studies, as well as for investigating the architecture and crosstalk of signaling pathways.[47] MS has developed into a key technique for site-specific protein phosphorylation analysis both for pinpointing phosphosites and for determining the degree of their phosphorylation.[48] In bottom-up proteomics, phosphorylation status data are extracted from molar ratios of a related peptide/phosphopeptide pair. This can be performed best by addition of a stable isotope-labeled peptide/phosphopeptide ratio standard (one-source standard), which provides an internal standard for both the phosphorylated and the unmodified peptide at defined molar ratio.[49] The production of one-source peptide/phosphopeptide ratio standards can be regarded as a combination of the procedures described in the two preceding sections. These standards are prepared by mixing dephosphorylation-interconnected peptide and phosphopeptide standard solutions at a known molar ratio, most often at a ratio of 1 : 1. A great advantage of one-source peptide/phosphopeptide standards is that they work for phosphorylation status determination even without quantification. This is because the molar ratio of these standards is connected only with two volumetric measurements so that precision and trueness of the ratio quantification can be kept below 2%. Numerous analyses of the dynamic activation behaviour of signaling proteins have been performed, which have proven the benefits of this concept in practical work. Measurements of protein phosphoform abundances of activation motifs with one[49] and two phosphorylation sites[50] have been demonstrated, which are characterized by their precision of <5%.

5.3 ICP-MS as an Absolute Quantification Method for Proteins

Numerous strategies have been developed for the heteroatom-tagging of proteins for absolute quantification *via* ICP-MS over the last few years.[51] Among the various ICP-tags available, only the use of selenium will be described and discussed here. A small group of naturally occurring selenoproteins contain genome-encoded selenocysteine (Sec). These proteins stoichiometrically incorporate Se, as defined by the Sec number, and can thus be quantified in a straightforward manner by ICP-MS. In contrast, natural SeMet-containing proteins contain variable amounts of Se since

L-SeMet is incorporated in place of L-Met in relation to the availability of L-SeMet in the pool of free amino acids. These L-SeMet-containing proteins can readily be detected, but not quantified by ICP-MS because, under normal circumstances, Se stoichiometry at the Met positions is unknown.[52–56] Stoichiometric introduction of selenium into proteins as an ICP-tag is an attractive concept for protein standard production. Generation of such SeMet-containing proteins can be achieved by cell-free protein synthesis in the presence of L-SeMet instead of L-Met since both amino acids are well accepted by the ribosomal protein synthesis machinery. This replacement is possible due to the high geometric and chemical similarity between Met and SeMet. Introduction of Se as an ICP-tag is superior to alternatives employing other elements. For instance, stoichiometric protein iodination is difficult to perform and, for other isotopes, such as ^{13}C or ^{15}N, the natural background is considerably higher. Furthermore, ICP-MS sensitivity for ^{13}C or ^{15}N is lower compared to that observed for selenium or iodine, which is <1 ppb.

5.3.1 Selenium-quantified Protein Standards

We developed the concept of double labeling of a protein standard by the introduction of SeMet in place of canonical Met and selected stable isotope-labeled amino acid analogues.[19] SeMet serves as tag for quantification by ICP-MS, as previously described. After digestion, all peptides that do not contain methionine can serve as internal standards for quantification of the corresponding target protein. SeMet-containing peptides of the standard cannot be used for accurate quantification of their Met-containing analogues. We named this type of protein standard RISQ (for recombinant isotope-labeled and selenium quantified). The structure of RISQ standards and their application in bottom-up proteomics is shown in Figure 5.5.

By performing cell-free protein synthesis in the presence of L-SeMet instead of L-Met, only the newly synthesized protein contains L-SeMet since

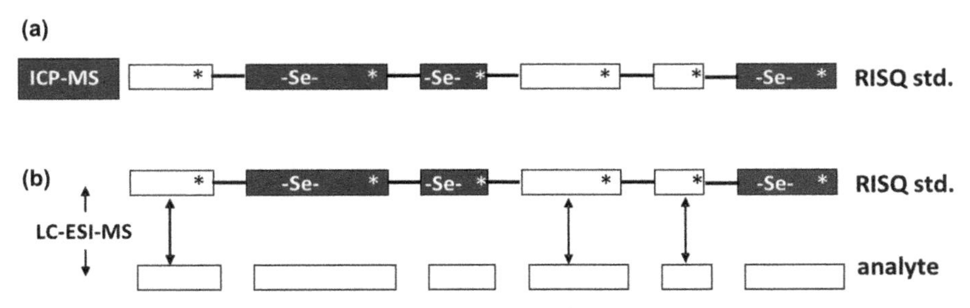

Figure 5.5 Structure and principle of RISQ standards. a) The intact RISQ standard contains SeMet and selected stable isotope-labeled amino acids, which are both introduced by cell-free protein synthesis; b) after enzymatic digestion of a RISQ standard in the presence of its non-labeled analyte protein, all SeMet-free peptides are observed as non-labeled/labeled peptide pairs, which can be used for relative quantification (*, stable isotope-labeled amino acid).

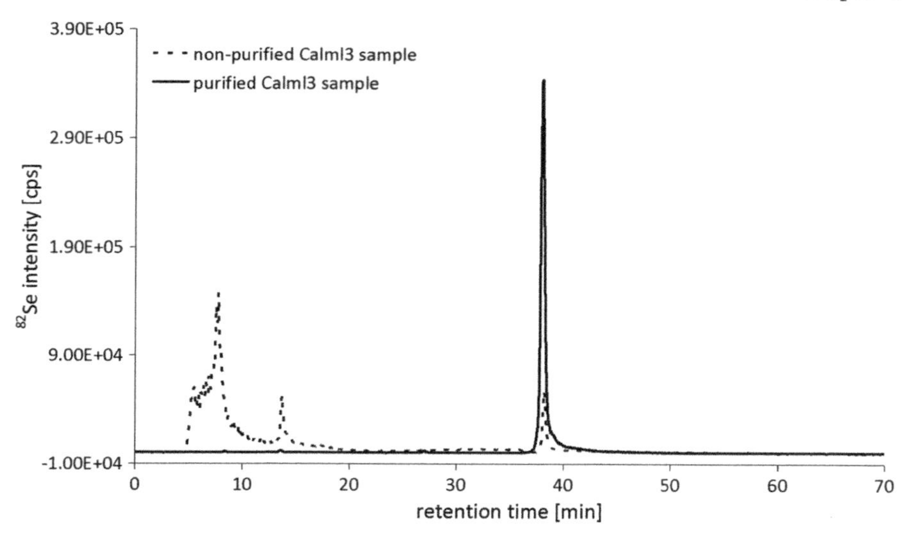

Figure 5.6 µLC-ICP-MS trace of the ^{82}Se isotope from CALML3 synthesized in a cell-free system containing L-SeMet, before and after purification by one-dimensional (1D) PAGE and electro-elution. After purification by one-dimensional PAGE and electro-elution of the CALML3 band, the RISQ standard is the only Se-containing species in the sample. Dashed line: µLC-ICP-MS trace of the crude cell-free synthesis reaction mixture; full line: µLC-ICP-MS trace of the CALML3 sample after 1D PAGE and electro-elution.

endogenous DNA transcription in the reaction mixture is inhibited. Analysis of the crude reaction mixture by µLC-ICP-MS with ^{82}Se detection mainly shows two peaks (Figure 5.6). One broad peak at a short retention time, which represents excess free SeMet, and one sharp peak at a longer retention time, representing the intact Se-containing protein standard, can be seen. After protein purification by one-dimensional (1D) PAGE and electro-elution of the protein from an excised gel band only the single large peak, indicating the intact protein, is observed. Figure 5.6 shows such a comparative analysis for the preparation of SeMet-labeled calmodulin-like protein 3, CALML3.

Once the RISQ protein has been verified as the sole Se-carrying constituent, as shown in Figure 5.6, it can be quantified directly by flow injection ICP-MS. Quantification is performed relative to an inorganic selenium standard calibration curve or by alternating sample and standard measurements. Table 5.1 summarizes Se-quantification data by ICP-MS recorded in quantification of serotransferrin labeled with SeMet. Averaged over 12 measurements, a concentration of 399 ng µL^{-1} (corresponding to 4.9 pmol µL^{-1}) was determined with a relative standard deviation of ~4%.

L-Selenomethionine is very similar to L-methionine. This similarity can be evidenced by the observation that both amino acids are incorporated into proteins. Their high functional similarity in Ca^{2+} activation has also been

Table 5.1 ICP-MS-based quantification of recombinant human SeMet-labeled serotransferrin produced by cell-free synthesis. The isotope, ^{82}Se, was monitored. Samples represent individual dilutions from the original batch of protein standard.

Sample number	*Concentration [ng μL^{-1}]*
1	397.4
2	382.9
3	385.4
4	397.0
5	409.0
6	395.9
7	387.6
8	411.7
9	419.9
10	418.9
11	415.5
12	367.2
average [ng μL^{-1}]	**399.0**
SD [ng μL^{-1}]	**16.4**
RSD [%]	**4.1**

observed for calmodulin highly substituted with SeMet and normal calmodulin.[57] However, an increased level of SeMet oxidation compared to Met oxidation has been observed for calmodulin recombinantly expressed in *E. coli*.[58] This effect was demonstrated by a stepwise change of the cell culture conditions from free L-Met to L-SeMet, which resulted in a mixture of normal and SeMet-labeled calmodulin. Analysis of this mixture by ESI-MS showed that the molecular ions of calmodulin only had a series of +16 Da satellites of minor abundance. In contrast, the SeMet-labeled calmodulin molecular ion signal was accompanied by a set of +16 Da satellite signals of roughly identical intensity among each other and to the unmodified SeMet-labeled calmodulin as well. Our finding in the analysis of a mixture of SeMet-tagged serotransferrin and normal serotransferrin confirmed the easier oxidative modification of SeMet compared to Met. A mixture of SeMet-tagged serotransferrin with its Met-analogue was digested by trypsin and analyzed by ESI-MS. The results obtained for the peptide LCMGSGLNLCEPNNK are shown in Figure 5.7. The oxidation status of the Met-peptide is lower compared to that of the SeMet-peptide. In addition, it can be recognized that the SeMet-peptide exhibits a longer retention time compared to the Met-peptide. This indicates that the substitution of S against Se results in an increased lipophilicity. For the LC retention time of the peptides hosting the oxidized peptide species, the reverse situation is observed as the oxSeMet peptide elutes before the oxMet

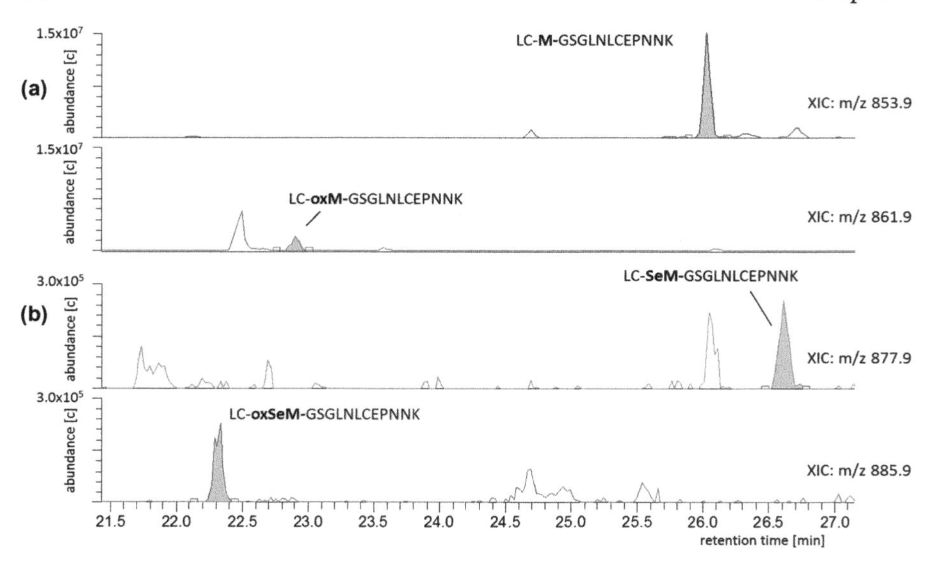

Figure 5.7 Differential LC-ESI-MS retention time and oxidation of a Met/SeMet pair of peptides within a single LC-MS run. Shown are the extracted ion chromatograms; a) elution of normal and oxidized peptide LC–M–GSGLNLCEPNNK; b) elution of normal and oxidized peptide LC–SeM–GSGLNLCEPNNK.

peptide. This indicates that oxidation of selenium generates a more pronounced increase in polarity compared to the corresponding oxidation of sulphur.

Figure 5.8 gives the doubly charged ion patterns of the peptide species annotated in Figure 5.7. The presence of SeMet can be directly recognized by the unusual isotopic pattern, which is caused by the presence of the multi-isotopic element selenium. As discussed above, there is evidence for the high structural and functional similarity between Met- and SeMet-containing proteins. However, analytical results also point towards a quantitatively different biochemical reactivity,[59] which may also cause a non-identical chemical reactivity of SeMet-containing standard proteins, in particular, during the analytical workflow of bottom-up proteomics or during storage. In studying digests of RISQ proteins mixed with non-labeled analytes (see Figure 5.5b) we observe systematic differences in the intensity ratios of non-labeled/labeled peptide pairs, which is probably caused by the presence of selenium. For Met/SeMet-free peptides, the non-labeled/labeled ratio was smaller compared to the corresponding ratio of peptide pairs containing Met/SeMet. Presumably, SeMet-specific side reactions are responsible for this observation. Therefore, to eliminate all differential effects caused by replacement of L-Met by L-SeMet, a strategy was developed that allows generation of selenium-quantified yet selenium-free protein standards.

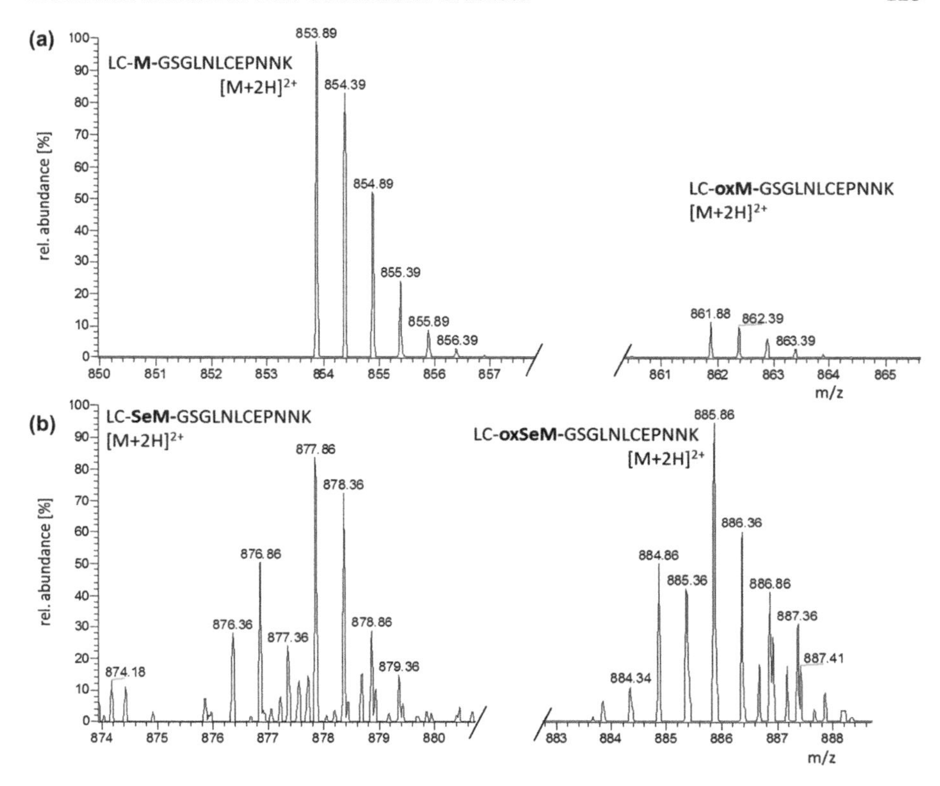

Figure 5.8 ESI mass spectra of the peptides annotated in Figure 5.7a and 5.7b. a) Ion patterns of the doubly charged species of the normal and oxidized peptide LC–M–GSGLNLCEPNNK; b) ion patterns of the doubly charged species of the normal and oxidized peptide LC–SeM–GSGLNLCEPNNK. These patterns can be used to determine the absence or presence of selenium due to the characteristic isotopic distribution.

5.3.2 Selenium-quantified but Selenium-free Protein Standards

The two types of labels present in a RISQ protein, the SeMet label and the stable isotope-labeled amino acids, can be separated into two protein species by adjusting the cell-free protein synthesis procedure for this purpose. As a result, two protein standard constructs are obtained, as shown in Figure 5.9. The SeMet containing protein species is quantified by ICP-MS with ^{82}Se detection, as described above (RSQ, recombinant and selenium quantified). Then, an aliquot of this RSQ standard is mixed with a standard species containing only the stable isotope-labeled amino acids (RIQ, recombinant isotope-labeled and quantified). When this mixture is subjected to enzymatic digestion, all Met-free peptide fragments form non-labeled/labeled pairs, which can be used for quantification of this protein species. In this way a secondary RIQ standard is created, which obtains its

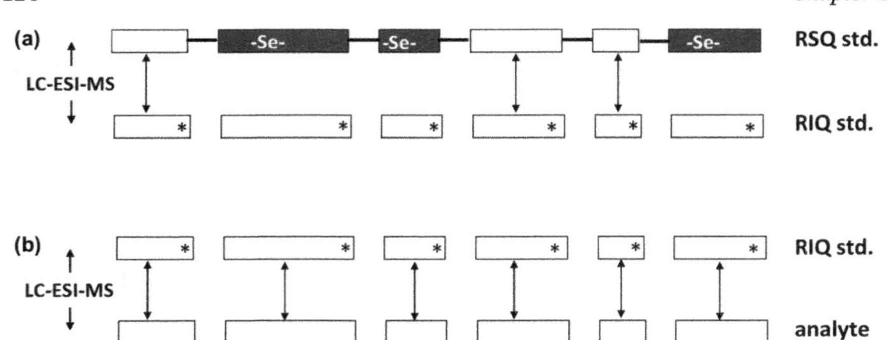

Figure 5.9 Principle of generation of selenium-quantified but selenium-free protein standards. a) The two types of labels present in a RISQ protein are distributed over two separate protein species, which are both generated by cell-free synthesis. Using protein digestion and LC-ESI-MS, the quantification value is transferred from the RSQ protein to the RIQ standard; b) the RIQ standard is then used for quantification of the analyte protein in a bottom-up experiment (*, stable isotope-labeled amino acid).

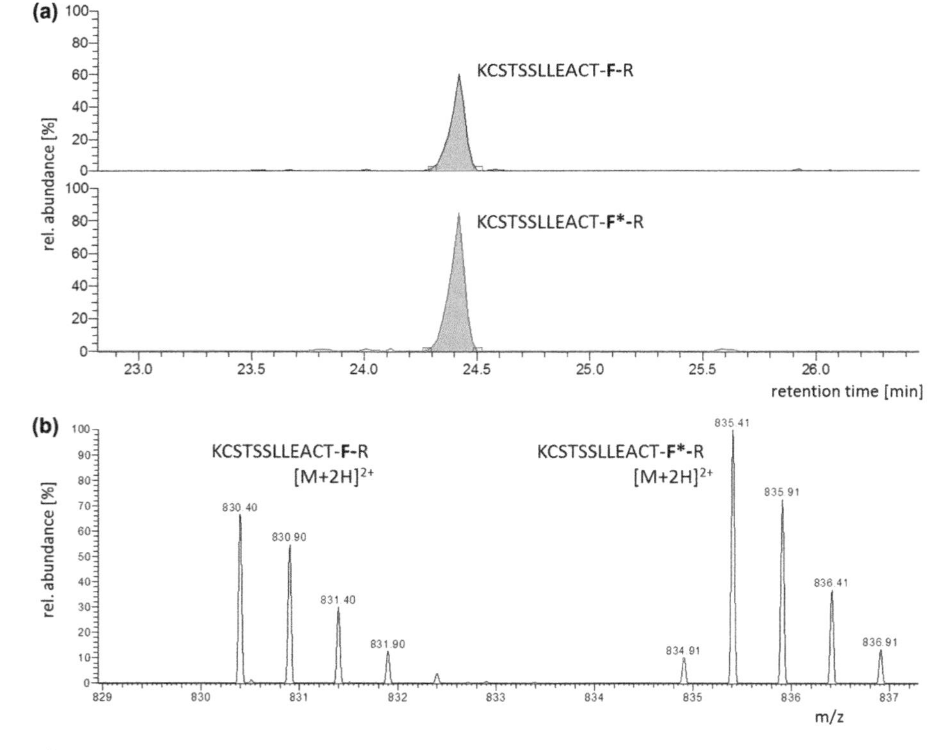

Figure 5.10 ESI-MS-based quantification transfer of an RSQ serotransferrin standard to a corresponding RIQ standard exemplified for a single tryptic peptide. a) Selected ion traces for the *m/z* values for the two $[M + 2H]^{2+}$ ions; b) mass spectrum obtained by averaging the MS spectra acquired over the complete LC peak. (F*, L-[$^{13}C_9$,^{15}N]-Phe; C, carbamidomethyl-Cys).

quantification from the RSQ standard, but which itself does not contain SeMet (see Figure 5.9a). Thus, the combined use of RSQ and RIQ standards has the advantage that, in the analyte protein quantification step on the peptide level, each proteolytic peptide can be used for (relative) quantification between the RIQ standard and the analyte protein, as depicted in Figure 5.9b. The "quantification transfer" between the RSQ and RIQ species can be performed with precision in the order of 1% in cases where several replicates are performed, and data are critically evaluated with respect to the isotope effects during the enzymatic digestion step.[60]

As an example for quantification transfer between an RSQ and an RIQ standard, Figure 5.10 gives a typical LC-ESI-MS trace observed in the elution of a non-labeled/labeled peptide pair and the corresponding MS spectrum averaged over the complete LC-peak. The extracted ion chromatogram of the two doubly charged ion signals in Figure 5.10a displays virtual co-elution of the non-labeled/labeled peptide pair. The mass spectrum in Figure 5.10b shows the clear separation of the two doubly charged ion species by 10 Da caused by the presence of one residue of [$^{13}C_9$,^{15}N]-phenylalanine.

5.4 Summary and Outlook

It has been demonstrated that, by using ICP-MS, μLC-[ICP/ESI]-MS, and μLC-ESI-MS in different combinations, a new class of more accurately quantified peptide, phosphopeptide and protein standards can be produced. All these standards incorporate stable isotope-labeled amino acids highly enriched in ^{13}C and ^{15}N. Peptides and phosphopeptides are generated by chemical synthesis and P is employed as the ICP-tag. In this way phosphopeptide standards are quantified directly by ^{31}P detection, whereas peptide standards are created *via* quantitative phosphopeptide standard dephosphorylation. Protein standards are generated by cell-free protein synthesis with stoichiometric L-SeMet incorporation in place of L-Met. Se serves as the ICP-tag for ICP-MS-based quantification of these standards. The standards described in this chapter represent a new class of proteomics standards characterized by their innovative methods of production and quantification. These methods ensure that their quantification will exhibit an outstanding level of trueness. We believe that the described methods will support current initiatives for the development of proteomics reference materials[61] and will thus improve comparability of quantitative proteomics data for different biological systems in different laboratories.

Acknowledgements

We are thankful to R. Pipkorn for expert phosphopeptide synthesis, N. Zinn, L. D'Alessandro, M. Schilling and U. Klingmüller for numerous valuable discussions, and to D. Winter for support in the LC-ESI-MS analysis. We gratefully acknowledge funding from the SBCancer network in the Helmholtz Alliance on Systems Biology and the German Federal Ministry of

Education and Research (BMBF)-funded MedSys-Network LungSys and the Virtual Liver Network. This research was undertaken within the project HLT05 (EMRP Researcher Grant). The EMRP is jointly funded by the EMRP participating countries within EURAMET and the European Union.

References

1. R. Phillips and R. Milo, *Proc. Natl Acad. Sci. USA*, 2009, **106**, 21465.
2. R. Picotti, B. Bodenmiller and R. Aebersold, *Nat. Methods*, 2013, **10**, 24.
3. S. E. Ong, L. J. Foster and M. Mann, *Methods*, 2003, **29**, 124.
4. M. Bantscheff, M. Schirle, G. Sweetmann, J. Rick and B. Kuster, *Anal. Bioanal. Chem.*, 2007, **389**, 1017.
5. J. R. Yates, C. I. Ruse and A. Nakorchevsky, *Annu. Rev. Biomed. Eng.*, 2009, **11**, 49.
6. V. Brun, C. Masselon, J. Garin and A. Dupuis, *J. Proteomics*, 2009, **72**, 740.
7. M. Bantscheff, S. Lemmer, M. M. Savitski and B. Kuster, *Anal. Bioanal. Chem.*, 2012, **404**, 939.
8. M. Monetti, N. Nagaraj, K. Sharma and M. Mann, *Nat. Methods*, 2011, **8**, 655.
9. J. V. Olsen, B. Blagoev, F. Gnad, B. Macek, C. Kumar, P. Mortensen and M. Mann, *Cell*, 2006, **127**, 635.
10. S. Singh, M. Springer, J. Steen, M. W. Kirschner and H. Steen, *J. Proteome Res.*, 2009, **8**, 2201.
11. S. A. Gerber, J. Rush, O. Stemman, M. W. Kirschner and S. P. Gygi, *Proc. Natl Acad. Sci. USA*, 2003, **100**, 6940.
12. D. S. Kirkpatrick, S. A. Gerber and S. P. Gygi, *Methods*, 2005, **35**, 265.
13. N. Zinn, B. Hahn, R. Pipkorn, D. Schwarzer and W. D. Lehmann, *J. Proteome Res.*, 2009, **8**, 4870.
14. R. J. Beynon, M. K. Doherty, J. M. Pratt and S. J. Gaskell, *Nat. Methods*, 2005, **2**, 587.
15. J. M. Pratt, D. M. Simpson, M. K. Doherty, J. Rivers, S. J. Gaskell and R. J. Beynon, *Nat. Protoc.*, 2006, **1**, 1029.
16. P. Brownridge, S. W. Holman, S. J. Gaskell, C. M. Grant, V. M. Harman, S. J. Hubbard, K. Lanthaler, C. Lawless, R. O'Cualain, P. Sims, R. Watkins and R. J. Beynon, *Proteomics*, 2011, **11**, 2957.
17. V. Brun, A. Dupuis, A. Adrait, M. Marcellin, D. Thomas, M. Court, F. Vandenesch and J. Garin, *Mol. Cell Proteomics*, 2007, **6**, 2139.
18. G. Picard, D. Lebert, M. Louwagie, A. Adrait, C. Huillet, F. Vandenesch, C. Bruley, J. Garin, J. Jaquinoda and V. Brun, *J. Mass Spectrom.*, 2012, **47**, 1353.
19. N. Zinn, D. Winter and W. D. Lehmann, *Anal. Chem.*, 2010, **82**, 2334.
20. M. I. Tyler, *Methods in Molecular Biology*, 2000, **159**, 1.
21. S. A. Cohen, *Methods in Molecular Biology*, 2000, **159**, 39.
22. S. M. Nelms (ed.), *ICP Mass Spectrometry Handbook*, Blackwell Publishing CRC Press, Oxford, 2005.
23. S. C. Shum, R. Neddersen and R. S. Houk, *Analyst*, 1992, **117**, 577.

24. J. Szpunar, R. Lobinski and A. Prange, *Appl. Spectrosc.*, 2003, **57**, 102A.
25. P. Giusti, D. Schaumlöffel, J. R. Encinar and J. Szpunar, *J. Anal. Atom. Spectrom.*, 2005, **20**, 1101.
26. J. Szpunar, *Analyst*, 2005, **130**, 442.
27. R. Lobinski, D. Schaumloffel and J. Szpunar, *Mass Spectrom. Rev.*, 2006, **25**, 255.
28. D. Profrock and A. Prange, *Appl. Spectrosc.*, 2012, **66**, 843.
29. J. Bettmer, M. Montes Bayón, J. Ruiz Encinar, M. L. Fernández Sánchez, M. del Rosario Fernández de la Campa and A. Sanz Medel, *J. Proteomics*, 2009, **72**, 989.
30. S. Mounicou, J. Szpunar and R. Lobinski, *Eur. J. Mass Spectrom.*, 2010, **16**, 243.
31. A. Pereira Navaza, J. Ruiz Encinar, A. Ballesteros, J. M. Gonzalez and A. Sanz-Medel, *Anal. Chem.*, 2009, **81**, 5390.
32. A. Pereira Navaza, J. Ruiz Encinar, M. Carrascal, J. Abian and A. Sanz-Medel, *Anal. Chem.*, 2008, **80**, 1777.
33. M. Wind, M. Edler, N. Jakubowski, M. Linscheid, H. Wesch and W. D. Lehmann, *Anal. Chem.*, 2001, **73**, 29.
34. B. O. Axelsson, M. Jornten-Karlsson, P. Michelson and F. Abou-Shakra, *Rapid Commun. Mass Spectrom.*, 2001, **15**, 375.
35. M. Wind and W. D. Lehmann, *J. Anal. Atom Spectrom.*, 2004, **19**, 20.
36. L. Tastet, D. Schaumloffel, B. Bouyssiere and R. Lobinski, *Talanta*, 2008, **75**, 1140.
37. D. Wesenberg, G. J. Krauss and D. Schaumloffel, *Int. J. Mass Spectrom.*, 2011, **307**, 46.
38. B. P. Jensen, C. J. Smith, C. Bailey, C. Rodgers, I. D. Wilson and J. K. Nicholson, *J. Chromatogr. B*, 2004, **809**, 279.
39. K. O. Amayo, A. Petursdottir, C. Newcombe, H. Gunnlaugsdottir, A. Raab, E. M. Krupp and J. Feldmann, *Anal. Chem.*, 2011, **83**, 3589.
40. D. Winter, J. Seidler, Y. Ziv, Y. Shiloh and W. D. Lehmann, *J. Proteome Res.*, 2009, **8**, 418.
41. J. Seidler, N. Zinn, E. Haaf, M. E. Boehm, D. Winter, A. Schlosser and W. D. Lehmann, *Amino Acids*, 2011, **41**, 311.
42. H. Steen, J. A. Jebanathirajah, M. Springer and M. W. Kirschner, *Proc. Natl Acad. Sci. USA*, 2005, **102**, 3948.
43. J. Gropengiesser, B. T. Varadarajan, H. Stephanowitz and E. Krause, *J. Mass Spectrom.*, 2009, **44**, 821.
44. J. Holzmann, P. Pichler, M. Madalinski, R. Kurzbauer and K. Mechtler, *Anal. Chem.*, 2009, **81**, 10254.
45. D. Winter, C. W. Hung, T. Jaskolla, M. Karas and W. D. Lehmann, *Proteomics*, 2012, **23–24**, 3470.
46. P. Cohen, *Trends Biochem. Sci.*, 2000, **25**, 596.
47. F. Marks, U. Klingmüller, K. Müller-Decker, *Cellular Signal Processing*, Garland Science, Oxford, 2009.
48. W. D. Lehmann, *Protein Phosphorylation Analysis by Electrospray Mass Spectrometry*, RSC Publishing, Cambridge, 2010.

49. B. Hahn, M. E. Boehm, V. Raia, N. Zinn, P. Möller, U. Klingmüller and W. D. Lehmann, *Proteomics*, 2011, **11**, 490.
50. B. Hahn, L. A. D'Alessandro, S. Depner, K. Waldow, M. E. Boehm, J. Bachmann, M. Schilling, U. Klingmüller and W. D. Lehmann, *J. Proteome Res.*, 2013, **12**, 637.
51. A. Sanz-Medel, M. Montes-Bayon, J. Bettmer, M. L. Fernandez-Sanchez and J. R. Encinar, *Trends Anal. Chem.*, 2012, **40**, 52.
52. P. Jitaru, G. Cozzi, R. Seraglia, P. Traldi, P. Cescon and C. Barbante, *Anal. Methods*, 2010, **2**, 1382.
53. G. Ballihaut, L. E. Kilpatrick and W. C. Davis, *Anal. Chem.*, 2011, **83**, 8667.
54. G. Ballihaut, L. E. Kilpatrick, E. L. Kilpatrick and W. C. Davis, *Metallomics*, 2012, **4**, 533.
55. I. L. Heras, M. Palomo and Y. Madrid, *Anal. Bioanal. Chem.*, 2011, **400**, 1717.
56. J. Bianga, G. Ballihaut, C. Pecheyran, Z. Touat, H. Preud'homme, S. Mounicou, L. Chavatte, R. Lobinski and J. Szpunar, *J. Anal. At. Spectrom.*, 2012, **27**, 25.
57. M. Zhang and H. J. Vogel, *J. Mol. Biol.*, 1994, **239**, 545.
58. G. Ballihaut, L. Tastet, C. Pecheyran, B. Boussiere, O. Donard, R. Grimaud and R. Lobinski, *J. Anal. At. Spectrom.*, 2005, **20**, 493.
59. D. T. Le, X. W. Liang, D. E. Fomenko, A. S. Raza, C. K. Chong, B. A. Carlson, D. L. Hatfield and V. N. Gladyshev, *Biochemistry*, 2008, **47**, 6685.
60. A. Konopka, M. E. Boehm, M. Rohmer, D. Baeumlisberger and W. D. Lehmann, *Anal. Bioanal. Chem.*, 2012, **404**, 1079.
61. (a) http://www.abrf.org/index.cfm/group.show/ProteomicsStandardsResearch Group.47.htm, (November 7, 2013); (b) http://www.fixingproteomics.org/advice/proteomics.asp, (November 7, 2013); (c) http://www.nibsc.ac.uk/products/biological_reference_materials/product_catalogue.aspx, (November 7, 2013); (d) http://irmm.jrc.ec.europa.eu/reference_materials_catalogue/Pages/index.aspx, (November 7, 2013); (e) http://www.nist.gov/bio-standard-reference-materials_pp.cfm, (November 7, 2013).

LABEL-FREE PROTEIN QUANTIFICATION

CHAPTER 6

Overview and Implementation of Mass Spectrometry-Based Label-Free Quantitative Proteomics

ERIK J. SODERBLOM,* J. WILL THOMPSON AND M. ARTHUR MOSELEY

Proteomics Core Facility, Institute for Genome Science & Policy, Duke University Medical Center, Durham, NC, USA
*Email: erik.soderblom@duke.edu

6.1 Introduction

Liquid chromatography coupled with mass spectrometry (LC-MS)-based quantitative proteomics has proven to be an invaluable tool to globally characterize proteins or peptides from complex biological systems.[1] Although many iterations of these experiments have been successfully employed over the past two decades, the most widely adopted strategy consists of four main components: (i) extraction and solubilization of proteins from a particular biological matrix; (ii) proteolytic digestion of solubilized proteins followed by LC separation and MS analysis of the resulting peptide mixtures; (iii) qualitative assignment of peptide fragment ion mass spectra through automated database searching and quantification of initial peptide abundances; and (iv) summarization of peptide data to the protein level and statistical analysis of the resulting expression changes across multiple

New Developments in Mass Spectrometry No. 1
Quantitative Proteomics
Edited by Claire Eyers and Simon J Gaskell

Published by the Royal Society of Chemistry, www.rsc.org

samples.[2] Through the use of modern LC-MS instrumentation, this centralized workflow can now routinely result in the simultaneous identification and quantification of up to thousands of proteins in a single experiment. Coupled with the increasing number of publically available and well-curated protein sequence databases, large scale proteomics has had a substantial impact on globally characterizing complex protein interaction networks, defining novel protein expression changes as a function of system perturbation, and identifying protein expression biomarkers correlated with human disease.[3]

Although there are a number of experimental variations within each of these core components, the methodology employed for MS-based protein quantification remains an area of significant research interest. Currently, the majority of LC-MS-based protein quantification strategies rely on the metabolic incorporation or covalent attachment of stable isotope labels. Metabolic labeling is performed at the intact protein level within actively growing cultured cells in the presence of ^{15}N-enriched media or in media containing isotope-labeled ^{13}C/^{15}N arginine and lysine residues, known as stable isotope labeling by amino acids in cell culture (SILAC).[4,5] Covalent labeling is most often performed at the peptide level by introducing a variety of different exogenous stable isotope-labeled reagents, such as ^{18}O,[6] isotope-coded affinity tags (ICAT),[7] isobaric tags for relative and absolute quantification (iTRAQ) reagents,[8] and tandem mass tags (TMT).[9] Regardless of how the stable isotope labels are introduced, each of these strategies results in a digested proteome containing multiple isotopically labeled versions of every peptide, the quantity of each uniquely corresponding to the amount of protein from which it was derived and particular to a different biological condition or sample, assuming equivalent labeling efficiency. As the labeled peptide groups within each experiment share the same chemical and physical properties, combining them prior to LC-MS analysis ensures that any technical variation incurred through the analytical process is shared equally between the species. This pre-combination may be powerful in experiments where multi-dimensional fractionation is utilized to increase depth of proteome coverage, or where subproteome enrichments, such as TiO_2 phosphopeptide enrichments, are employed as these may represent sources of large quantitative variation unless rigorous analytical protocols and appropriate QC samples are included in the experimental design. Another advantage of a labeling strategy is the multiplexed nature of the resulting sample as a single LC-MS analysis can yield quantitative information on peptides and proteins originating from multiple unique biological conditions, albeit from a multiplexed sample where any one sample is diluted by the combination of the samples.

While labeling strategies have distinct benefits in minimizing technical error within the analytical measurement and increasing throughput, there are inherent limitations of these approaches, which reduce their effectiveness for certain experimental designs. For example, metabolic labeling strategies require the biological system to be amenable to the labeling procedure as normal endogenous protein expression machinery within actively growing cells is required to incorporate the isotope label. This limitation was partially overcome by the Mann group through the introduction of whole SILAC-labeled mice and the Super SILAC approach in which SILAC ratios between native peptides derived from a tissue sample and heavy SILAC peptides derived from a representative cultured cell line are generated.[10] However, the overall approach remains non applicable to many commonly used animal model systems, many human tissues, and nearly all biofluids for which a representative cultured cell line cannot be generated.[10,11] There are also restrictions to the number of unique biological conditions (up to three in a SILAC experiment) that can optimally be analyzed in a single LC-MS experiment with the proteome of two samples of interest being analyzed concurrently with a reference proteome. Covalent labeling strategies provide more flexibility in the nature of the biological sample that can be studied as they are almost exclusively performed after the proteolytic digestion. However, they are also limited to the total number of unique isotope labels (a maximum of eight for current iTRAQ reagents), are expensive to implement over a large number of samples, and add technical variation to the experiment from the additional chemical reaction steps.[12] In addition, the quantitative information from many chemical labels, including iTRAQ, are derived from MS/MS spectra and their quantitative accuracy and precision may be compromised if a correction factor is not applied due to compression of reporter ion intensities.[13–15]

In the past decade there has been growing interest in the use of label-free approaches to globally characterize complex proteomes. In contrast to labeling approaches, each sample in a label-free quantification experiment is individually processed and analyzed by LC-MS in their natural isotopic state. Quantification is then performed post-data acquisition using either spectral counting or ion intensity measurements. The most attractive features of these experiments are the straightforward nature of the sample preparation and the technically limitless flexibility in experimental design, both in the nature of the samples being analyzed, as well as the number of unique samples or biological conditions. Although the overall strategy of a label-free quantitative proteomic experiment is relatively straight forward, early iterations of these experiments were relatively limited in the dynamic range of quantification and the accuracy and precision of quantification was notably less than that of isotope labeling approaches, likely because of the lack of sophisticated software and the prevalence of low-resolution mass spectrometers. However, in recent years, there have been a number of advances in both data acquisition strategies and bioinformatic solutions, which have reduced technical variation and increased the accuracy of quantification so that they are at least equivalent with covalent labeling strategies, such as iTRAQ.[16,17] The focus of this chapter is to provide an overview of the various label-free quantification acquisition and informatics strategies, and highlight general procedural considerations in label-free experimental designs.

6.2 Label-free LC-MS Quantification Strategies

Label-free quantification can generally be categorized into two different strategies: spectral counting or ion intensity-based measurements. Although these approaches were both introduced over a decade ago, the use of spectral counting was initially more widely utilized for complex biological systems. This is likely due the compatibility of the approach with low-resolution mass spectrometers (*i.e.*, ion-traps), flexibility and compatibility with existing data acquisition workflows. Furthermore, sophisticated bioinformatics tools were not required to extract the quantitative information from the resulting datasets. In recent years ion intensity-based quantification strategies, most notably through the generation of extracted ion chromatograms (XICs) of eluting precursor ions, have become an increasingly popular alternative. Not surprisingly, this has paralleled the trend of many laboratories to adopt the use of high resolution chromatographic separations coupled with high resolution and high mass accuracy MS (*i.e.*, orbitrap and QToF instruments) analysis. In general, ion intensity-based measurements are more computationally intensive, but result in a more accurate description of protein abundance across a wider dynamic range than spectral counting that are not maintained for lower abundant proteins. Section 6.2.1 provides a technical overview of these two strategies.

6.2.1 Spectral Counting and Derived Indices for Label-free Quantification

The general workflow for a spectral counting quantification experiment is straightforward and does not require any specific sample handling or preparation procedures other than proteolytic digestion of each sample followed by a data-dependent LC-MS/MS analysis of the resulting peptides. As there are a number of analytical variables in a data-dependent acquisition (DDA), including the number of precursor ions that are selected for MS/MS per survey scan, the intensity thresholds that trigger MS/MS and dynamic exclusion times that can alter the number of peptides identified within a single analysis, quantification by spectral counting requires that all user-defined analytical variables are kept constant across multiple sample acquisitions. All acquired MS/MS spectra are then subjected to database searching using one of a number of suitable algorithms, and spectra that meet appropriate scoring thresholds are annotated and assigned to a specific gene product. The quantification is based on the rationale that proteins that are in higher abundance generally yield a higher number of uniquely identified peptides, a higher percent sequence coverage, and higher total assigned MS/MS spectra (*i.e.*, spectral count).[18] To increase both the number of uniquely identified proteins and the average number of peptides identified per protein, this approach has been commonly coupled with multidimensional LC fractionation approaches, such as strong-cation exchange (SCX) followed by reversed phase LC (RPLC).[18]

One of the first demonstrations of spectral counting in a complex system was from Liu *et al.*, in which a six protein standard mixture was spiked into a *S. cerevisiae* lysate at four different levels ranging from 25% to 0.25% relative to the background. The samples were subjected to a twelve-fraction SCX-RPLC-MS/MS analysis in triplicate using an ion-trap mass spectrometer.[19] Spectra were subjected to database searching using the *Pep_Prob* algorithm and spectral counts were compared for the six proteins across the spiking concentrations. Their results indicated that the spectral count metric yielded an accurate description of the relative protein abundance with R^2 linear fit values between 0.9967 and 0.9995 over a dynamic range of approximately two orders of magnitude. Their data were independently interrogated by Zhang *et al.*, in which five different statistical tests were applied to the Liu *et al.* spiked protein dataset. Using a Fisher's exact test, the results confirmed the ability to statistically differentiate large fold changes (at least five fold) in protein quantity from singlicate measurements, but required three replicates to confidently quantify a two-fold change.[20]

To improve the quantitative accuracy of this approach, spectral normalization strategies, such as total spectral count normalization (TSpC), normalized spectral abundance factor (NASF) and normalization to selected proteins (NSP), were introduced to account for the differences in protein sizes and technical variations occurred across multiple data acquisitions. In TSpC normalization, each protein's spectral count (SpC) value is normalized to the highest SpC value within a set of technical replicates and an average value is calculated. The averaged values across multiple samples are then normalized to the sample with the highest technical replicate subjected to TSpC before a comparison of relative protein abundance is made across multiple samples.[21] The NASF approach was first described by Zybailov *et al.* and accounted for differences in protein sizes by first dividing the SpC for each protein by the number of amino acids in that protein (L) to yield a spectral abundance factor (SAF). Each protein's SAF value is then normalized to the sum of all SAF values within the sample to yield the NASF, which could be used to compare a proteins relative abundance both within a sample and across multiple samples.[22] In contrast to TSpC and NASF approaches, the NSP approach normalizes the SpC of each protein to that of the internal protein standard(s) and assumes that the total SpC for the spiked internal standard(s) at the same concentration should be constant within replicates and across multiple samples. These three normalization approaches were investigated by Gokce *et al.* in a GeLC-MS/MS analysis of *Magnaporthe oryzae* spiked with equine myoglobin and chicken ovalbumin and analyzed on an LTQ-Orbitrap XL mass spectrometer. In a scatter plot comparison of each protein's SpC value *versus* its averaged SpC value across two technical replicates, applying either NSpC or NASF normalizations to the data significantly improved the deviation from an ideal correlation (1.0) from 0.099 in the unnormalized data to 0.0093 or 0.004, respectively. Although NSP improved the correlation between replicates, it was not as effective as the NSpC or NASF approaches.[23]

In addition to relative protein quantification across multiple samples, spectral counting approaches have also been applied to absolute protein quantification through the development of spectral indices. The protein abundance index (PAI) is defined as the number of observed MS/MS spectra from a protein divided by the number of all peptides potentially observable (*i.e.*, within *m*/*z* ranges) following proteolytic digestion.[24] A later revision of this approach exponentially modified the PAI index score to emPAI, which more accurately estimates the abundance of a protein within a complex mixture.[25] In the initial description of the emPAI approach, Ishihama *et al.* performed absolute quantification of 46 unique proteins by spiking a corresponding stable isotope-labeled synthetic peptide into a mouse whole cell lysate ranging from 30 fmol μL^{-1} to 1800 fmol μL^{-1}. The calculated emPAI concentrations of those interrogated proteins were consistent with the actual values with $R^2 = 0.89$. Similarly, the absolute protein expression (APEX) index estimates a protein's concentration per cell by introducing a correction factor, which makes the fraction of the expected peptides observed in an experiment proportional to the fraction of peptides actually observed in the experiment.[26] The correction factor, referred to as the Qi value, is calculated by first acquiring a separate highly fractionated dataset of the proteome being studied. A machine learning classification algorithm based on peptide length and amino acid composition then uses this training dataset to calculate Qi. In a standard ten-protein mixture analyzed by SCX-LC-MS/MS on an LCQ Deca ion-trap mass spectrometer alone or in the presence of a digested yeast background, APEX values for each protein correlated very well with R^2 values of 0.98 over approximately three orders of magnitude.[26]

A variation of spectral counting, which includes the use of summed ion intensities of fragment ion spectra, has been described by Griffin *et al.* in which they use a spectral index (SI), which is the cumulative product ion intensity for each significantly identified peptide (including all its spectra) for an individual protein.[27] In their evaluation of the approach they performed experiments in which either BSA (2.65 μg) or a complex plasma membrane fraction (40 μg) was spiked with a mixture of nineteen protein standards across a wide dynamic range (0.5–50 000 fmol). The normalized SI for each of the standard proteins was calculated and plotted as a function of protein load with $R^2 = 0.9239$. More recently, Colaert *et al.* described an approach called RIBAR, in which the fragment ion intensities of each unique peptide (including those containing differential post-translation modifications) belonging to a protein are summed.[28] A peptide ratio for each unique modified peptide is then calculated using these abundance estimations and then a $\log_2$ ratio for that protein is obtained by averaging the calculated $\log_2$-modified peptide ratios for a protein. In the xRIBAR method an extra ratio for each protein is obtained by calculating the ratio of the average summed fragment ion intensities in all spectra belonging to the protein. The results from this technique have outperformed both NASF and emPAI approaches in minimizing technical variation and more accurately estimating a protein's abundance when applied to two publically available datasets from CPTAC (Clinical Proteomic Technology Assessment for Cancer) Network and the 2009 Proteome Informatics Research Group (iPRG) of the Association of Biomolecular Resource Facilities.[29,30]

Although these spectral counting approaches have been shown to be useful in certain experimental designs, their implementation to accurately and reproducibly quantify proteins across the entire identifiable proteome has been relatively limited. Even with the application of various normalization approaches, saturation effects (*i.e.*, when no additional peptides are identified even at higher protein concentrations) are commonly observed.[19,26] Signal saturation is especially problematic for samples that have a high dynamic range of expressed proteins, such as serum, where a relatively small number of unique proteins account for the majority of the total number of identified spectra and lower abundant proteins are identified by fewer unique peptides. The main reason for this signal saturation is the inherent serendipity of a data-dependent analysis in which a precursor ion is selected for fragmentation and MS/MS analysis based on its relative abundance in a full MS scan compared to other co-eluting peptides. While the serendipity of precursor selection in a DDA experiment is less affected in the case of higher abundant precursor ions, it represents a considerable problem for lower abundant ions, which may indeed be present in the full MS scan but are simply not selected for subsequent MS/MS analysis.[31] This serendipity, in part, results in a reduced quantitative accuracy (unable to measure less than a three-fold change in abundance) for proteins identified by less than four spectral counts, which may make up a considerable portion of an identified proteome.[32] Although the linear quantifiable range of the proteome can be extended by coupling these experiments with multidimensional fractionation (for example, SCX-rpLC-MS/MS), these fractionation techniques add additional time to the overall experiment and will also result in a new population of proteins identified by less than four spectral counts.

6.2.2 Ion Intensity-based Label-free Quantification

As an alternative to spectral counting, a precursor ion's intensity or signal response within a mass spectrometer can also be used to quantify its relative abundance across multiple samples. In combination with high resolution chromatographic separations and high resolution mass spectrometers, this approach generally results in a more accurate and precise quantification of relative protein abundances across the entire identifiable dynamic range in comparison to spectral counting approaches. The overall workflows employed for these experiments are similar to other label-free experiments in that samples are individually prepared and are then subjected to an LC-MS/MS analysis. Qualitative identifications of eluting precursor ions are generated through normal database searches of fragmentation spectra and those passing certain confidence thresholds are subsequently assigned to the corresponding precursor ion at a particular retention time. Unlike spectral counting strategies, the quantification is performed from precursor MS spectra by generating extracted ion chromatograms (XIC), either of all detectable precursor ion *m/z* values or exclusively those that were qualitatively identified, using automated software algorithms. The quantification is based on the fact that an individual precursor ion's measured intensity linearly correlates with its concentration (*i.e.*, abundance) during the electrospray ionization process.[19,33] Since a peptide's chemical and physical properties remain constant from run to run, it is therefore possible to use the peak area under the curve (AUC) or peak height measurement from XICs of the same precursor ion to yield relative quantification information across multiple samples (see also Chapter 7 for MS1 label-free quantification using Skyline and its application in the area of cancer phenotyping).

This approach was first described in 2002 by Chelius *et al.* as a method for quantifying proteins within complex proteomic samples, where tryptically digested horse myoglobin (10 fmol–100 pmol) was analyzed on a low-resolution LTQ-Deca ion-trap mass spectrometer. Across the four orders of magnitude concentration range they found a linear response ($R^2 = 0.991$) in the XIC peak areas of five different myoglobin peptides.[34] The group extended their analysis to include a complex proteomic sample by spiking digested myoglobin into a human serum background at 250 fmol and 500 fmol. Using the XIC peak areas of three different myoglobin peptides, the measured ratio of protein abundance between the two samples was 1.91 after a normalization factor was applied, with a 5% error from the expected two-fold change.[34]

Although this work successfully demonstrated the ability to quantify a single well-characterized protein within a complex proteome, extrapolating the approach to provide quantitative information for all qualitatively identified proteins within a complex sample becomes challenging. This is in part due to unavoidable analytical fluctuations in the LC-MS platform, such as drift in precursor ion retention times or measured mass accuracy from injection to injection. In the context of a typical complex proteome, which may contain several near-isobaric precursor ions eluting within some retention time window, it quickly becomes very difficult to ensure that the exact same precursor ion is being compared across the multiple injections. This is compounded when considering all detectable or all identified precursor ions within a single LC-MS experiment. To help address these challenges, a number of XIC-based label-free workflows have utilized more sophisticated and comprehensive informatics solutions, which typically consist of the following four core components: (i) peak detection and filtering; (ii) LC-MS data alignment; (iii) data normalization and quantification; and (iv) assignment of the qualitative identifications to corresponding peaks (Figure 6.1).

Peak detection is typically an automated process in which a software package (see section 6.3) determines which features of the MS spectrum are going to be included for further interrogation. This is an important step as low quality precursor ions, which may not be derived from peptides, may

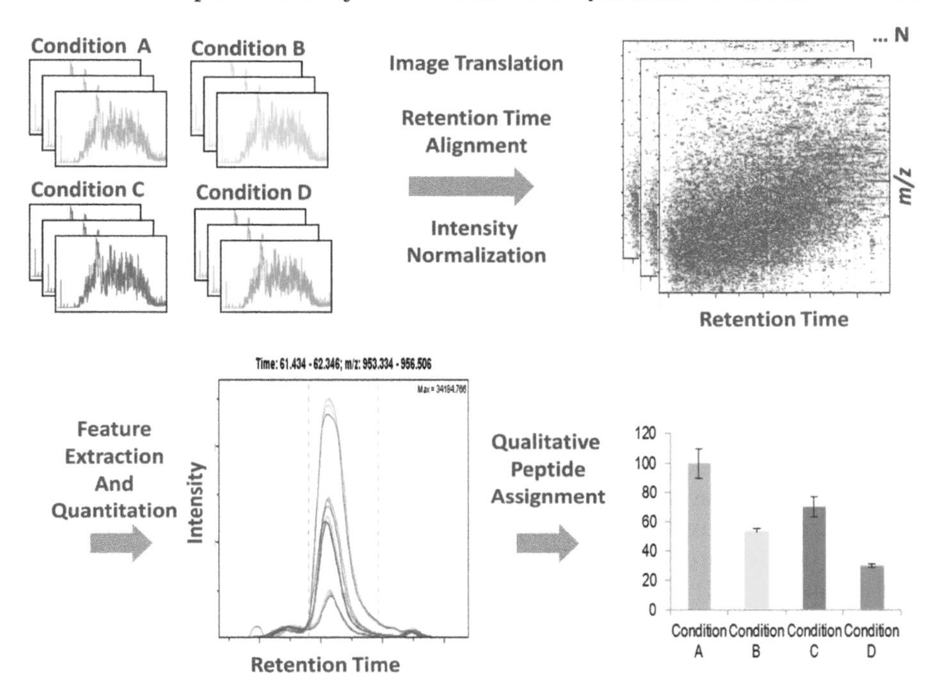

Figure 6.1 An overview of an extraction ion chromatogram (XIC)-based label-free quantitative proteomics workflow. Following independent LC-MS acquisition of any number of unique samples, raw data files are typically subjected to accurate mass and retention time alignment (AMRT) and intensity normalization. All detectable isotopomer features are subjected to extraction ion chromatogram generation and area-under-curve measurements are generated following peak integration across all individual LC-MS data files. Qualitative identifications are generated and independently scored using database searching algorithms and are then assigned to the aligned precursor ion groups from which the identification was initially made. To yield a protein level expression value, the same set of isotopomers belonging to a particular protein are summed or averaged before comparing across all samples.

interfere with the subsequent data alignment. These filters often include absolute intensity thresholds, restrictions on precursor ion charges or *m*/*z* ranges, chromatographic peak shapes and peak width, and normal isotope distribution of precursor ions. Modern high resolution mass spectrometers offer a distinct advantage to efficiently detect and differentiate precursor ions as compared with low resolution ion traps; baseline resolution can easily be achieved at charge states much higher than typically observed for tryptic peptide ions generated by electrospray ionization. In many software packages this resolution allows peak detection at the individual isotopomer (*i.e.*, monoisotopic peak, natural ^{13}C isotope, $^{13}C_2$ isotope *etc.*) level. Depending on the signal-to-noise of the precursor ions, typically three to six isotopomers (or "features") can be included in one precursor ion isotope group. Although this may increase total computational time for peak detection, it also allows for multiple XICs to be generated for each precursor peptide and results in a more robust analytical measurement. It also permits quantification on the ^{13}C isotopomer if the ^{12}C isotopomer is saturating.

Following peak detection, it is critical to ensure that the same precursor ion is being considered across the multiple samples before relative quantification can be performed. To allow precursor ion grouping to be performed comprehensively, alignment of all precursor ions within the raw LC-MS files is typically performed first across the entire data set based on their accurate mass (*m/z* and charge state) and retention time (AMRT) (Figure 6.2). An example alignment is shown in Figure 6.2 in which XIC traces of an eluting peptide in a dataset containing 40 raw data files (upper left panel) were subjected to AMRT alignment within Rosetta Elucidator®. The resulting overlap XIC traces (upper right panel) showed excellent alignment and subsequently allowed the software to more easily integrate the same peak areas across the multiple samples. The average amount of time for each of the 140 000 features within this particular dataset to shift into an aligned state (lower panel) was typically less than 30 s for the majority of the chromatographic separation time. In addition to increasing the confidence in qualitative identifications by assigning replicate search results correlated to the same precursor ion, this alignment also allows for the projection of qualitative identifications to samples in which the peptides were not confidently identified. Another important benefit of this alignment is a reduction in the consequence of a serendipitous data-dependent acquisition because each peptide does not necessarily need to be identified in every sample, as is often the case for lower abundant proteins or proteins that may have higher expression in a unique experimental condition.

The earliest description of the use of AMRT-aligned datasets was from Smith *et al.*, who created a database of high confidence peptide mass tags (PMT), which associated high confidence MS/MS qualitative information of peptides to their precursor accurate mass and retention time.[35] Since that work, a number of strategies have subsequently been described to align multiple LC-MS data files, including warping,[36] vectorized peaks,[37] non-linear regression methods,[38] statistical alignment,[39] and clustering[40] strategies. One common feature between all of these alignment approaches is the increased effectiveness arising as a result of using high resolution chromatographic separations, such as those obtained by ultra-performance LC (UPLC) systems coupled to high resolution, high mass accuracy mass spectrometers. The use of these systems, in combination with commercially available capillary chromatography hardware, typically results in highly reproducible separation with retention time peak widths of less than 20 s at the base for most eluting peptides. Ambiguity in assigning the same precursor ion across multiple samples to the same group is also significantly reduced through the use of high resolution, high mass accuracy mass spectrometers, mainly orbitrap or QToF mass analyzers. In combination

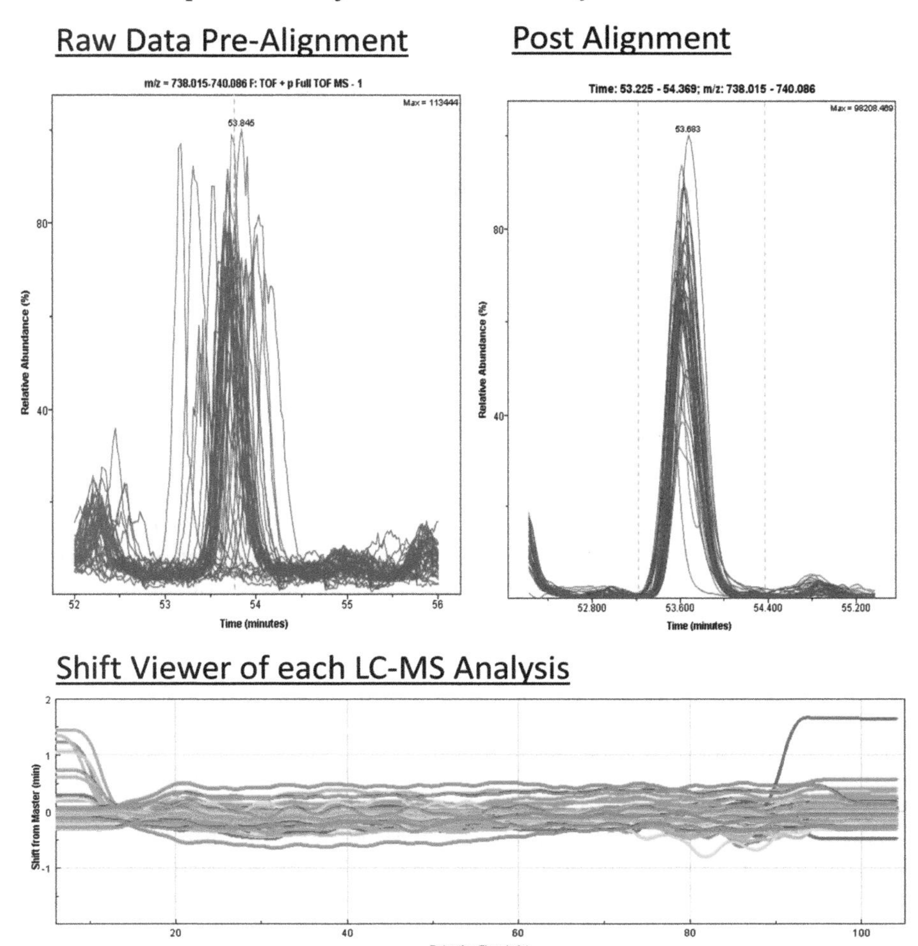

Figure 6.2 Utility of AMRT data alignment in an XIC-based label-free proteomic analysis. Prior to peak integration, an example feature (*m/z* 739.045) within the raw LC-MS files (upper left panel) from a 40-injection quantitative study was subjected to AMRT within Rosetta Elucidator®. The resulting AMRT aligned data (upper right panel) required that the apex of these XIC traces be shifted as much as 55 s, but resulted in a peak grouping that was able to be more accurately integrated. The average shift in retention time for all features across the 40 injections was, in most cases, less than 30 s, which is reflective of the highly reproducible nature of nanoscale capillary chromatography.

with real-time or post-acquisition mass correction based on a known calibrant, these instruments can routinely achieve a mass accuracy of less than ten parts per million at resolutions greater than 20 000. To ensure the accuracy of data alignment procedures, many software packages incorporate quality control visualizations to allow users to manually inspect random peak alignments at various intensities or assign and subsequently filter on peak alignment scores based on goodness of fit.

Following peak alignment, data are then typically subjected to intensity normalization to improve quantification. These normalizations adjust for slight variations in the total sample loaded on the column, which are determined by upstream total protein assays (Bradford, BCA, *etc.*) or from technical variations that may arise during injection, subtle changes in electrospray efficiency, or ion detector fluctuations. A number of different strategies for normalization have been described in the literature, including total ion current of all detectable signals, summed ion current from only those precursor ions qualitatively identified, summed feature intensity of all signals belonging to an endogenous protein not expected to change in abundance, or the summed feature intensity of an exogenously spiked protein standard (reviewed in ref. 41). Wang *et al.* found that, to increase the effectiveness of global normalization in complex proteomes, a robust normalization excluding the lowest portion of the detectable signals was more effective due to missing precursor ion intensity data when approaching instrument detection limits.[42] Once these various normalization strategies are performed, area-under-curve (AUC) or the peak heights of the integrated peak groups can be calculated. As these initial XIC measurements are typically performed at the individual isotopomer or feature level, a summarization of the data must be performed to report relative abundance changes at the peptide or protein level. Although this can be performed by either summing or averaging, it is important to note that it is always performed on the same set of features, thus enabling an accurate quantitative comparison across all samples.

Although various LC-MS acquisition methods can be employed for XIC intensity-based label-free studies, it is important that an adequate number of data points from the precursor MS scans are acquired over the entire peptide elution profile. For typical peak widths achieved by current nanoscale capillary chromatography (approximately 20 s at the base), a data point every 1.5 s (thirteen points across a 20 s wide peak) is generally desired to define a Gaussian distribution. An acquisition approach that is particularly well suited for XIC-based label-free quantification is data-independent acquisition (DIA), such as the MS^E approach used by the Waters Corporation. These acquisitions alternate between a low-energy MS scan and a high-energy scan during which all precursor ions are simultaneously fragmented and detected (see Chapter 8 for further information). Software is then used during spectral processing to assign specific high energy product ions to their respective precursor ions prior to database searching. The benefit for quantification comes from the high duty cycle of the precursor MS scan (every $\sim$0.5 s) with a constant frequency between each of the precursor ions scans, allowing for accurate generation of precursor ion XIC.

Over the past several years there have been an increasing number of XIC-based label-free quantitative proteomic experiments described in the literature (recently reviewed in ref. 43 and 44). The majority of these experiments have been applications to clinical studies from human tissue or biofluids. By using modern LC-MS instrumentation and more comprehensive software solutions, these experiments are capable of achieving a higher degree of quantitative accuracy and reproducibility than spectral-counting approaches. For example, a study by Nagaraj and Mann analyzed the variability in relative protein expression levels within the human urine proteome using an XIC-based label-free LC-MS approach with data analysis performed in MaxQuant.[42] Their workflow resulted in a median coefficient of variation in individual protein intensities ($n = >600$ proteins) of 18% within 21 technical replicates, which spanned approximately four orders of magnitude in intensity. Recently, Ralston-Hooper *et al.* utilized a label-free quantitative approach to measure protein expression changes in fathead minnow livers as a function of an oestrogen synthesis inhibitor drug.[43] As part of a quality control to assess analytical platform variability, a QC pool was created from equal parts of each of the 33 unique samples and was run periodically throughout the entire acquisition window. Following label-free quantification using the Rosetta Elucidator® software package, the technical variation across all proteins identified within the QC pool ($n = 782$ unique proteins) was an average of 8.9% RSD across four orders of magnitude. This level of reproducibility over a large dynamic range is not solely limited to unbiased protein expression studies. In recent work performed in our laboratory to assess quantitative changes in global peptide level phosphorylation directly from red blood cell membranes of sickle cell patients (eight unique treatment groups), an average technical coefficient of variation in peak intensity of 19.8%, including variation from independent TiO_2-based phosphopeptide enrichments, was achieved over three orders of magnitude.[44] This level of reproducibility was achieved, in part, by identifying and optimizing several key upstream variables during the enrichment process, such as resin-to-input material ratios and quantity of enrichment modifier used, which significantly reduced overall technical variation.[45]

6.3 Software Packages for Label-free Quantitative Proteomics

As with other MS-based quantification approaches, the use of appropriate software packages to accurately analyze and interpret the acquired data is critical for success. There are now several software packages, both commercial and open-source, available for interpreting and generating quantitative measurements from label-free datasets.[46] Although most instrument vendors offer software packages that fully integrate with data generated from their specific line of instruments, there has recently been a growing trend toward open source solutions that are instrument vendor neutral and can accept a variety of different standardized file formats.

For quantitative datasets acquired using spectral counting approaches, much of the data processing can be performed manually using simple spreadsheet programs following the exporting of database search results. Commonly used database searching programs, such as Mascot, automatically calculate emPAI values, which are spectral counting indexes used to calculate absolute protein expression values. Other open-source programs, such as PANORAMICS,[47] PepC,[48] and IDPicker,[49] offer more easily navigable graphical user interfaces and can accept common data file formats, such as pepXML, and incorporate additional features, such as the ability to easily calculate error statistics or apply differential expression tests. One of the more comprehensive data integration and visualization programs for label-free data is Scaffold, developed by Proteome Software. Although a commercial license is required to initially build Scaffold files, a free to download viewer is available to read the resulting Scaffold files so that they can be shared easily. In addition to accepting search results from the major database algorithms, such as Mascot, Sequest, and X! Tandem, the newest versions of Scaffold (v3.6 +) offer a number of built-in quantitative options. For normal spectral counting, Scaffold takes the sum of all the unweighted spectral counts for each sample, normalizes that sum across all samples, and then outputs a normalized spectral count for each protein. The total ion current (TIC) of the product ion spectra belonging to a protein can also be used for quantification and users have the option to choose the average of those TIC values, the sum total of those TIC values, or the summed total of TIC values from the three most intense precursor ions of a unique peptide sequence.

For ion intensity-based quantitative datasets, a number of open-source solutions are available, such as MSight,[50] PEPPeR,[51] MSInspect,[52] and MSQuant.[53] These programs incorporate alignment algorithms and data normalizations options to perform XIC-based quantitative measurements across a number of samples. These programs typically each have a set of unique features, such as being able to incorporate multiple reaction monitoring data (MSInspect) or isotope-labeled datasets (MSInspect and MSQuant). Another open-source solution is Skyline, developed by the MacCoss laboratory (University of Washington), which has the capability of importing data files from all major instrument manufacturers, as well as standardized data formats.[54] Although much of Skyline's functionality is directed toward targeted MRM data interpretation, the program also allows for the global generation of XICs based on a list of proteins input into the program. The graphical user interface within Skyline easily allows the user to visualize the XIC peak area of peptides across multiple samples in addition to tracking retention times and predicted *versus* experimental isotope distributions (see Chapter 7 for further information). In addition to open-source solutions there are also a number of full featured commercial ion intensity-based software solutions. SIEVE (Thermo Scientific) processes raw data files from Thermo instruments and uses an algorithm called ChromAlign to perform chromatographic retention time alignment. Progenesis LC-MS (NonLinear Dynamics) is capable of accepting common raw data file formats from Waters, Agilent, and Thermo instruments, and uses a peak-modeling algorithm to handle very large datasets, as well as an AMRT alignment algorithm, which is compatible with single- or multi-dimensional LC separations. Progenesis has a powerful graphical user interface with many visualization tools to identify and manipulate differentially expressed features through PCA and multi-dimensional clustering.

6.4 Label-free Experimental Design Considerations

The use of modern high resolution chromatography and MS analysis coupled with the increasing availability of commercial or open-source software solutions has been a driving force behind an increasing trend of employing label-free approaches for LC-MS protein quantification. Although these advances greatly enhance the utility of these workflows, the overall accuracy and precision of the protein quantification is still highly dependent on the protocols employed for upstream sample handling and downstream LC-MS analysis, primarily because each sample is processed and subjected to LC-MS analysis independently and has the potential to incur a significant amount of technical variation. Historically, this has been one of the most cited disadvantages of a label-free experimental approach as compared with, for example, a SILAC metabolic labeling approach in which any variation in sample handling and data acquisition procedures is realized equally by isotopically heavy and light versions of the protein. To address this potential challenge, a number of laboratories, including our own, have incorporated "best-practice" standard operating protocols into nearly all aspects of an LC-MS-based label-free workflow. The use of these protocols, in addition to a number of integrated quality control experiments and quality control samples addressing both sample handling and data acquisition procedures, routinely result in a highly reproducible and accurate peptide or protein level quantification (typically <10% at the protein level). The remaining portion of this section will highlight several of these best-practice standard operating protocols applicable to any quantitative MS-based experiment, with a particular focus on the application to XIC-based label-free quantitative experiments.

6.4.1 General Sample Handling Guidelines for Label-free Quantitative Experiments

Although it may seem very straight forward, arguably one of the most critical steps in a label-free LC-MS experiment is the protocol employed for initial protein solubilization and digestion across a sample cohort. Regardless of whether the starting biological matrix is cultured cell lines or various tissues or biological fluid samples, an effective and reproducible strategy needs to be employed to generate soluble protein prior to proteolytic digestion. If sample processing is not performed consistently from sample to sample, there is the possibility of broadly selecting for or against classes of proteins based on physical protein properties, such as hydrophobicity, net charge, or molecular weight. One of the first considerations in minimizing this variation is to start, wherever possible, with roughly the same quantity (*i.e.*, number of cells or wet weight of tissue) of intact biological material from all unique samples. Ideally, all non-MS compatible components, such as non-volatile salts and surfactants, should then be removed through multiple washes with a MS-compatible buffer, such as ammonium bicarbonate. Performing this at the level of the intact tissue or cell pellet prior to protein solubilization often pre-empts the requirement for additional downstream sample cleanup procedures, which may add unnecessary technical variation. Although there are many very effective solubilization protocols that can be employed, the use of MS-compatible surfactants, such as RapiGest™ (Waters Corporation) or PPS (Protein Discovery), combined with mechanical disruption, such as burst sonication, is often preferred because the surfactants can be removed easily through acid hydrolysis and centrifugation. To account for different quantities of starting biological material between each sample, a constant ratio of solubilization reagent volume to the number of cells or wet weight of the tissue should be used. It is also best to avoid reagents that can covalently modify amino acid residues. Although urea is a very effective chaotropic agent, it is known to carbamylate primary amines (*i.e.*, lysine residues) at temperatures above (37 °C). This often results in non-specific partial derivatization of many lysine-containing peptides, which subsequently increases the complexity of the proteome.

Following protein solubilization, insoluble material, such as cell debris and lipids that could potentially interfere with downstream protein quantification assays, should be removed by high-speed centrifugation. Protein assay (Bradford or BCA) calibration curves should be generated each time an assay is performed to account for any reduction in the reagent activity over time. All calibration curve samples and test samples should be run at least in duplicate, and a buffer blank should always be used to assess the background from the buffer constituents. Unexpectedly high background values in these assays may suggest an interferant and an orthogonal measurement approach should be considered. Once a total protein quantity is determined, samples should be concentration-normalized and digested under the exact same conditions. In many workflows a purified protein standard from a different species is spiked into each sample at a constant quantity per amount of lysate. This internal quality control can be assessed downstream following LC-MS analysis and serves to ensure consistent reduction, alkylation, and digestion across all samples.

For specialized sample preparation protocols, such as the immunodepletion of serum samples or phosphopeptide enrichment from various biological matrices, the same core principles of normalization and consistency of protocols should be employed (Figure 6.3). Through our experience across several clinical biomarker discovery studies, our laboratory routinely employs three independent QC metrics to assess the immunodepletion process: 1) a protein assay QC based on Bradford assays of both the

Immunodepletion QCs

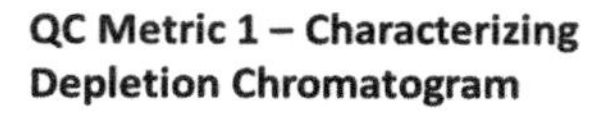

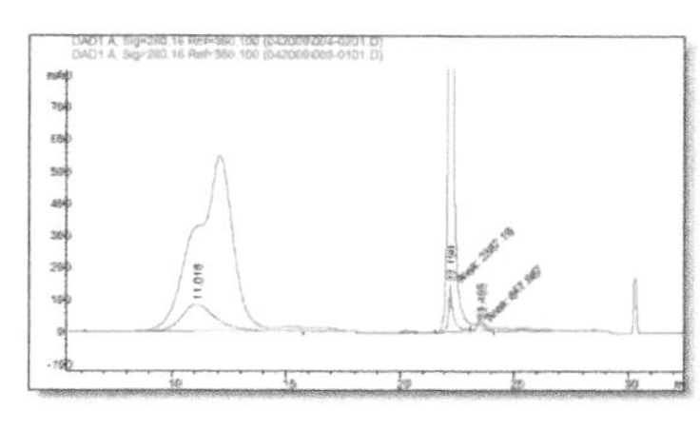

QC Metric 2 – Bradford Assay

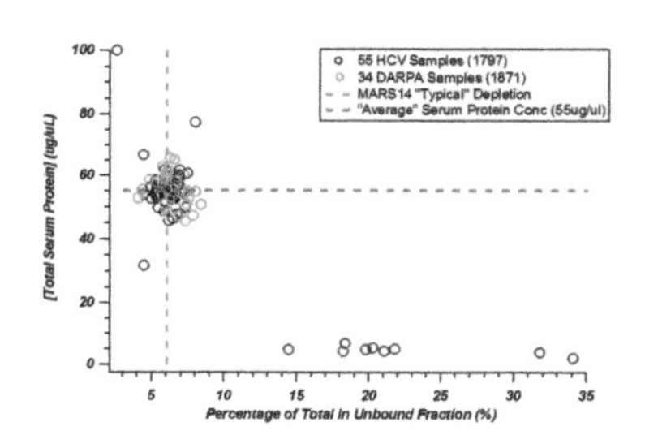

QC Metric 3 – SDS-PAGE Gel Image Analysis

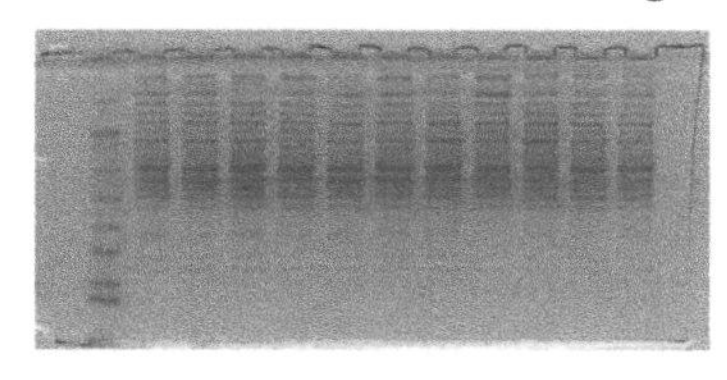

Phosphopeptide Enrichment QCs

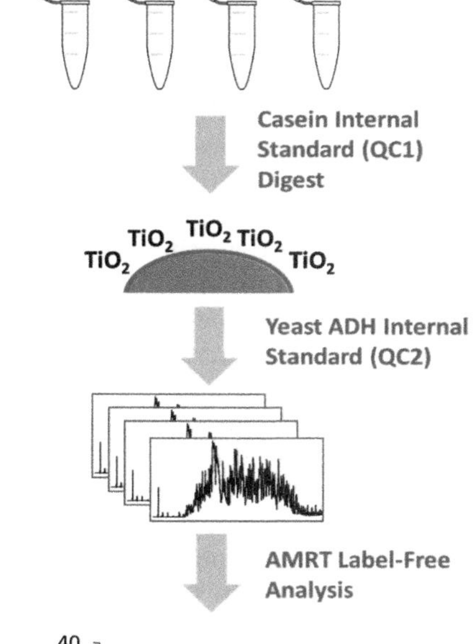

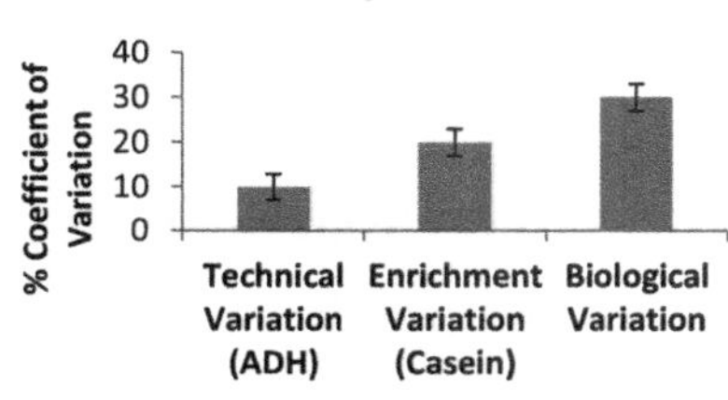

Figure 6.3 Overview of quality control metrics integrated into immunodepletions and phosphopeptide enrichments prior to label-free quantitative analysis. (Left) A three component quality control is employed for all immunodepletions performed in our laboratory: A) protein assay QC based on Bradford assays of both the pre-depleted sample and flow-through (non-depleted) fractions; B) measurement of peak areas (A_{280}) for unbound and bound fractions to visualization sample outliers; and C) SDS-PAGE and coomassie blue staining of each sample to visualize the most abundant proteins and the gross detection of outliers. (Right) Two independent quality control samples are utilized for phosphopeptide enrichment samples: A) a spike of a known highly phosphorylated casein protein prior to enrichment to permit measurement of enrichment variation, and B) spike a pre-digested standard protein prior to LC-MS analysis to measure system variation. The combination of these two quality control samples is important for assessing and troubleshooting potential variation in sample preparation procedures when coupled to a label-free LC-MS analysis.

pre-depleted sample and flow-through (non-depleted) fraction; this allows measurement of the initial column loading and final protein yield and, importantly, the overall fraction of the sample that was removed during depletion. 2) For LC column-based immunodepletions, the peak areas of the absorbance at 280 nm (A_{280}) for unbound and bound fractions are recorded and a graphical display of the bound *versus* unbound AUC allows easy visualization of sample outliers. 3) A quantitative analysis of each sample by SDS-PAGE and coomassie blue staining is performed using image densitometry. This allows for a high-level visualization of the most abundant proteins and the gross detection of outliers. For instance, hemolysis during sample isolation will yield an intense hemoglobin band. This approach is also useful for detecting proteins differentially depleted over the course of a study due to non-specific binding to the immunodepletion column. Similarly, for phosphopeptide enrichments, it is important to keep constant the amount of input material *versus* the binding capacity of the resin, as well as the final concentration of enrichment-modifying compounds (*e.g.*, glycolic acid or dihydroxybenzoic acid). For any enrichment-based protocol, it is also best to spike a known quantity of an exogenous phosphorylated protein into the lysate prior to sample processing. Our practice is to spike bovine alpha-casein at 25 fmol μg^{-1} total lysate prior to sample digestion. Quantitative analysis of this protein within the biological background across the sample cohort provides an internal standard to measure the digestion and enrichment reproducibility.

6.4.2 LC-MS System Qualification

Prior to any quantitative study it is important to ensure both the LC separation and mass spectrometer are operating within normal performance tolerances. This is especially critical for label-free studies, where each sample is run separately and a reduction in chromatographic or MS performance will reduce the effectiveness of accurate mass and retention time alignment and add technical variation in peak areas across multiple samples. To assess instrument performance prior to any quantitative study, our laboratory routinely performs a system suitability analysis of a standard protein digest. Ideally the standard protein digest will be pre-aliquoted from a stock single batch digestion or purchased from a commercial supplier. The same quantity of this standard should be analyzed using a standardized LC-MS/MS method and metrics, such as average chromatographic peak width, mass accuracy, peak intensity, individual peptide retention time, *etc.*, should be measured. Two software solutions that perform well for assessing and storing these metrics over time are the open-source Skyline and MassQC (Proteome Software), which is a subscription-based program that runs locally on the acquisition workstation. Both of these software solutions offer the ability to quickly compare a current system suitability QC sample against a previously run QC sample and determine if the LC column hardware needs to be replaced (*via* a shift in retention time or widening of peak widths), the ion optics need to be cleaned (total signal intensity decreased), or the instrument needs to be recalibrated (through poor mass accuracy based on expected values).

6.4.3 Quality Control and Data Outlier Determination

When designing a label-free quantification study, it is important to incorporate quality control checks such that an accurate assessment of the technical variation across the multiple samples can be measured. These metrics could, for example, be used to determine how much of a fold change is statistically required to claim a protein expression change due to biology, *versus* simply being within analytical variation. After performing a number of these types of studies, an LC-MS system's historical technical reproducibly metrics (which should be independent of the biological sample itself) can be tracked and used to determine if the system was operating at optimal performance. There have been numerous approaches to incorporate these quality controls described in the literature and include strategies such as single internal protein spikes or spiking in a mixture of proteins at different quantities.[55,56] Our laboratory routinely incorporates two different quality controls into all label-free LC-MS experiments. The first is to spike predigested yeast alcohol dehydrogenase (ADH) into each unique sample at a constant quantity per amount of lysate for use as a surrogate standard. This is typically performed immediately prior to transferring samples in an autosampler vial such that no other sample handling variations upstream of this step can incurred. Following the LC-MS data acquisition and label-free data analysis portion of the experiment, the variation in peak areas of each qualitatively identified yeast ADH peptide is measured and a percent coefficient of variation (% CV) is calculated. The second QC is to generate a new unique QC pool sample from equal portions of all samples. This QC pool is typically the first sample to be run following system qualification and is then run periodically throughout the study, typically a minimum of once every 12 h. Since these QC pool samples are derived from all of the samples in the study, they can be aligned with the individual unique samples. After the label-free data analysis has been performed, a % CV of both peptide level and protein level variations is calculated across all qualitatively identified proteins within the sample. Although this measure of system variability is not made within every unique sample, it provides a much more comprehensive assessment of the variation within all of the identified proteins across the entire dynamic range of the analysis. As all of the data is aligned in a single experiment, it is then possible to determine a technical variation for each individual protein in the entire dataset and plot their individual variations within the QC pools. Our laboratory uses these % CV distributions to ensure that 80% of the identified proteins within the dataset have technical variations less than 20% and failure to meet these specifications indicates sub-optimal system performance.

In addition to assessing the technical variation within a label-free experiment, a global characterization of the protein expression profile allows potential sample outliers to be flagged or potentially removed from the dataset. Two approaches that can effectively visualize outliers are principle component analysis (PCA) and two-dimensional agglomerative clustering based on the intensity of each peptide or protein within the dataset. Both of these strategies group samples together based on the levels of similarity of the overall protein expression profile. As the dynamic range in the intensity of identified proteins can be as large as four orders of magnitude, it is often easier to identify outlier samples by first transforming each protein expression value to a Z-score, which allows the PCA or clustering algorithms to use the significance of change in expression values rather than absolute values. If QC pools are included with the data alignment, then both of these visualizations should result in the grouping of all QC pool samples together as these samples will contain only technical variation without any additional biological variation. It is also particularly informative to investigate these grouping algorithms as a function of when the sample was acquired within the context of the entire acquisition window. This visualization is often helpful in determining if a systematic degradation in system performance, for example, a decrease in chromatographic resolution, occurred over the course of the experiment, requiring repeat analysis to be performed.

6.5 Conclusions

Label-free quantitative proteomic approaches offer a high degree of flexibility in experimental design, both in the nature of the sample being analyzed and the number or complexity of the overall experimental design. Two general strategies have been employed: spectral counting approaches and ion intensity-based measurements. Spectral counting and the associated normalization approaches work reasonably well across a wide range of LC-MS instrumentation and can generally be implemented in most laboratories with current workflows. It is possible to achieve reasonable quantification from a spectral counting experiment given an adequate number of spectral counts and replicate measurements. Ion intensity-based quantification has the potential to be more accurate across the entire dynamic range, especially for lower abundant proteins; our laboratory routinely achieves technical variation of <10% at the protein level. Moving forward, as more laboratories incorporate high resolution LC separations and high resolution MS instrumentation, we anticipate a movement towards the application of more XIC-based label-free quantification solutions to larger clinical sized human cohorts.

Acknowledgment

The authors would like to thank Dr. Brenna Richardson for critical review of this manuscript.

References

1. A. Pandey and M. Mann, *Nature*, 2000, **405**, 837.
2. R. Aebersold and M. Mann, *Nature*, 2003, **422**, 198.
3. M. Gstaiger and R. Aebersold, *Nat. Rev. Genet.*, 2009, **10**, 617.
4. Y. Oda, K. Huang, F. R. Cross, D. Cowburn and B. T. Chait, *Proc. Natl Acad. Sci. USA*, 1999, **96**, 6591.
5. S. E. Ong, B. Blagoev, I. Kratchmarova, D. B. Kristensen, H. Steen, A. Pandey and M. Mann, *Mol. Cell Proteomics*, 2002, **1**, 376.
6. K. J. Reynolds, X. Yao and C. Fenselau, *J. Proteome Res.*, 2002, **1**, 27.
7. S. P. Gygi, B. Rist, S. A. Gerber, F. Turecek, M. H. Gelb and R. Aebersold, *Nat. Biotechnol*, 1999, **17**, 994.
8. C. Evans, J. Noirel, S. Y. Ow, M. Salim, A. G. Pereira-Medrano, N. Couto, J. Pandhal, D. Smith, T. K. Pham, E. Karunakaran, X. Zou, C. A. Biggs and P. C. Wright, *Anal. Bioanal. Chem.*, 2012, **404**, 1011.
9. A. Thompson, J. Schafer, K. Kuhn, S. Kienle, J. Schwarz, G. Schmidt, T. Neumann and R. Johnstone, A. K. Mohammed and C. Hamon, *Anal. Chem.*, 2003, **75**, 1895.
10. T. Geiger, J. Cox, P. Ostasiewicz, J. R. Wisniewski and M. Mann, *Nat. Methods*, 2010, **7**, 383.
11. S. Zanivan, M. Krueger and M. Mann, *Methods Mol. Biol.*, 2012, **757**, 435.
12. A. J. Thompson, M. Abu and D. P. Hanger, *Amino Acids*, 2012, **43**, 1075.
13. T. S. Collier, S. M. Randall, P. Sarkar, B. M. Rao, R. A. Dean and D. C. Muddiman, *Rapid Commun. Mass Spectrom.*, 2011, **25**, 2524.
14. V. J. Patel, K. Thalassinos, S. E. Slade, J. B. Connolly, A. Crombie, J. C. Murrell and J. H. Scrivens, *J. Proteome Res.*, 2009, **8**, 3752.
15. A. M. Falick, W. S. Lane, K. S. Lilley, M. J. MacCoss, B. S. Phinney, N. E. Sherman, S. T. Weintraub, H. E. Witkowska and N. A. Yates, *J. Biomol. Tech.*, 2011, **22**, 21.
16. H. Wang, S. Alvarez and L. M. Hicks, *J. Proteome Res.*, 2012, **11**, 487.
17. M. P. Washburn, D. Wolters and J. R. Yates III, *Nature Biotechnol.*, 2001, **19**, 242.
18. H. Liu, R. G. Sadygov and J. R. Yates III, *Anal. Chem.*, 2004, **76**, 4193.
19. B. Zhang, N. C. VerBerkmoes, M. A. Langston, E. Uberbacher, R. L. Hettich and N. F. Samatova, *J. Proteome Res.*, 2006, **5**, 2909.
20. M. Q. Dong, J. D. Venable, N. Au, T. Xu, S. K. Park, D. Cociorva, J. R. Johnson, A. Dillin and J. R. Yates III, *Science*, 2007, **317**, 660.
21. B. Zybailov, A. L. Mosley, M. E. Sardiu, M. K. Coleman, L. Florens and M. P. Washburn, *J. Proteome Res.*, 2006, **5**, 2339.
22. E. Gokce, C. M. Shuford, W. L. Franck, R. A. Dean and D. C. Muddiman, *J. Am. Soc. Mass Spectrom.*, 2011, **22**, 2199.
23. J. Rappsilber, U. Ryder, A. I. Lamond and M. Mann, *Genome Res.*, 2002, **12**, 1231.
24. Y. Ishihama, Y. Oda, T. Tabata, T. Sato, T. Nagasu, J. Rappsilber and M. Mann, *Mol. Cell Proteomics*, 2005, **4**, 1265.

25. P. Lu, C. Vogel, R. Wang, X. Yao and E. M. Marcotte, *Nature Biotechnol.*, 2007, **25**, 117.
26. N. M. Griffin, J. Yu, F. Long, P. Oh, S. Shore, Y. Li, J. A. Koziol and J. E. Schnitzer, *Nat. Biotechnol*, 2010, **28**, 83.
27. N. Colaert, K. Gevaert and L. Martens, *J. Proteome Res.*, 2011, **10**, 3183.
28. S. Ghaemmaghami, W. K. Huh, K. Bower, R. W. Howson, A. Belle, N. Dephoure, E. K. O'Shea and J. S. Weissman, *Nature*, 2003, **425**, 737.
29. W. M. Old, K. Meyer-Arendt, L. Aveline-Wolf, K. G. Pierce, A. Mendoza, J. R. Sevinsky, K. A. Resing and N. G. Ahn, *Mol. Cell Proteomics*, 2005, **4**, 1487.
30. R. D. Voyksner and H. Lee, *Rapid Commun. Mass Spectrom.*, 1999, **13**, 1427.
31. D. Chelius and P. V. Bondarenko, *J. Proteome Res.*, 2002, **1**, 317.
32. R. D. Smith, G. A. Anderson, M. S. Lipton, L. Pasa-Tolic, Y. Shen, T. P. Conrads, T. D. Veenstra and H. R. Udseth, *Proteomics*, 2002, **2**, 513.
33. J. T. Prince and E. M. Marcotte, *Anal. Chem.*, 2006, **78**, 6140.
34. C. A. Hastings, S. M. Norton and S. Roy, *Rapid Commun. Mass Spectrom.*, 2002, **16**, 462.
35. B. Fischer, J. Grossmann, V. Roth, W. Gruissem, S. Baginsky and J. M. Buhmann, *Bioinformatics*, 2006, **22**, e132.
36. P. Wang, H. Tang, M. P. Fitzgibbon, M. McIntosh, M. Coram, H. Zhang, E. Yi and R. Aebersold, *Biostatistics*, 2007, **8**, 357.
37. J. C. Silva, R. Denny, C. A. Dorschel, M. Gorenstein, I. J. Kass, G. Z. Li, T. McKenna, M. J. Nold, K. Richardson, P. Young and S. Geromanos, *Anal. Chem.*, 2005, 77, 2187.
38. S. Cappadona, P. R. Baker, P. R. Cutillas, A. J. Heck and B. van Breukelen, *Amino Acids*, 2012, **43**, 1087.
39. P. Wang, H. Tang, H. Zhang, J. Whiteaker, A. G. Paulovich and M. McIntosh, *Pac. Symp. Biocomput.*, 2006, 315.
40. M. D. Filiou, D. Martins-de-Souza, P. C. Guest, S. Bahn and C. W. Turck, *Proteomics*, 2012, **12**, 736.
41. J. Martinez-Aguilar, J. Chik, J. Nicholson, C. Semaan, M. J. McKay and M. P. Molloy, *Proteomics Clin. Appl.*, 2013, 7, 42.
42. N. Nagaraj and M. Mann, *J. Proteome Res.*, 2011, **10**, 637.
43. K. J. Ralston-Hooper, M. E. Turner, E. J. Soderblom, D. Villeneuve, G. T. Ankley, M. A. Moseley, R. A. Hoke and P. L. Ferguson, *Environ. Sci. Technol.*, 2013, **47**, 1091.
44. E. J. Soderblom, J. W. Thompson, E. A. Schwartz, E. Chiou, L. G. Dubois, M. A. Moseley and R. Zennadi, *Clin. Proteomics*, 2013, **10**, 1.
45. E. J. Soderblom, M. Philipp, J. W. Thompson, M. G. Caron and M. A. Moseley, *Anal. Chem.*, 2011, **83**, 3758.
46. L. N. Mueller, M. Y. Brusniak, D. R. Mani and R. Aebersold, *J. Proteome Res.*, 2008, **7**, 51.
47. B. Cooper, J. Feng and W. M. Garrett, *J. Am. Soc. Mass Spectrom.*, 2010, **21**, 1534.

48. N. L. Heinecke, B. S. Pratt, T. Vaisar and L. Becker, *Bioinformatics*, 2010, **26**, 1574.
49. Z. Q. Ma, S. Dasari, M. C. Chambers, M. D. Litton, S. M. Sobecki, L. J. Zimmerman, P. J. Halvey, B. Schilling, P. M. Drake, B. W. Gibson and D. L. Tabb, *J. Proteome Res.*, 2009, **8**, 3872.
50. P. M. Palagi, D. Walther, M. Quadroni, S. Catherinet, J. Burgess, C. G. Zimmermann-Ivol, J. C. Sanchez, P. A. Binz, D. F. Hochstrasser and R. D. Appel, *Proteomics*, 2005, **5**, 2381.
51. J. D. Jaffe, D. R. Mani, K. C. Leptos, G. M. Church, M. A. Gillette and S. A. Carr, *Mol. Cell Proteomics*, 2006, **5**, 1927.
52. M. Bellew, M. Coram, M. Fitzgibbon, M. Igra, T. Randolph, P. Wang, D. May, J. Eng, R. Fang, C. Lin, J. Chen, D. Goodlett, J. Whiteaker, A. Paulovich and M. McIntosh, *Bioinformatics*, 2006, **22**, 1902.
53. J. S. Andersen, C. J. Wilkinson, T. Mayor, P. Mortensen, E. A. Nigg and M. Mann, *Nature*, 2003, **426**, 570.
54. B. MacLean, D. M. Tomazela, N. Shulman, M. Chambers, G. L. Finney, B. Frewen, R. Kern, D. L. Tabb, D. C. Liebler and M. J. MacCoss, *Bioinformatics*, 2010, **26**, 966.
55. T. A. Addona, S. E. Abbatiello, B. Schilling, S. J. Skates, D. R. Mani, D. M. Bunk, C. H. Spiegelman, L. J. Zimmerman, A. J. Ham, H. Keshishian, S. C. Hall, S. Allen, R. K. Blackman, C. H. Borchers, C. Buck, H. L. Cardasis, M. P. Cusack, N. G. Dodder, B. W. Gibson, J. M. Held, T. Hiltke, A. Jackson, E. B. Johansen, C. R. Kinsinger, J. Li, M. Mesri, T. A. Neubert, R. K. Niles, T. C. Pulsipher, D. Ransohoff, H. Rodriguez, P. A. Rudnick, D. Smith, D. L. Tabb, T. J. Tegeler, A. M. Variyath, L. J. Vega-Montoto, A. Wahlander, S. Waldemarson, M. Wang, J. R. Whiteaker, L. Zhao, N. L. Anderson, S. J. Fisher, D. C. Liebler, A. G. Paulovich, F. E. Regnier, P. Tempst and S. A. Carr, *Nat. Biotechnol*, 2009, **27**, 633.
56. P. Lescuyer, A. Farina and D. F. Hochstrasser, *Trends Biotechnol.*, 2010, **28**, 225.

CHAPTER 7

MS1 Label-free Quantification Using Ion Intensity Chromatograms in Skyline (Research and Clinical Applications)

BIRGIT SCHILLING,[a] BRENDAN X. MACLEAN,[b] ALEXANDRIA D'SOUZA,[a] MATTHEW J. RARDIN,[a] NICHOLAS J. SHULMAN,[b] MICHAEL J. MACCOSS[b] AND BRADFORD W. GIBSON*[a]

[a] Buck Institute for Research on Aging, 8001 Redwood Boulevard, Novato, California 94945, USA; [b] University of Washington, Department of Genome Sciences, Foege Building S113, 3720 15th Ave NE, Seattle, WA 98195, USA
*Email: bgibson@buckinstitute.org

7.1 Introduction—Label-free Quantification Using MS1 Filtering

Over the last several years, label-free quantification strategies have become more widely used in various types of mass spectrometric proteomics experiments.[1,2] Software solutions to process label-free data sets, however, are in many cases either expensive commercial software products, or programs created in academic laboratories around the experimental workflow of an

New Developments in Mass Spectrometry No. 1
Quantitative Proteomics
Edited by Claire Eyers and Simon J Gaskell

Published by the Royal Society of Chemistry, www.rsc.org

individual research group. In addition, commercial software programs are often designed for processing the data from instruments of a specific vendor, further limiting their utility. Although the HUPO Proteomics Standards Initiative has made great efforts to generate uniform data formats, such as mzML, many software tools do not support these standardized formats. From a user's perspective, it is noticeable that open-source software in this area is often poorly documented with minimal commitment to ongoing support. With recent improvements in mass spectrometric technology and instrumentation, as well as in chromatographic reproducibility, label-free quantitative workflows have become increasingly attractive. These advances have created an even more critical need for comprehensive software solutions to support label-free quantitative proteomics experiments.

In 2012, a new open-source software solution was introduced for MS1-based label-free quantification, referred to as Skyline MS1 Filtering.[3] Skyline had already been well established for quantifying targeted mass spectrometric data, in particular, for selected reaction monitoring (SRM) assays,[4] and it remains freely available. Features, such as vendor independence, ease of use, the ability to import raw data without prior conversion, graphical tools to assess data quality visually, and tools to manually adjust automatically chosen peak integration boundaries, have made Skyline an essential tool for performing quantitative proteomics experiments in many laboratories. There have been over 15 000 installations of Skyline worldwide and the Skyline program is initiated for use by scientists over 4500 times each week (status 04/2013).

The label-free Skyline MS1 Filtering workflow, as originally described, is reviewed in Figure 7.1.[3] Briefly, data-dependent acquisitions (DDA) are directly imported into Skyline as raw data. MS1 scans are then filtered to yield extracted ion chromatograms (XICs) for target peptides. Typically, all MS/MS spectral data is independently interrogated using one of several compatible search algorithms. Skyline supports importing results from a wide variety of peptide search pipelines into a spectral library that may be used to direct MS1 peak picking and peak integrations from extracted MS1 ion chromatograms. In Skyline chromatogram plots, 'ID' lines indicate exact elution times of underlying MS/MS spectra by peptide and replicate. Particular emphasis during development of the original MS1 Filtering method was placed on visual tools to assess data quality and peak integration, to probe for interferences, and to provide quality controlled, quantitative MS1 peak area reports. Two major advantages of the Skyline MS1 Filtering technology were its open-source licence, which makes it freely-available, and support for the native data formats of five major instrument vendors, due to the internal use of the ProteoWizard data access library.[5] As demonstrated in the original report,[3] MS1 Filtering was shown to have a high degree of linearity and reproducibility across multiple acquisitions in a variety of sample workflows, suggesting that this approach could be broadly applicable to many types of experimental designs. Development for MS1 Filtering is ongoing and the Skyline web site (http://proteome.gs.washington.edu/software/skyline)

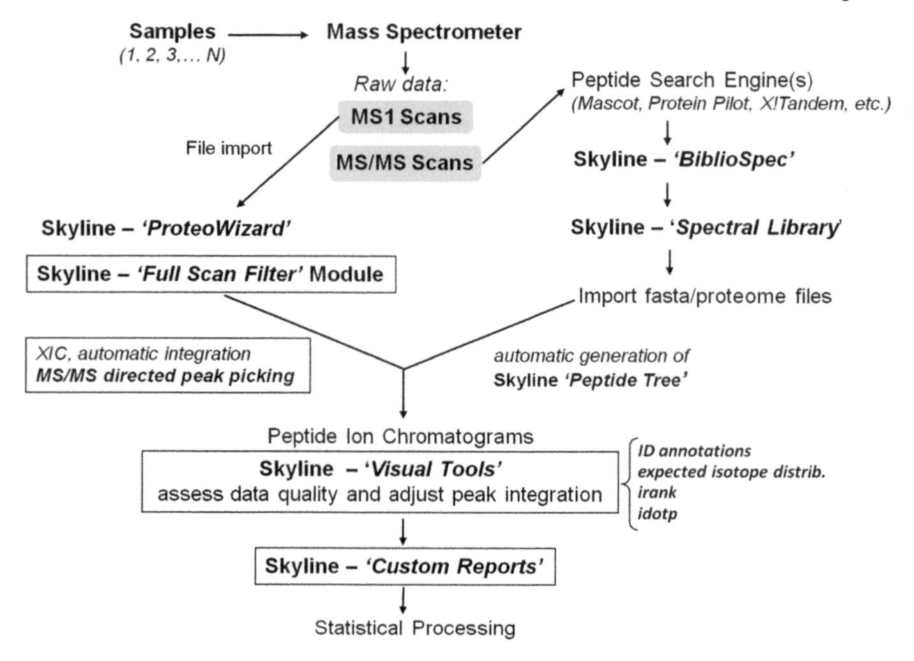

Figure 7.1 Skyline MS1 Filtering for quantification of extracted MS1 ion intensities. MS1 Filtering was implemented as a new module into Skyline extracting ion chromatograms (XICs) for precursor ions from full MS1 scans from data-dependent acquisitions (see text for details). This figure was originally published in *Mol. Cell Proteomics*. B. Schilling *et al.*, *Mol. Cell Proteomics*, 2012, **11**, 202. © the American Society for Biochemistry and Molecular Biology.

provides access to extensive documentation and active technical support of both new and old features alike.

Although Skyline MS1 Filtering has gained widespread use since its introduction, there are a number of other software packages and approaches available for both label and label-free proteomic quantification.[1,2] These methods encompass both metabolic and post-metabolic chemical labeling approaches, such as SILAC,[6] iTRAQ,[7] and TMT,[8] as well as several label-free methods, such as MaxQuant,[9] and Elucidator® (Rosetta). Skyline MS1 Filtering does, however, provide unique features as reviewed in this article and can readily be applied to essentially any data-dependent acquisition experiment generated from multiple vendors' instrumentation. For example, performing MS1 Filtering in Skyline allows for very fast information transfer and linking to other quantification methods that are also supported by the Skyline software, such as SRM, or full scan MS2 product ion-based quantification approaches (*i.e.*, MS/MS^{ALL} with SWATH,[10] MS^{E} [11]). This is especially important as several of these quantitative approaches work on only a single vendor's instrumentation and, therefore, do not allow cross-platform comparisons or experimental designs involving multiple instrument types. Technological improvements in mass spectrometric instruments (*i.e.*, high resolution MS1 scans, new MS detector developments), as well as for chromatographic systems (*i.e.* ultra high pressure UPLC systems with good retention time stability), also contribute to improved, high quality data sets, which results in increased accuracy when employing label-free methods. Label-free quantification is certainly cost effective compared to metabolic (*i.e.*, SILAC[6]) and post-metabolic (*e.g.*, iTRAQ[7] or TMT[8]) labeling, although these latter chemical labeling methods do offer the ability to multiplex sample analysis. MS1 Filtering does have a clear advantage over labeling methods in that one can directly process samples from human and other mammalian sources, something that is restrictive or impossible for SILAC-type experiments. Lastly, as Skyline MS1 Filtering is a non-targeted quantification approach it does not require any prior assay development, which can be especially time consuming for SRM experiments, making it more flexible and rapid to employ.

This article describes experiments from both recently published and unpublished research and clinical studies using MS1 Filtering quantification and further documents the robustness of the methodology. Included is a study investigating a known cancer target, receptor tyrosine-protein kinase (ErbB2),[12] and experiments designed to identify new breast cancer subtype-specific glycoprotein biomarker candidates originally described by Drake *et al.*[13] We also review recent advancements developed for improving the accuracy of peak picking and integration, which greatly improve experimental throughput in research and clinical applications. In addition, we report new support for 'point-and-click' integration of externally developed statistical processing tools with Skyline MS1 Filtering.

7.2 Applications, Improved Features and Applied Statistical Processing

Skyline has been widely used in quantitative proteomics studies extracting chromatograms from selected reaction monitoring (LC-SRM-MS) experiments.[14,15] This broad familiarity facilitates a transition to using Skyline for other quantitative methods, such as MS1 Filtering, which use chromatograms extracted from full-scan mass spectra. Recent studies that have adopted this new label-free MS1 methodology in Skyline have demonstrated its robustness and high throughput capabilities. Rardin *et al.* described a large study designed to identify substrates of mitochondrial deacetylase sirtuin 3 (SIRT3) by comparing the levels of over 2000 acetyllysine-containing peptides from over 500 proteins in liver mitochondria from wild-type and knockout mice using MS1 Filtering.[16,17] A recent review by Gonzalez-Galarza *et al.* describing both free, open-source, and commercial software for quantitative proteomics also referred to Skyline's label-free, full-scan filtering approaches.[1] Compared with other open-source software, there are several attributes of Skyline and MS1 Filtering that are noteworthy, including responsive technical support, extensive documentation, thorough beta-testing 'in the field' by a diverse group of laboratories, open user meetings at international conferences (*i.e.*, ASMS, HUPO), import of MS raw data formats, export of native MS instrument methods (for follow-up experiments), and overall support of diverse laboratory pipelines to generate, search, and process mass spectrometric data for proteomics investigation.

7.2.1 Improved and Expanded Features for Quantification Using Skyline MS1 Filtering

Since the first report of Skyline MS1 Filtering,[3] significant progress has been made that improves and expands upon previously released features (Figure 7.2). As indicated, new developments include the addition of support for full-scan raw data from Bruker instruments to the previously supported data formats from AB SCIEX, Agilent, Thermo, and Waters, as well as improvements for raw data import speed using 64-bit computers. As shown in Figure 7.2B, the variety of database search pipelines that can be integrated with Skyline MS1 Filtering has been greatly expanded, largely based on community demand, demonstrating the diversity of data formats supported

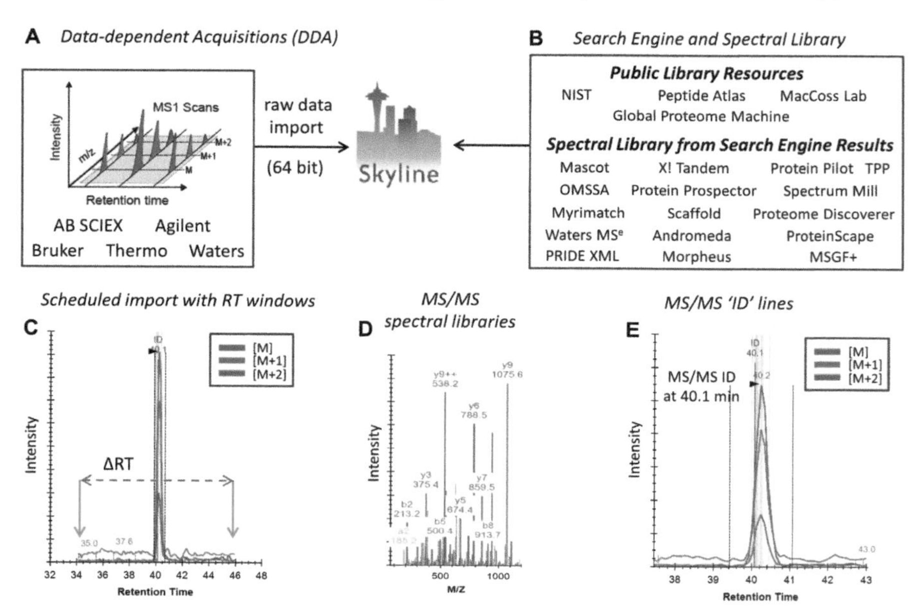

Figure 7.2 Improved and new features for MS1 Filtering. A) Data-dependent acquisitions (DDA) and expanded vendor support. B) List of resources and search engines for spectral library building. C) Scheduled import of raw files defining retention time windows for MS1 chromatograms (△RT). Example shown for acetyllysine-containing peptide LVESANE^{3071}**K$_{Ac}$**AER from mouse histone-lysine *N*-methyltransferase (Q6PDK2, MLL2) with precursor ion M at *m/z* 644.33 and an elution time of 40.1 min. D) Display of corresponding MS/MS from spectral libraries. E) The MS/MS 'ID' line from the current acquisition is indicated with a black solid line while other replicate/sample acquisitions are projected and displayed with thin blue lines.

and flexibility provided. Attention to preserving scan source file and retention time (RT) information through search pipelines and into spectral libraries allows this information to be used extensively with MS1 Filtering data sets for retention time alignment, peak picking, and the display of MS/MS 'ID' lines in chromatogram plots, indicating underlying MS/MS peptide search engine results. All of these features are now supported for the search engines listed in Figure 7.2B; although Mascot conversion of the raw data files to generic (MGF) files may require the use of ProteoWizard MSConvert tools to preserve source file and RT information.

This broad support of peptide search pipelines combined with the ability to import raw data from many instrument vendors allows investigators to process data sets outside their typical laboratory formats and workflows, such as those available for downloading from data repository servers as part of peer-reviewed publications. Another new feature in Skyline MS1 Filtering is the 'retention time scheduled file import' with dynamic ID-based retention time windows that the user can define based on observed HPLC retention time reproducibility (Figure 7.2C). Scheduled import of a selected RT window (ΔRT) from the total HPLC gradient greatly reduced import time and Skyline file size. For example, importing and extracting MS1 data for 450 de-glycosylated peptides from 40 QqTOF files across the entire 2 h HPLC gradient on a 64-bit computer took 123 min, while a corresponding scheduled import around the RT of each peptide (ΔRT = 5 min) required only 22 min with a reduction in Skyline file size of ~10 fold from 2.0 GBytes (full import) to 0.2 GBytes (scheduled import). Similar reductions in import time (from 220 to only 41 min) and file size (from 4.9 to 0.51 GBytes) were observed using the scheduled import of ~2600 peptides from ten raw data files generated on a Thermo Orbitrap with a 2 h HPLC gradient.

In addition, a key feature in Skyline has been the availability of underlying MS/MS spectra from redundant spectral libraries that allows the user to visualize MS/MS from each replicate acquisition showing the time of sampling in the MS1 trace (Figure 7.2D). New features expand upon the peptide 'ID' line or 'RT indicator', not only showing sampled MS/MS in the active replicate (dark vertical 'ID' lines) but RT-aligned MS/MS 'ID' lines (projected blue vertical lines) from other replicate acquisitions that now provide greater confidence in MS1 peak picking (Figure 7.2E). The latter feature led to the development of enhanced, automated MS1 peak integration in an acquisition where a given peptide was not sampled for MS/MS. Skyline peak picking now uses not only the retention times of identified MS/MS scans that are displayed as 'ID' lines in chromatogram plots (Figure 7.3A), but also retention times projected from other replicate acquisitions displayed as thin blue lines in plots where no MS/MS was acquired in the active sample (also referred to as RT alignment) (Figure 7.3B).

This retention time alignment feature dramatically improved the accuracy of automatic peak picking (see, for example, retention time replicate view in Figure 7.3C). Thorough evaluation of automatic MS1 peak picking performance using these new features found false MS1 peak picking rates to be

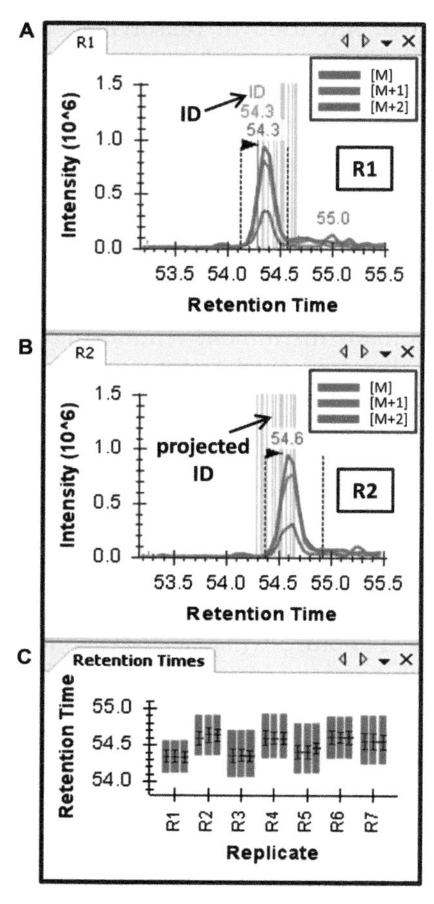

Figure 7.3 Retention time alignment and directed peak picking. A) MS1 Filtering peptide peak picking is directed by underlying MS/MS per acquisition replicate as indicated by solid, red vertical 'ID' lines (shown for replicate R1). B) For acquisition replicates (*i.e.*, replicate R2) that did not sample specific peptide MS/MS, blue vertical lines within the XIC indicate projected MS/MS ID lines as they were observed in other replicates. These projected ID lines are used for retention time alignment directed peak picking to improve peak picking accuracy. The peptide displayed for selected acquisitions is an acetyllysine-containing peptide HLQEM$_{ox}$GL^{330}**K**$_{ac}$FFVK from mouse serine-pyruvate aminotransferase (O35423, SPYA) with precursor ion M observed at *m*/*z* 512.27. C) Acquisition replicates R1 to R7 are displayed in the retention time replicate view demonstrating retention time stability and repeatability (here showing a 2 min retention time window zoomed in from the 2 h total runtime of the HPLC gradient).

between 5 and 7%, which substantially reduced the amount of time that had to be spent manually correcting automatic peak integration. Lastly, we recently added support for annotating replicates after file import that allows sample conditions and sampling groups to be annotated within the Skyline document, *i.e.*, knock-out (KO) *vs.* wild-type (WT), cancer *vs.* control *etc.*, which can be later exported with Skyline reports for use in downstream statistical analysis. In summary, the improved feature set in Skyline MS1 Filtering is comprised of faster data import, reduced file size, retention time aligned MS/MS ID annotations across replicates, and more robust peak picking to provide a high throughput and automated laboratory tool for label-free peptide quantification with greater confidence in peak integration and minimal need for user adjustments.

7.2.2 Development of Statistical Tools for Processing MS1 Filtering Data Sets

To complete the integration of an MS1-based quantification program into laboratory research pipelines, several statistical tools were recently developed and made freely available that initiate post-Skyline statistical processing of precursor peak areas obtained from biological experiments and replicate acquisitions.

As shown in Figure 7.4, to obtain quantitative results after processing data with the Skyline MS1 Filtering software, peak areas can be exported to the recently developed statistical program 'MS1Probe',[18] a Python-coded bioinformatics tool. MS1Probe allows users to enter sample information and requested statistics, such as ratios for differentiating conditions (*i.e.*, KO *vs.*

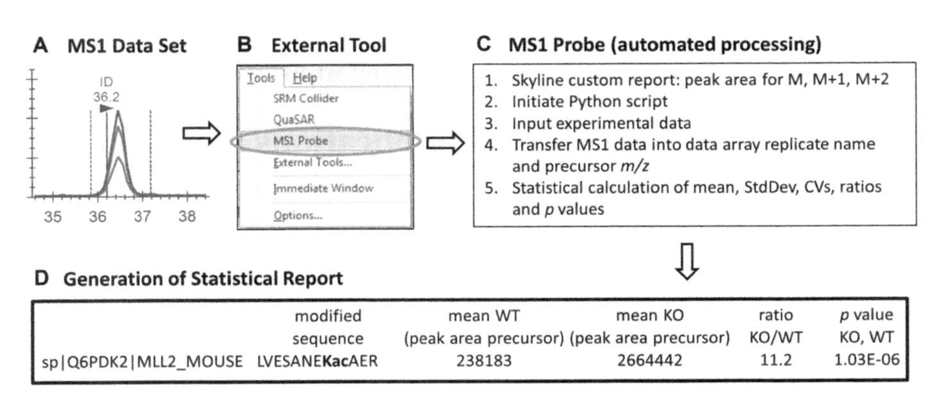

Figure 7.4 Skyline statistical results processing using external bioinformatics tool 'MS1 Probe'. Skyline MS1 Filtering data sets (A) are subjected to further statistical processing by connecting with an externally developed tool, referred to as 'MS1Probe' (B) that was added to the Skyline 'Tools' menu (see text). (C) Once the 'MS1Probe' option is selected, subsequent steps are automatically triggered, as shown in the inset box, including simple queries to the user *via* command prompts requesting replicate group information (*i.e.*, knock-out (KO), wild-type (WT) *etc.*) to define statistical details that subsequently will be calculated (D), such as means, standard deviations, CVs, and ratios between different groups/conditions, as well as significance measurements (p values).

WT), which will then be automatically generated for MS1 precursor ions. To use the MS1Probe software, installation of the following open-source programs is required: the programming language Python 2.7.3, and corresponding extensions 'NumPy' and 'SciPy', which provide support for large, multi-dimensional arrays, matrices, and mathematical tools. The Python program 'MS1Probe', corresponding tutorials, and detailed instructions on how to integrate MS1Probe as an external tool in Skyline can be downloaded from http://www.gibsonproteomics.org/resources/MS1Probe. Installation of new external tools into Skyline, such as the one described here, can be conveniently achieved using 'SkylineRunner', a module available within Skyline. Once MS1Probe is installed, it appears as an option in the Skyline tools menu, allowing for easy point-and-click access to this externally developed program from within Skyline. Choosing the MS1Probe option in Skyline will initiate all steps to automatically generate a statistical MS1 Filtering report that summarizes the results for a given biological experiment. In addition to calculating Student's t-test and associated p values (unadjusted for multiple testing) to assess statistical significance within the data sets, the 'MS1Probe' tool can also further determine q values, which have been proposed as a false discovery rate (FDR)-based measure of significance in large multiple testing data sets typical of genomics and proteomics studies.[19] Additional statistical analysis programs for interfacing with Skyline MS1 Filtering will be made available at the same resource website (listed above), including 'MS1-InterferenceFilter' for applying MS1-precursor peak area CV filters to assess technical sample reproducibility and flag potential outliers or interferences. The currently available statistical tool, MS1Probe, can facilitate and accelerate MS1 quantitative data processing within laboratories.

7.3 Research and Clinical Applications

Applications for research projects using label-free MS1-based quantification are broad and extensive. As MS1 Filtering may be applied to any discovery-type data set that typically would be acquired in DDA mode, MS1 quantification can quickly provide an overview of the experimental status, sample quality, or quantitative comparisons between samples. This unbiased MS1-based quantitative approach supports monitoring for unexpected experimental observations and fast screening of conditions without prior labeling chemistry or quantitative assay development that are typically necessary when using (label-free) SRM-based workflows.

7.3.1 MS1-based Label-free Quantification Tools to Assess and Optimize Sample Preparation Workflows

Results from two research projects are presented as examples that emphasize the benefits of MS1 Filtering for assessing sample preparation and reproducibility. Recently, MS1 Filtering was used to investigate the effectiveness and reproducibility of using several types of protease digestion regimens that were

Table 7.1 MS1 Filtering for workflow reproducibility assessment and optimization of ErbB2 digestion using different proteases: MS1 Filtering was used to determine peak areas for technical and process replicates for all individual proteolytic ErbB2 peptides that were generated during digestion. The percent coefficient of variation (% CV) for the peak areas across replicates was calculated for each peptide. Median % CV and % CV quartile distribution are reported.

		MS Replicates[a] (% CV)		*Process Replicates[b] (% CV)*	
Protease	*# of peptides quantified*	*Median[c]*	*Lower–Upper[d] Quartile*	*Median[c]*	*Lower–Upper[d] Quartile*
Trypsin	140	6.1	3.2–12.0	15.1	8.2–26.5
AspN	95	4.6	2.9–9.1	14.1	8.7–21.3
Chymotrypsin	146	9.1	5.3–14.2	22.9	14.9–41.1
AspN + Chymo	63	14.6	10.4–19.2	40.8	27.5–56.9

[a]MS replicates: each individually digested sample was analyzed in triplicate.
[b]Process replicates: each enzyme digestion was performed in triplicate.
[c]Median % CV as determined for all proteolytic peptides quantified.
[d]% CV distribution for all proteolytic peptides quantified (the range comprises data from the second and third quartile, 50% of the data).

designed to target and increase coverage of a specific breast cancer receptor, ErbB2.[12] While trypsin digestions are typically assumed to have good reproducibility, proteases, such as AspN and chymotrypsin, or conditions where more than one protease is used are often avoided in quantitative studies, mostly due to concerns of introducing greater digestion variability. Table 7.1 shows the coefficient of variation (% CV) of individual ErbB2 peptides obtained from digestions using different proteases or combinations of proteases when analyzing three process/digestion replicates and three technical MS replicates.[12]

ErbB2 proteolytic peptide reproducibility for the various conditions was monitored using the unbiased MS1-based approach. Results were promising for the different single protease digestions, especially for trypsin and AspN, with median CVs of 6.1% and 4.6 % for technical MS reproducibility, and 15.1% and 14.1 % for process reproducibility (where the entire workflow was examined). However, variance in process reproducibility was unacceptably high with a median CV of 40.8% for double enzyme digestions (see Table 7.1).[12] This ErbB2 study would have been unfeasible or significantly more difficult to execute if LC-SRM-MS had been used instead, mostly due to time-intensive SRM assay development, unexpected/non-specific or missed enzyme cleavages, and resulting high multiplexing levels needed for proper implementation.

MS1 Filtering was further evaluated to assess the reproducibility of complex sample preparation for glycopeptide analysis in the context of a previously published study. In this case, the available published mass spectrometric data was re-analyzed using MS1 Filtering. The original study by Drake *et al.*[13] investigated lectin chromatography/mass spectrometry discovery workflows for identification of putative biomarkers of aggressive breast cancers and aimed at identifying glycopeptides from conditioned media (CM) of various basal and luminal breast cancer cell lines (for project details see section 7.3.2). The sample workup included a complex series of sample processing steps, such as preparation and lysis of conditioned media, trypsin digestion, lectin chromatography enrichment for glycopeptides containing specific sugars or epitopes, PNGase F treatment to remove the glyco-moiety for easier mass spectrometric analysis of the 'de-glycosylated' peptides, and finally HPLC-ESI-MS/MS analysis. For each of the ten breast cancer cell lines that were investigated, two lectin chromatography 'process replicates' and two mass spectrometric 'technical replicates' were acquired. The lectin used was *Aleuria aurantia* lectin (AAL), which binds to and enriches fucose-containing carbohydrate structures. Here, we subjected the original high resolution QqTOF data sets to the Skyline MS1 Filtering processing and quantification pipeline for a thorough reproducibility assessment.

Figure 7.5A shows MS1 peak areas displayed for the two process and technical replicates of the de-glycosylated peptide GSNYSEILDK (follistatin-related protein 1) for the CM of ten individual cancer cell lines (lines A–L). Although, as expected, differences are observed between the CM obtained from several biologically diverse cancer cell lines, strong MS1 peak area reproducibility was observed within each individual line. For example, the coefficient of variation across four replicates for the de-glycosylated peptide GSNYSEILDK was 10.5% in cell line J (MDAMB231), 12.2% in cell line K (BT549) and 9.3% in cell line L (HS5787). Individual extracted MS1 chromatograms for CM replicates from cell line L showed similar abundance and yielded comparable peak areas, as displayed in the peak area replicate view (Figure 7.5B–C). In addition, the replicate ID lines contribute to high confidence that the correct peak was integrated (Figure 7.5B). Expanding beyond the specific de-glycosylated peptide example shown here, we similarly calculated lectin workflow reproducibility for larger sets of peptides, *i.e.*, the median coefficient of variation across four technical/process replicates was 16.0% in cell line J (for 85 investigated de-glycosylated peptides), and 17.3% in cell line K (for 113 investigated de-glycosylated peptides). In summary, using MS1 Filtering we were able to rapidly assess both workflow reproducibility in our process replicates, and MS instrument stability in our technical replicates, which are required for any quantification study. This example also demonstrates that publicly available datasets can be re-analyzed in a relatively straightforward manner using MS1 Filtering. For this purpose, Skyline vendor independence is particularly advantageous as separately installed third-party vendor software is not necessary in order to process data sets acquired on MS instruments from any of the five vendors supported by Skyline.

7.3.2 MS1-based Label-free Quantification to Investigate Subtype-Specific Breast Cancer

Many quantitative proteomics studies have attempted to discover new cancer biomarkers for a wide variety of clinical applications, including pre-symptomatic disease prognosis, monitoring of disease progression,

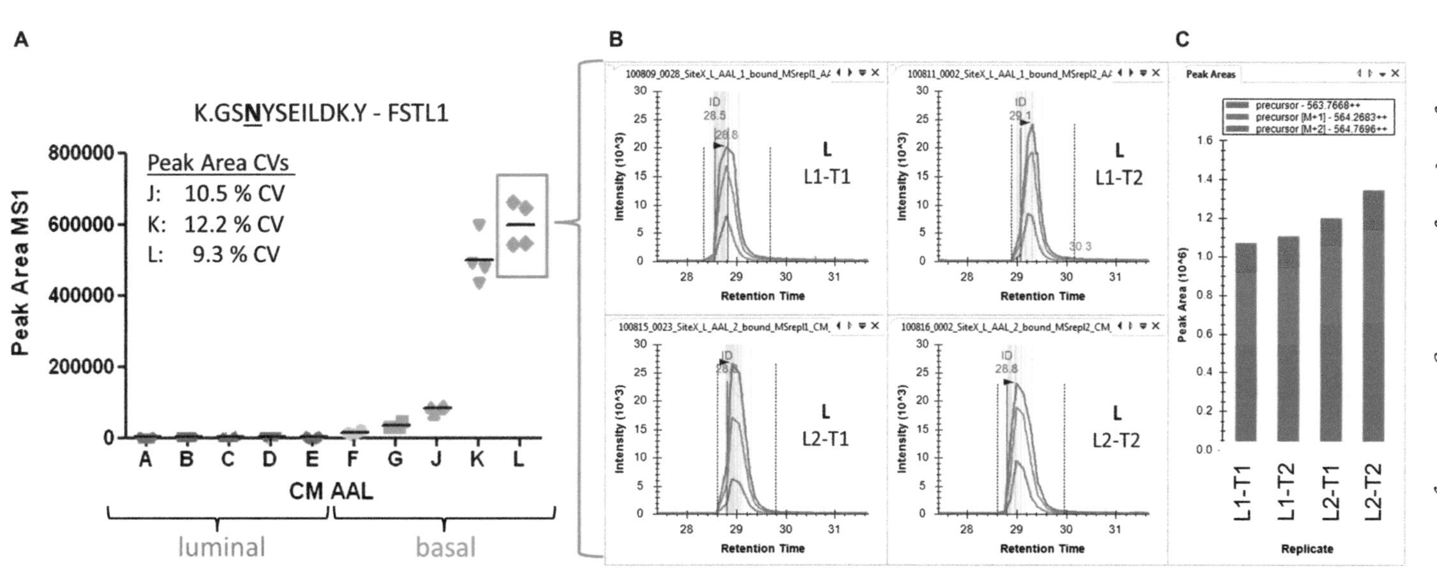

Figure 7.5 Assessment of process and technical reproducibility during AAL lectin enrichment. MS1 Filtering of de-glycosylated peptide GSNYSEILDK from follistatin-related protein 1 (Q12841, FSTL1) with precursor ion M observed at m/z 563.77. A) MS1 peak area was extracted for all ten cell line CM samples (five luminal, A–E, and five basal, F–L). Each CM sample is displayed with four data points (two AAL lectin processes and two technical MS replicates). B) Replicates from CM cell line 'L' are shown in more detail with their precursor intensity chromatograms showing precursor ions M, M + 1 and M + 2. Solid vertical lines show the ID line indicating sampled MS/MS for the actively displayed sample; thin blue lines indicate projected MS/MS sampling lines from other replicate acquisitions. C) MS1 peak area replicate view for CM from basal cell line 'L' (HS5787) showing precursor ions M, M + 1 and M + 2. Replicate annotation for panels B) and C): samples L1 and L2 represent process replicates assessing variation in tryptic digestion, lectin enrichment, PNGase F de-glycosylation (for cell line HS5787); T1 and T2 indicate technical MS replicates (repeat acquisitions of the same sample on the LC-MS/MS system).

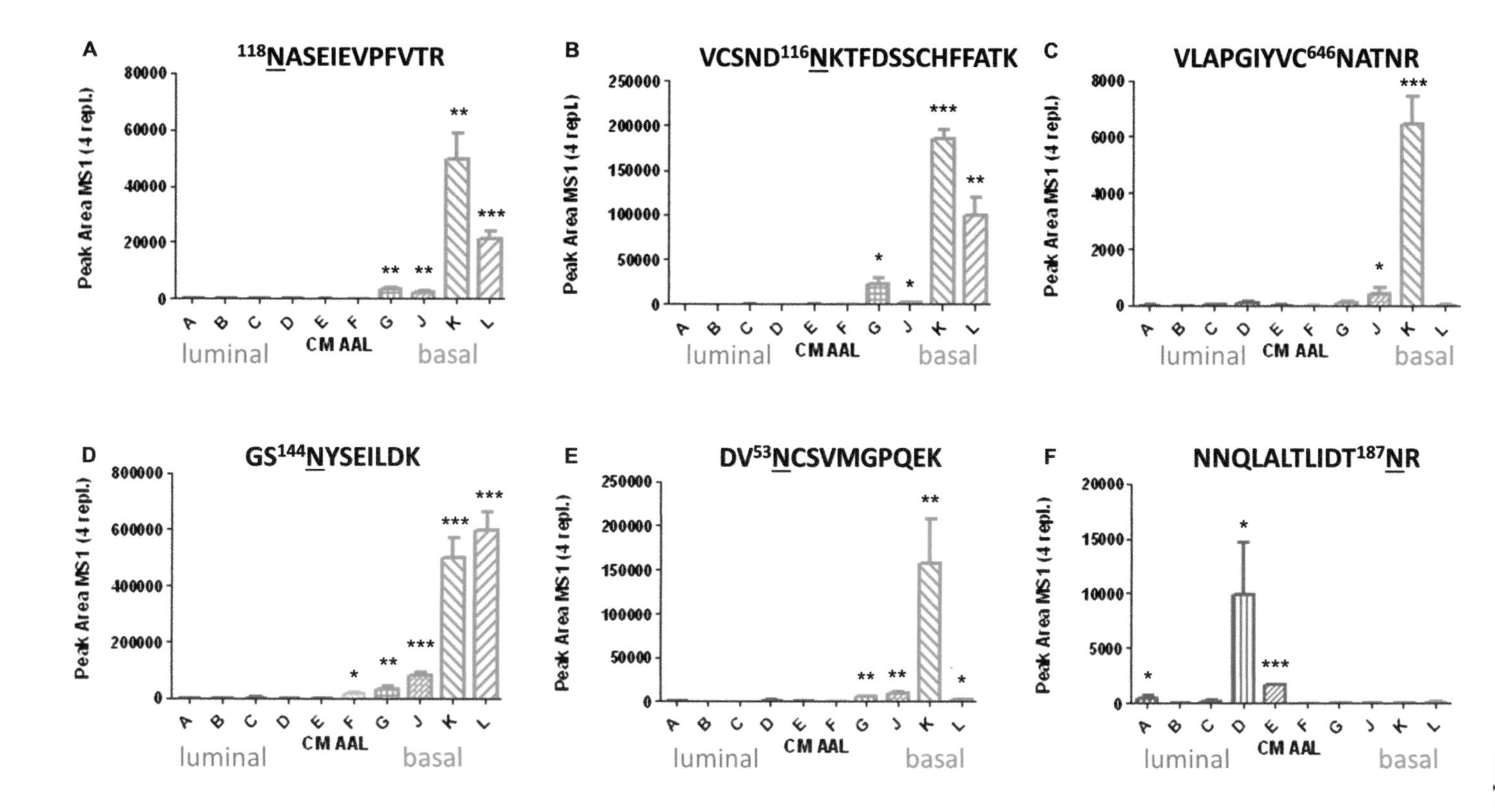
A
118NASEIEVPFVTR
B
VCSND116NKTFDSSCHFFATK
C
VLAPGIYVC646NATNR
D
GS144NYSEILDK
E
DV53NCSVMGPQEK
F
NNQLALTLIDT187NR
Peak Area MS1 (4 repl.)
A B C D E F G J K L
luminal
CM AAL
basal

treatment efficiency, and cancer recurrence.[20,21] The Cancer Genome Atlas Network[22] recently published a report on breast-cancer heterogeneity and breast cancer tumor subtype specificities in an extensive study of primary breast cancers in 825 patients. As breast cancer is a particularly heterogeneous disease, determining cancer subtype specificity is extremely relevant for supporting a patient-specific choice of therapy. Currently, therapeutic approaches are chosen based on cancer type: i) the estrogen receptor (ER)-positive group that is likely to be responsive to endocrine therapies;[23,24] ii) the HER2, or ERBB2, amplified group[25] that is typically treated with therapeutics targeting HER2; and iii) triple-negative breast cancers, lacking expression of ER, progesterone receptor (PR) and HER2, also referred to as basal-like breast cancers[26] that typically require chemotherapy.

The goal of the biomarker discovery study by Drake *et al.* was to determine subtype-specific glycoproteins that were secreted into CM from various luminal and more aggressive basal breast cancer cell lines.[13] In the original report triple negative subtype-specific biomarker candidate selection was performed based on sampling of MS/MS spectra predominantly in the CM from the basal breast cancer cell lines and subsequent non-parametric statistical testing and permutation analysis. Re-analysis of this publicly available data set with Skyline MS1 Filtering provided us with an opportunity to obtain both greater insight into glycopeptide abundance levels within subcategories and individual cell lines, and also proper quantitative assessments, independent from the MS/MS sampling that was used previously for candidate selection. As shown in Figure 7.6, label-free MS1 Filtering for

Figure 7.6 MS1 Filtering of breast cancer cell line conditioned media (CM): luminal *vs.* basal cancer. MS1 Filtering of peptides obtained after AAL lectin enrichment (for fucose-containing glycopeptides) of CM from ten breast cancer cell lines, followed by subsequent PNGase F de-glycosylation. For each de-glycosylated peptide, the peak area for the highest ranked precursor ion is extracted from five luminal (cell lines A–E) and five basal (cell lines F–L) CM samples. Each sample is acquired as four replicates (two process and two technical replicates). The following peptides were monitored: A) $^{118}\underline{\mathbf{N}}$ASEIEVPFVTR from glia-derived nexin (P07093) with the doubly charged precursor ion M observed at *m*/*z* 681.85, B) VCSND$^{116}\underline{\mathbf{N}}$KTFDSSCHFFATK from SPARC/osteonectin (P09486) with the doubly charged precursor ion M + 1 observed at *m*/*z* 567.49, C) VLAPGIYVC$^{646}\underline{\mathbf{N}}$ATNR from intercellular adhesion molecule 5 (Q9UMF0, ICAM5) with the doubly charged precursor ion M observed at *m*/*z* 774.90, D) GS$^{144}\underline{\mathbf{N}}$YSEILDK from follistatin-related protein 1 (Q12841, FSTL1) with the doubly charged precursor ion M observed at *m*/*z* 563.77, E) DV$^{53}\underline{\mathbf{N}}$CSVMGPQEK from dipeptidyl peptidase 1/Cathepsin C (P53634, CATC) with the doubly charged precursor ion M observed at *m*/*z* 682.79, and F) NNQLALTLIDT$^{187}\underline{\mathbf{N}}$R from receptor tyrosine-protein kinase ErbB-2 (P04626, ERBB2) with the doubly charged precursor ion M observed at *m*/*z* 743.90. $\underline{\mathbf{N}}$ indicates a de-glycosylated asparagine residue (after PNGase F treatment). Statistical significance from two-tailed Student's t-tests are marked as * (P<0.05), ** (P<0.01), and *** (P<0.001). CM analyzed from luminal cell lines MDA361 (A), Sum52PE (B), MDAMB175 (C), UACC812 (D), SKBR3 (E), and from basal cell lines MDAMB 468 (F), HCC38 (G), MDAMB231 (J), BT549 (K), and HS5787 (L).

precursor peak areas and quantification provides detailed information about glycoprotein secretion levels detected in five individual luminal cell line CM (indicated as A–E) and five basal cell line CM (indicated as F–L).

Many of these glycopeptides were derived from glycoproteins previously suggested as potential biomarker candidates, such as glia-derived nexin, SPARC/osteonectin, and intercellular adhesion molecule 5.[13] Using MS1 Filtering, we were able to provide quantitative confirmation of the early results that indicated these glycopeptides may be 'basal-specific' candidates (Figure 7.6A–C). However, MS1 Filtering provided greater insight into the abundance of glycosylated peptides and relative quantitative levels in basal *vs.* luminal CM, as well as differences between different basal cell subtypes themselves. In addition to those previously reported we were able to identify several new basal specific biomarkers, including glycopeptides from follistatin-related protein 1 and cathepsin C (Figure 7.6D and E), demonstrating the reliability of peak area-based quantification using MS1 Filtering (Table 7.2). A case of luminal preference can be found in Figure 7.6F, where a significantly higher level of secretion of the receptor tyrosine-protein kinase ErbB-2 into CM was seen in the luminal 'ErbB2 positive' cell lines D and E, which provide a convincing positive control for the workflow.

As the ultimate goal of this work was to identify blood-based biomarkers, we compared the candidates with the ones reported in the human Plasma Atlas,[27,28] which is available at http://www.peptideatlas.org/hupo/hppp/ (see Table 7.2). Interestingly, the glycopeptides/glycoproteins listed in Table 7.2 were also reported in the human Plasma Atlas and, therefore, could in principal serve as starting points to verify and validate these glycopeptides as plasma biomarker candidates for differentiating breast cancer subtypes.

As demonstrated above, one aspect that makes MS1 Filtering somewhat unique is its ability to re-investigate and re-process published data sets from previous publications, regardless of the instrument platform or file format. Data sets acquired on various vendor platforms can readily undergo quantitative re-processing using the existing set of Skyline tools. Indeed, as more published mass spectrometric data sets from a variety of research groups and a variety of instrumentation are uploaded to public data storage repositories, the ability to mine such data sets in a common open-source software tool as flexible as Skyline becomes relatively straight-forward. With quantitative pipelines like MS1 Filtering, that work on a diverse set of file formats, these data repositories become a truly dynamic community resource and not just a static archive of large data sets.

7.3.3 Laboratory Research Pipelines Connecting MS1-based Label-free Quantification with Targeted, Data-independent Quantification Methods

A typical pipeline for research applications that started with data-dependent discovery experiments processed using label-free MS1-based quantification is often followed up using an independent quantitative method for

Table 7.2 MS1 Quantification of de-glycosylated peptides with breast cancer subtype specific preferences.

De-glyco peptide, N-glyco site	*Protein accession, name*	*Ratio*[a] *(p) Bas F : Lum*	*Ratio (p) Bas G : Lum*	*Ratio (p) Bas J : Lum*	*Ratio (p) Bas K : Lum*	*Ratio (p) Bas L : Lum*
NASEIEVPFVTR N-118	P07093 | GDN, Glia-derived nexin	0.5 (0.020)	**22 (0.007)**	**18 (0.002)**	**336 (0.002)**	**148 (0.0005)**
VCSNDNKTFDSSCHFFATK N-116	P09486 | SPRC, SPARC	4.4 (0.310)	**353 (0.012)**	**31 (0.011)**	**2900** (3.4E-5)	**1557 (0.003)**
VLAPGIYVCNATNR N-646	Q9UMF0 | ICAM5, Intercell. adhesion molecule 5	0.7 (0.329)	2.2 (0.051)	**9.1 (0.021)**	**128 (0.001)**	0.7 (0.354)
GSNYSEILDK N-144	Q12841 | FSTL1 Follistatin-related protein 1	**10 (0.012)**	**21 (0.005)**	**49 (0.0005)**	**297 (0.0008)**	**356 (0.0003)**
DVNCSVMGPQEK N-53	P53634 | CATC, Dipeptidyl peptidase 1	0.6 (0.271)	**10 (0.002)**	**17 (0.002)**	**290 (0.009)**	**3.4 (0.017)**
		Basal : Lum A	Basal : Lum B	Basal : Lum C	Basal : Lum D	Basal : Lum E
NNQLALTLIDTNR N-187	P04626 | ERBB2, Receptor tyr-protein kinase	**1:6.4 (0.015)**	1:0.9 (0.722)	1:2.8 (0.101)	**1:123 (0.025)**	**1:21 (3.8E-5)**

[a]MS1 Precursor peak area ratios were calculated comparing CM from an individual basal cell line with the mean of all luminal cell lines (top), as well as comparing CM from an individual luminal cell line with the mean of all basal cell lines (bottom). All ratios >7 are rounded.

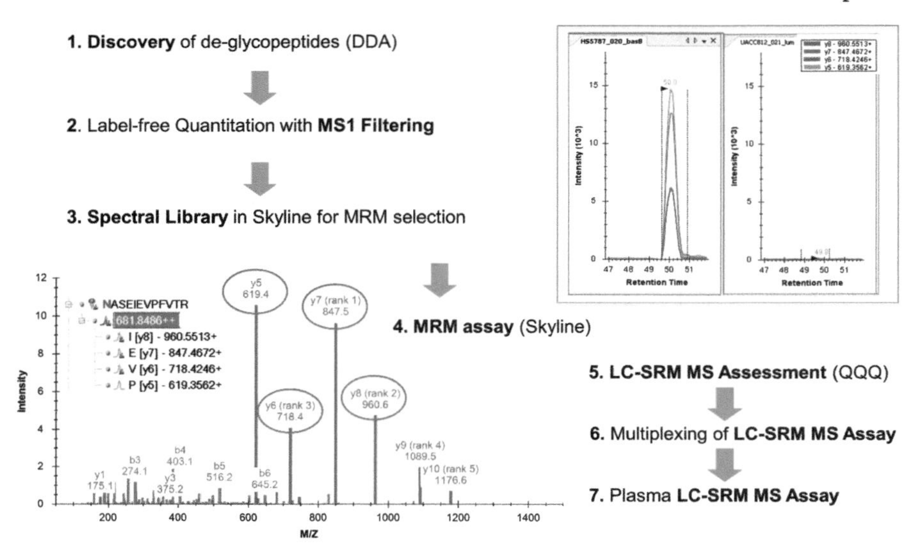

Figure 7.7 Implementation of MS1 Filtering quantification into laboratory pipelines. De-glycosylated peptides were identified in CM from basal and luminal breast cancer cell lines during data-dependent acquisitions. Label-free quantification using MS1 Filtering indicated the de-glyco peptide, ^{118}NASEIEVPFVTR, from glia-derived nexin as basal-specific. Product ions from the corresponding MS/MS spectra available in the Skyline spectral library were used for automatic SRM transition ranking in Skyline. Selected highest ranked transitions are displayed in the tree view for direct export of the SRM transition lists for triple quadrupole instruments. LC-SRM-MS acquisitions on a hybrid triple quadrupole linear ion trap (4000 QTRAP) were performed on protein digestions of CM to confirm candidates. Using this workflow, additional SRM assays for human plasma analysis can be easily designed in Skyline.

confirmation, such as selected reaction monitoring (SRM) (see Chapter 4 for further information). As described for the glycopeptide-based cancer biomarker candidate discovery workflow, potential biomarkers suggested by MS1 Filtering can be developed into targeted quantitative assays using existing features of the Skyline environment (Figure 7.7).

Based on data-dependent MS/MS acquisitions and the resulting spectral libraries, LC-SRM assays can be generated in Skyline, as shown for a potential basal candidate de-glycosylated peptide NASEIEVPFVTR, which is a glycosite in glia-derived nexin. LC-SRM-MS data acquired from CM of cell lines 'L' (basal) and 'D' (luminal) confirmed the anticipated quantitative trend. As shown in Figure 7.7, further higher multiplexed SRM assays can be developed for studies in human plasma as a cancer diagnostic.

It is intriguing to anticipate how additional follow-up experiments, such as some of the newer types of data-independent acquisition (DIA) or SRM-like targeted MS/MS acquisitions can be supported by the Skyline software. Indeed, preliminary experience using some of these additional acquisition strategies suggests that Skyline MS1 Filtering can be seamlessly integrated with these other workflows, all inside Skyline. We have already shown that one can readily transfer results obtained from MS1 Filtering quantification to develop LC-SRM assays in Skyline for triple quadrupole instruments. Similar relations can also be established to move from MS1 Filtering to MS2-based quantification workflows. For example, some commercial mass spectrometers are capable of acquiring targeted, high resolution MS2 full scans in a cycle repeated throughout the entire gradient (also known as '*pseudo* SRM' or 'MRM-HR'). Alternatively, some of these same high resolution, fast-scanning instruments can acquire spectra in a DIA mode, such as that recently popularized as MS/MS^{ALL} with SWATH,[10] MS^E,[11] and multiplexed-DIA.[29]

7.4 Conclusions and Future Outlook

With many technical improvements made in recent years for mass spectrometric instrumentation and chromatographic equipment, label-free high throughput quantitative data acquisition and processing pipelines have become increasingly attractive. In parallel, recent software developments, including Skyline MS1 Filtering, have offered significant improvements in more accurate label-free quantification and more feasible high throughput processing. The field of mass spectrometry and proteomics has shifted to great degrees toward standardizing data formats that do not require vendor-specific software and are uniformly accessible (HUPO Proteomics Standards). At the same time, ProteoWizard, accessed from within Skyline, has established a software library capable of reading the native data formats of five major mass spectrometer vendors (AB SCIEX, Agilent Technologies, Bruker, Thermo, and Waters), as well as standard open formats.[30] As more large mass spectrometric data repositories become publicly available, the availability of vendor-independent software solutions, such as Skyline MS1 Filtering, provides an attractive opportunity for broader data processing and interrogation of existing data sets. In addition, data sharing of raw data or processed data has become more relevant with increased inter-laboratory collaborations that are more than likely to have been generated by different commercial instrumentation. In that context, Skyline MS1 Filtering offers a robust label-free processing solution that supports most major instrument vendors and peptide search engine formats. Investigators using Skyline can also readily extend their workflows from initial MS1 label-free quantification into more specialized quantitative workflows, such as SRM-MS, and data-independent or targeted MS/MS acquisitions, such as *pseudo* SRM, MRM-HR, SWATH-MS2, and other full-scan MS2 quantification methods.

In summary, data were presented here that demonstrate how typical discovery-based proteomic data sets can be mined for quantitative information using MS1 Filtering. The rigorous processing of MS1 ion intensity data into highly quantitative measurements can be achieved across multiple samples and replicates, and even from multiple laboratories with different instrumentation. MS1 Filtering can also be used to assess proper sample handling and workflow reproducibility. For both clinical and research applications, the high reproducibility of MS1 Filtering affords the opportunity to identify specific sets of candidates that can be readily integrated into various types of targeted quantitative workflows. MS1 Filtering also offers opportunities for re-mining published data sets across vendors and data pipelines by providing the tools to interrogate existing data repositories and resources of published scientific data sets. Future directions for MS1 Filtering and label-free quantification can be anticipated as new Skyline features are implemented, such as the integration of quantitative results into the server-based, web-accessible database system known as Panorama,[31] which can be interfaced directly with Skyline for web-based data sharing and archiving data processed with Skyline.

Acknowledgements

This work was supported by grants from the NIH, including R24 DK085610 and R21 CA138308 (BWG), and R01 RR032708 (MJM). We acknowledge the support of instrumentation from the NCRR shared instrumentation grant S10 RR024615 (BWG) and AB SCIEX for evaluation of the TripleTOF 5600 at the Buck Institute. We thank Daniel Broudy and Julian Chytrowski for help with Skyline External Tools features.

References

1. F. F. Gonzalez-Galarza, C. Lawless, S. J. Hubbard, J. Fan, C. Bessant, H. Hermjakob and A. R. Jones, *OMICS*, 2012, **16**, 431.
2. K. A. Neilson, N. A. Ali, S. Muralidharan, M. Mirzaei, M. Mariani, G. Assadourian, A. Lee, S. C. van Sluyter and P. A. Haynes, *Proteomics*, 2011, **11**, 535.
3. B. Schilling, M. J. Rardin, B. X. MacLean, A. M. Zawadzka, B. E. Frewen, M. P. Cusack, D. J. Sorensen, M. S. Bereman, E. Jing, C. C. Wu, E. Verdin, C. R. Kahn, M. J. MacCoss and B. W. Gibson, *Mol. Cell Proteomics*, 2012, **11**, 202.
4. B. MacLean, D. M. Tomazela, N. Shulman, M. Chambers, G. L. Finney, B. Frewen, R. Kern, D. L. Tabb, D. C. Liebler and M. J. MacCoss, *Bioinformatics*, 2010, **26**, 966.
5. D. Kessner, M. Chambers, R. Burke, D. Agus and P. Mallick, *Bioinformatics*, 2008, **24**, 2534.
6. S. E. Ong, B. Blagoev, I. Kratchmarova, D. B. Kristensen, H. Steen, A. Pandey and M. Mann, *Mol. Cell Proteomics*, 2002, **1**, 376.
7. P. L. Ross, Y. N. Huang, J. N. Marchese, B. Williamson, K. Parker, S. Hattan, N. Khainovski, S. Pillai, S. Dey, S. Daniels, S. Purkayastha, P. Juhasz, S. Martin, M. Bartlet-Jones, F. He, A. Jacobson and D. J. Pappin, *Mol. Cell Proteomics*, 2004, **3**, 1154.

8. L. Dayon, A. Hainard, V. Licker, N. Turck, K. Kuhn, D. F. Hochstrasser, P. R. Burkhard and J. C. Sanchez, *Anal. Chem.*, 2008, **80**, 2921.
9. J. Cox and M. Mann, *Nat. Biotechnol.*, 2008, **26**, 1367.
10. L. C. Gillet, P. Navarro, S. Tate, H. Rost, N. Selevsek, L. Reiter, R. Bonner and R. Aebersold, *Mol. Cell Proteomics*, 2012, **11**, O111 016717.
11. C. E. Doneanu, A. Xenopoulos, K. Fadgen, J. Murphy, S. J. Skilton, H. Prentice, M. Stapels and W. Chen, *MAbs*, 2012, **4**, 24.
12. J. M. Held, B. Schilling, A. K. D'Souza, T. Srinivasan, J. B. Behring, D. J. Sorensen, C. C. Benz and B. W. Gibson, *Int. J. Proteomics*, 2013, **2013**, 791985, http://dx.doi.org/10.1155/2013/791985.
13. P. M. Drake, B. Schilling, R. K. Niles, A. Prakobphol, B. Li, K. Jung, W. Cho, M. Braten, H. D. Inerowicz, K. Williams, M. Albertolle, J. M. Held, D. Iacovides, D. J. Sorensen, O. L. Griffith, E. Johansen, A. M. Zawadzka, M. P. Cusack, S. Allen, M. Gormley, S. C. Hall, H. E. Witkowska, J. W. Gray, F. Regnier, B. W. Gibson and S. J. Fisher, *J. Proteome Res.*, 2012, **11**, 2508.
14. S. E. Abbatiello, D. R. Mani, B. Schilling, B. MacLean, L. J. Zimmerman, X. Feng, M. P. Cusack, N. Sedransk, S. C. Hall, T. Addona, S. Allen, N. G. Dodder, M. Ghosh, J. M. Held, V. Hedrick, H. D. Inerowicz, A. Jackson, H. Keshishian, J. W. Kim, J. S. Lyssand, C. P. Riley, P. Rudnick, P. Sadowski, K. Shaddox, D. Smith, D. Tomazela, A. Wahlander, S. Waldemarson, C. A. Whitwell, J. You, S. Zhang, C. R. Kinsinger, M. Mesri, H. Rodriguez, C. H. Borchers, C. Buck, S. J. Fisher, B. W. Gibson, D. Liebler, M. MacCoss, T. A. Neubert, A. Paulovich, F. Regnier, S. J. Skates, P. Tempst, M. Wang and S. A. Carr, *Mol. Cell Proteomics*, 2013, **12**, 2623.
15. T. Bock, H. Moest, U. Omasits, S. Dolski, E. Lundberg, A. Frei, A. Hofmann, D. Bausch-Fluck, A. Jacobs, N. Krayenbuehl, M. Uhlen, R. Aebersold, K. Frei and B. Wollscheid, *J. Proteome Res.*, 2012, **11**, 4885.
16. M. J. Rardin, J. C. Newman, J. M. Held, M. P. Cusack, D. J. Sorensen, B. Li, B. Schilling, S. D. Mooney, C. R. Kahn, E. Verdin and B. W. Gibson, *Proc. Natl Acad. Sci. USA*, 2013, **110**, 6601.
17. M. Rardin, B. Schilling, M. Cusack, D. Sorensen, B. MacLean, M. J. MacCoss, C. R. Kahn, E. Verdin and B. W. Gibson, *Proceedings of the 60th Annual ASMS Conference on Mass Spectrometry & Allied Topics*, Vancouver, Canada, May 20–24, 2012.
18. A. K. D'Souza, B. Schilling, J. Chytrowski, B. MacLean, D. Broudy, N. J. Shulman, M. J. MacCoss and B. W. Gibson, *Proceedings of the 61st Annual ASMS Conference on Mass Spectrometry & Allied Topics*, Minneapolis, USA, June 8–13, 2013.
19. J. D. Storey and R. Tibshirani, *Proc. Natl Acad. Sci. USA*, 2003, **100**, 9440.
20. H. T. Tan, Y. H. Lee and M. C. Chung, *Mass Spectrom. Rev.*, 2012, **31**, 583.
21. L. Anderson, *Clin. Chem.*, 2012, **58**, 28.
22. T. C. G. A. Network, *Nature*, 2012, **490**, 61.
23. S. Paik, S. Shak, G. Tang, C. Kim, J. Baker, M. Cronin, F. L. Baehner, M. G. Walker, D. Watson, T. Park, W. Hiller, E. R. Fisher,

D. L. Wickerham, J. Bryant and N. Wolmark, *N. Engl. J. Med.*, 2004, **351**, 2817.

24. L. J. van 't Veer, H. Dai, M. J. van de Vijver, Y. D. He, A. A. Hart, M. Mao, H. L. Peterse, K. van der Kooy, M. J. Marton, A. T. Witteveen, G. J. Schreiber, R. M. Kerkhoven, C. Roberts, P. S. Linsley, R. Bernards and S. H. Friend, *Nature*, 2002, **415**, 530.
25. D. J. Slamon, G. M. Clark, S. G. Wong, W. J. Levin, A. Ullrich and W. L. McGuire, *Science*, 1987, **235**, 177.
26. C. M. Perou, *Oncologist*, 2011, **16**(Suppl 1), 61.
27. T. Farrah, E. W. Deutsch and R. Aebersold, *Methods Mol. Biol.*, 2011, **728**, 349.
28. T. Farrah, E. W. Deutsch, G. S. Omenn, D. S. Campbell, Z. Sun, J. A. Bletz, P. Mallick, J. E. Katz, J. Malmstrom, R. Ossola, J. D. Watts, B. Lin, H. Zhang, R. L. Moritz and R. Aebersold, *Mol. Cell Proteomics*, 2011, **10**, M110 006353.
29. J. D. Egertson, A. Kuehn, G. E. Merrihew, N. Bateman, B. X. MacLean, Y. S. Ting, J. D. Canterbury, D. M. Marsh, M. Kellmann, V. Zabrouskov, C. C. Wu and M. J. MacCoss, *Nature Methods*, 2013, **10**, 744.
30. M. C. Chambers, B. MacLean, R. Burke, D. Amodei, D. L. Ruderman, S. Neumann, L. Gatto, B. Fischer, B. Pratt, J. Egertson, K. Hoff, D. Kessner, N. Tasman, N. Shulman, B. Frewen, T. A. Baker, M. Y. Brusniak, C. Paulse, D. Creasy, L. Flashner, K. Kani, C. Moulding, S. L. Seymour, L. M. Nuwaysir, B. Lefebvre, F. Kuhlmann, J. Roark, P. Rainer, S. Detlev, T. Hemenway, A. Huhmer, J. Langridge, B. Connolly, T. Chadick, K. Holly, J. Eckels, E. W. Deutsch, R. L. Moritz, J. E. Katz, D. B. Agus, M. MacCoss, D. L. Tabb and P. Mallick, *Nat. Biotechnol.*, 2012, **30**, 918.
31. V. Sharma, B. MacLean, J. Eckels, A. B. Stergachis, J. D. Jaffe and M. J. MacCoss, *Proceedings of the 60th Annual ASMS Conference on Mass Spectrometry & Allied Topics*, Vancouver, Canada, May 20–24, 2012.

CHAPTER 8

Label-free Quantification of Proteins Using Data-Independent Acquisition

YISHAI LEVIN

Proteomics Unit, Israel National Center for Personalized Medicine, Weizmann Institute of Science, Rehovot 76100, Israel
Email: yishai.levin@weizmann.ac.il

8.1 Introduction

The most common approach for identification of proteins is by 'bottom-up' proteomics using liquid chromatography (LC) coupled to tandem mass spectrometry (MS/MS).[1,2] Over a decade ago it was established that peptides in complex mixtures could be identified and even quantified using this approach.[2,3] For reliable identification, peptide ions need to be subjected to dissociation (fragmentation), making use of the generated product ions for sequence determination. Performing MS/MS fragmentation manually for individual ions during an MS acquisition is not practical, even for simple mixtures. Thus, an automated acquisition mode was devised to facilitate selection and fragmentation of thousands, and in some cases tens of thousands, of ions in a single acquisition. This mode was termed data-dependent acquisition (DDA).[2] Regardless of the instrument being used, a DDA experiment is a serial process of peptide ion selection and fragmentation. The cycle starts by acquisition of an MS survey scan, which records all

New Developments in Mass Spectrometry No. 1
Quantitative Proteomics
Edited by Claire Eyers and Simon J Gaskell

Published by the Royal Society of Chemistry, www.rsc.org

the precursor ions generated at a given point in time. The instrument control software then reviews the acquired mass spectrum and decides automatically which peptide ions to isolate and fragment, typically on the basis of signal intensity. The survey scan is then followed by selection of precursor ions for fragmentation that may or may not be at the chromatographic peak apex. Selected precursor ions are isolated serially for MS/MS acquisition for a defined time period or until a specific ion current is breached. This cycle of MS and MS/MS acquisitions continues throughout the period of chromatographic separation. With the technological improvements in mass spectrometers in terms of sensitivity, mass accuracy and speed, the approach was applied to increasing levels of sample complexity. A landmark in this journey was the introduction of the multidimensional peptide identification technology (MuDPIT) approach, which enabled a dramatic increase in the number of peptides and proteins identified in a single experiment by virtue of combining two orthogonal modes of chromatographic separation prior to tandem MS analysis.[4]

Several years following the application of bottom-up proteomics using DDA for analysis of complex samples it was recognized that there existed an inherent issue of poor peptide identification reproducibility.[5] The root of the problem lies in the bias of the DDA towards fragmentation of abundant peptide ions, hand-in-hand with the fact that fragmentation of these ions is conducted in series, not in parallel. This results in a low duty cycle and under-sampling of the peptide flux. The approach was further hampered by the fact that the ion transmission window, set on the quadrupole, is typically ± 1.5 to 3.0 *m*/*z* units in order to capture the complete isotopic distribution of an ion of interest. In a complex sample, other co-eluting and isobaric peptide ions will frequently be co-isolated and transmitted into the collision cell.[6] These contaminating ions also undergo fragmentation and will produce a multitude of aberrant product ions. Therefore, the MS/MS spectra will contain product ions derived from multiple precursors and compromise the selectivity of the resulting MS/MS data. These additional product ions can, and often do, lead to low scoring identifications. The end result is often irreproducible identification and inaccurate quantification in the case of isobaric tagging methods.[5]

In recent years alternative data acquisition modes have been devised.[25,26] Two of these are both referred to as data-independent acquisition (DIA). One DIA approach, termed SWATH™, is to scan the required mass range by setting a wide isolation window of the quadrupole, typically between 10 and 25 *m*/*z* units, and then progressing this window along the mass range and cycling the isolation windows throughout the run time.[7,8] In this approach all peptide ions that are stably transmitted to the collision cell within the SWATH window are fragmented. Identification is achieved by aligning product ion spectra with data in spectral libraries, generated in previous experiments by shotgun proteomics.[8] The fact that previous experiments are required in order to identify the proteins is a major limitation of this approach.

The second approach, termed MS^E, facilitates measurement of the *m*/*z* values of both the intact peptide ions, at a high sampling rate, and their fragmentation products.[6,9] In this method the quadrupole is used as an ion guide without isolating any of the precursor ions.[9] Proprietary software is then used post-acquisition to match the product ions with the precursor ions based on a number of factors, including the chromatographic elution profile and accurate mass measurement of both product ions and precursors.

The concept was introduced in 2002 and has since been recognized as a powerful approach for identification and, especially, quantification of proteins in various applications.[10–16] The fundamentals of this unique method, including the technical details, the combination of ion mobility and MS^E and expected performance will be discussed in this chapter.

8.1.1 Fundamentals

The concept behind an MS^E acquisition is simple. Instead of isolating each precursor ion using the quadrupole, all ions pass through it and straight in to the collision cell. The collision energy is alternated between low and elevated energy at constant intervals throughout the acquisition. In the low-energy scans, intact ions are measured. In the elevated-energy scans, peptide ions are subjected to collision-induced dissociation. Since the switching of low and elevated energies is done rapidly, relative to the elution time of the peptides, the product ions have the same apex retention time as their corresponding precursors. Thus, it is possible to align them in time and match fragments to precursors. An illustration of this is shown in Figure 8.1.

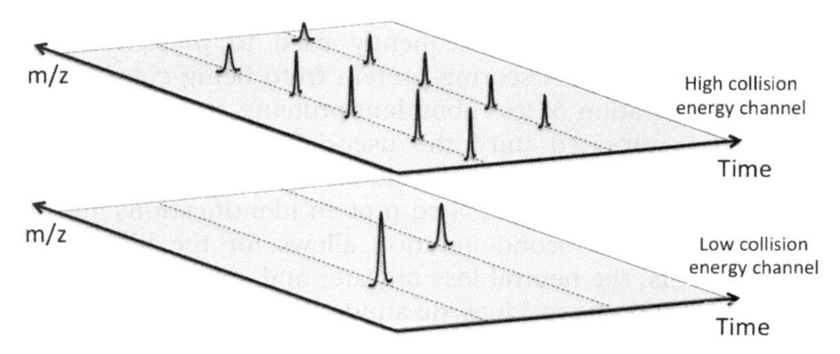

Figure 8.1 A data structure illustration of an LC-MS^E acquisition. The low collision energy scans, which include measurement of the precursor ions, are stored in one channel (bottom). The high collision energy scans, which include the measurement of all product ions, are stored in another channel (top). Based on the retention time and other parameters the software matches the product ions to each corresponding precursor.

8.2 Identifying Proteins

The raw data structure of an MS^E acquisition is non-standard and thus requires tailor-made software to process it. Currently, the only software able to process such data is $Identity^E$, part of ProteinLynx Global Server (PLGS, Waters Corp., Milford MA, USA). The software utilizes a peak detection algorithm called Apex3D and sequence database search algorithm named Ion Accounting, both of which are described in detail by Geromanos *et al.*[9]

The initial stages of the Apex3D algorithm include ion detection, lock mass correction, de-isotoping and charge-state reduction for both low and elevated scans. It creates lists for precursor mass and retention time clusters in order to match fragments to precursors. The basis of the time-alignment algorithm is that elevated ions with apex retention times that are within plus or minus one tenth the chromatographic peak width of a precursor ion in the low energy scans are associated with that particular precursor ion (peptide).

In the case of co-eluting peptides the same product ions will be initially assigned to multiple precursors. Thus, the initial peak lists would not be searchable by standard search engines, such as Mascot, because the list of product ions is redundant.

The final assignment of product ions to precursors is done by the Ion Accounting algorithm, which employs a hierarchal database search strategy, including the use of a reverse or random decoy database. Tryptic peptides are tentatively identified according to initial search parameters. These tentative identifications are ranked and scored by how well they conform to a number of predetermined models of physicochemical attributes. These identifications are then used to refine the retention time and fragmentation models and used to further score the identified peptides. All tentative peptides are collapsed into their parent proteins utilizing only the highest scoring peptides that contribute to the total protein score.

A depletion algorithm is subsequently used to prevent the peptide detections from the highest scoring protein from being considered in the subsequent identification of less abundant proteins.

The process is repeated until the user-defined false positive rate is obtained (based on identification of random or reverse proteins). A subset database is created from the validated protein identifications and the next iteration initiated. The second iteration allows for the identification of in-source fragments, the neutral loss of water and ammonia, missed cleavages, oxidized methionine residues, de-amidation, *N*-terminal alkylation and any other user-defined modifications.

8.3 Quantifying Proteins

The MS^E acquisition performs high sampling of precursor peptide ions. Thus, accurate and label-free quantification is achievable, assuming sample preparation is consistent and reproducible. One can rely on the peak detection of the PLGS software, which facilitates exporting of all spectra, peak lists and identifications for quantification of proteins in a small-scale experiment.

In cases where large-scale experiments are conducted, time alignment of precursor peptide peaks is required. There are several commercial software packages that can import raw MS^E data, as well as the identifications, directly from PLGS. Such software includes Progenesis LC-MS® (Nonlinear Dynamics, UK), Elucidator® (Rossetta Biosoftware, USA) and Expressionist® (Genedata). The advantage of using such software is the ability to perform quantitative analysis based on detected peaks, irrespective of peptide identification. Furthermore, identifications can be propagated from sample to sample based on the aligned mass and retention time, significantly increasing proteomic coverage (similar to the 'match between runs' option in MaxQuant[17]).

One key feature of an MS^E acquisition is the ability to obtain an assessment of the absolute concentration of proteins based on a single spike-in protein.[18] Silva *et al.* have described an approach to move from peptide-level quantification to the protein level by what is known as the Hi-3 method,[18] which has been widely accepted in the field. In this approach the mean intensity of the three most abundant unique peptides per protein is calculated to generate a quantitative index at the protein level. This has been shown to produce accurate quantification in both simple and complex samples.[19]

8.4 Ion Mobility and MS^E

Ion mobility (IM) spectrometry is a gas-phase separation method where ions are separated based on their charge state, shape and, to some degree, their mass. It is applicable to a wide range of chemical and biochemical substances and provides a fast analytical strategy for the separation of components of complex biological samples.[20] Moreover, IM can address chimeric events, whereby peptide ions are co-fragmented and produce mixed product ion spectra.[21,22] Recently, IM was coupled with liquid chromatography and orthogonal acceleration time-of-flight (oa-ToF) mass spectrometry.[23] In this hybrid instrument, IM separation takes place on the millisecond timescale and is used to separate eluted peptide ions before or after collision-induced dissociation; whereas oa-ToF samples the mobility domain efficiently.[23] The introduction of this orthogonal IM separation into the analytical workflow can significantly reduce ion interference and increases the overall peak capacity of the analytical platform by adding an additional separation space. The combination of MS^E and ion mobility has been shown to significantly improve the performance of the method.[23]

When acquiring data in IM-MS^E mode, fragmentation is performed in a collision cell that is placed after the IM separation chamber. Thus, it can be observed that not only do product ions from a given precursor peptide have the same chromatographic retention time but also the same IM drift time

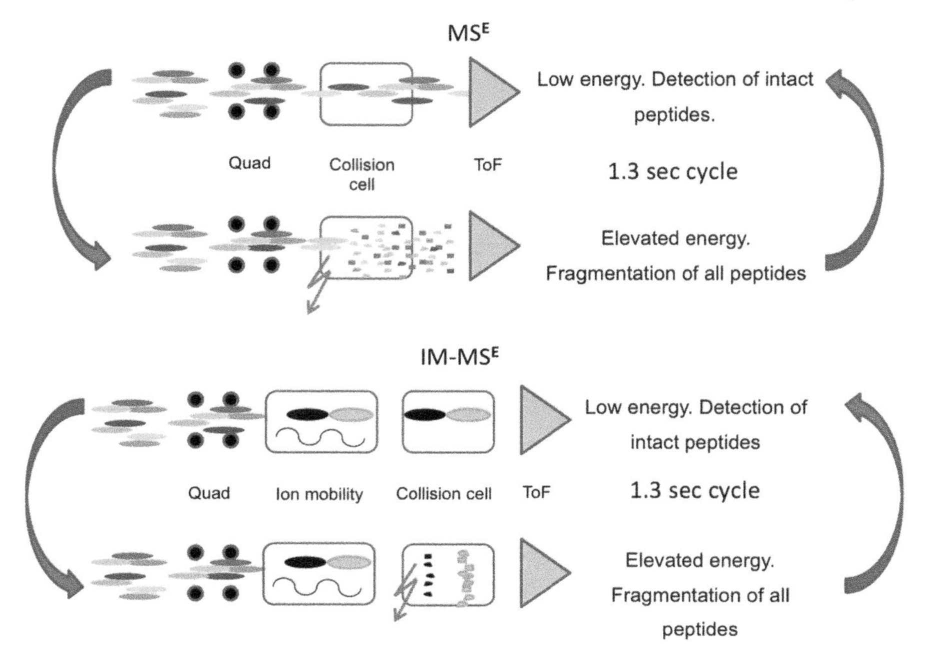

Figure 8.2 Schematic representation of MS^E and IM-MS^E acquisitions. In an analysis of a complex sample, peptides often co-elute making the task of matching fragments to precursor peptides more challenging. However, when ion mobility is enabled, the task becomes much easier due to the additional separation dimension.

(as illustrated in Figure 8.2). This considerably reduces the number of product ions assigned to each parent precursor. The increased selectivity in the alignment process further reduces the number of co-generated product ions presented to the database search algorithm, producing a significant increase in the number of peptide and protein identifications.[23] This is illustrated in Figure 8.3, where an *Escherichia coli* tryptic digest was analyzed by both MS^E and IM-MS^E modes. Comparisons of the two modes were conducted at different on-column loadings. It can be seen that, irrespective of gradient length and loading, IM-MS^E produces significantly more identifications.

8.5 Performance

Among the criteria for evaluating the performance of an analytical platform for large-scale quantification of proteins in complex samples, are quantification accuracy, reproducibility and proteomic coverage. For clinical applications, an additional criterion is the ability to analyze dozens of samples in a single experiment. Inclusion of multiple biological replicates enables the

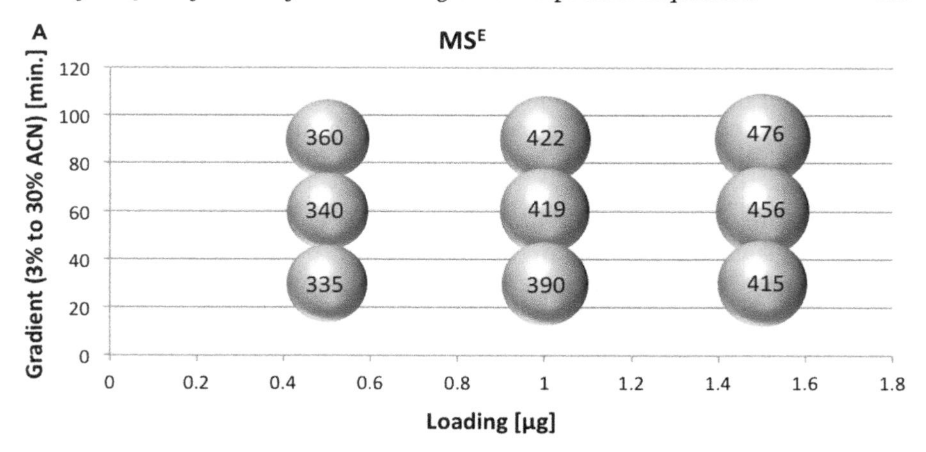

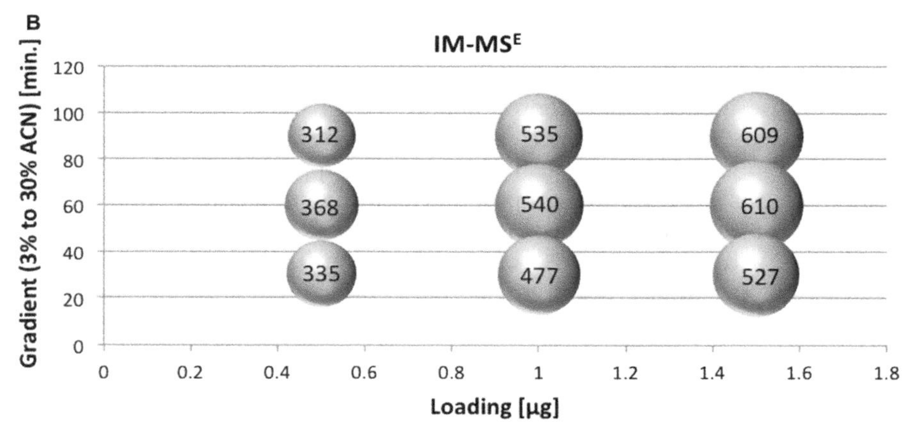

Figure 8.3 Comparison between an MS^E and IM-MS^E analysis of tryptic peptides from a cytosolic *E. coli* extract using one-dimensional chromatography and a Synapt G2 instrument. Shown is the number of proteins identified (minimum two peptides per protein, maximum protein false discovery rate of 1%). It can be seen that, independent from the loading or gradient, the IM-MS^E acquisition produces more identifications. The data have been deposited in the ProteomeXchange Consortium http://proteomecentral.proteomexchange.org *via* the PRIDE partner repository[27] with the dataset identifier PXD000150.

analysis to be statistically powered. This is critical for the success of any biomarker discovery study, regardless of what is being measured.

In a recent publication using a theoretical simulation we showed that, in an experiment that includes four biological replicates per group and a total abundance variation of 40%, the minimum fold change would be four fold at a power of 0.8.[24] This highlights the fact that most shotgun proteomic experiments are, in fact, underpowered because not enough replicates are included in the experiment. However, the MS^E label-free approach poses no theoretical limit on the number of samples that can be included in a single experiment, unlike labeling methods where one is limited by the number of tags.

Evaluation of quantification accuracy of a proteomic platform or method can best be determined using standards, which are spiked into complex samples in known concentrations to simulate a differential expression. The samples can then be analyzed by the platform and used to determine how well the simulated expression difference was detected. In a recent paper we conducted such an experiment to evaluate the performance of the Hi-3 methodology. Four proteins were spiked into serum samples and rat brain tissue digests. It was shown that, using the MS^E approach, a differential expression as low as 50% change (or 1.5 fold change) could be detected with a mean error of 23%.[19] The protein standards were spiked at concentrations as low as 10 fmol μL^{-1}.[19] This quantitative accuracy can be attributed mainly to the acquisition mode, which enables measurement of intact peptides at a high sampling rate, as previously discussed.

The second factor for evaluating a proteomic platform is reproducibility. While there are many factors influencing the reproducibility of an LC-MS platform, the identification reproducibility relies solely on the acquisition mode and sample complexity. Here we present two typical examples of the identification reproducibility using the IM-MS^E approach. Figure 8.4 shows Venn diagrams of the number of proteins identified in technical replicates of two types of complex samples. It can be seen that identification is very consistent, with 85% and 86% of proteins identified in two out of three replicates of the mouse brain and porcine epithelial samples, respectively. Furthermore, the analysis of the mouse brain samples produced 1196

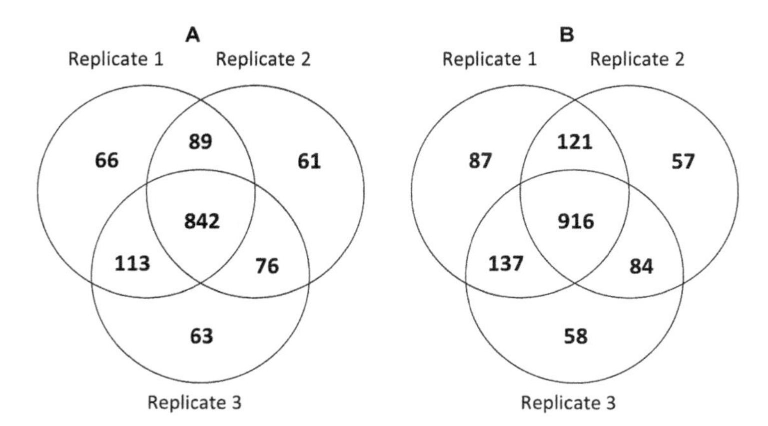

Figure 8.4 Venn diagrams comparing protein identifications from replicate analyses. A) An analysis of a mouse whole brain lysis, where 85% of proteins were identified in two out of three replicates. Data was acquired using a Synapt G2-S loading 300 ng. B) An analysis of a porcine epithelial cell line, where 86% of proteins were identified in two out of three replicates. Data acquired using a Synapt G2 instrument, loading 10 μg.

protein identifications with at least two peptides using 300 ng of digested protein, a 90 min gradient and one-dimensional chromatography. While this is not a comprehensive coverage, it does suggest that utilizing two-dimensional chromatography, in addition to IM separation of the gas-phase ions, reaches significant proteomic coverage.

8.6 Conclusions

In this chapter we reviewed MS^E acquisition for the analysis of proteins in complex samples. This data-independent acquisition technique resolves some of the limitations of the traditional analysis mode of DDA. The extent to which this approach deals with the complexity of identifying and quantifying proteins was presented. It can be concluded that MS^E enables accurate quantification of proteins and the identification is significantly more reproducible than other approaches.

Acknowledgements

The Albert Scholars Program for New Scientists is kindly acknowledged for their generous financial support of our work.

References

1. M. L. Yarmush and A. Jayaraman, *Annu. Rev. Biomed. Eng.*, 2002, **4**, 349–373.
2. M. Quadroni and P. James, *Electrophoresis*, 1999, **20**, 664–677.
3. M. Mann, P. Hojrup and P. Roepstorff, *Biol. Mass Spectrom.*, 1993, **22**, 338–345.
4. D. A. Wolters, M. P. Washburn and J. R. Yates III, *Anal. Chem.*, 2001, **73**, 5683–5690.
5. R. Aebersold, *Nat. Methods*, 2009, **6**, 411–412.
6. G. Z. Li, J. P. Vissers, J. C. Silva, D. Golick, M. V. Gorenstein and S. J. Geromanos, *Proteomics*, 2009, **9**, 1696–1719.
7. J. D. Venable, M. Q. Dong, J. Wohlschlegel, A. Dillin and J. R. Yates III, *Nat. Methods*, 2004, **1**, 39–45.
8. L. C. Gillet, P. Navarro, S. Tate, H. Rost, N. Selevsek, L. Reiter, R. Bonner and R. Aebersold, *Mol. Cell. Proteomics*, 2012, **11**, O111 016717.
9. S. J. Geromanos, J. P. Vissers, J. C. Silva, C. A. Dorschel, G. Z. Li, M. V. Gorenstein, R. H. Bateman and J. I. Langridge, *Proteomics*, 2009, **9**, 1683–1695.
10. B. Reidel, J. W. Thompson, S. Farsiu, M. A. Moseley, N. P. Skiba and V. Y. Arshavsky, *Mol. Cell. Proteomics*, 2011, **10**, M110 002469.
11. L. Pizzatti, C. Panis, G. Lemos, M. Rocha, R. Cecchini, G. H. Souza and E. Abdelhay, *Proteomics*, 2012, **12**, 2618–2631.
12. Y. Levin, L. Wang, E. Schwarz, D. Koethe, F. M. Leweke and S. Bahn, *Molecular Psychiatry*, 2010, **15**, 1088–1100.

13. V. Stelzhammer, B. Amess, D. Martins-de-Souza, Y. Levin, S. E. Ozanne, M. S. Martin-Gronert, S. Urday, S. Bahn and P. C. Guest, *Proteomics*, 2012, **12**, 3386–3392.
14. H. A. Saka, J. W. Thompson, Y. S. Chen, Y. Kumar, L. G. Dubois, M. A. Moseley and R. H. Valdivia, *Mol. Microbiol.*, 2011, **82**, 1185–1203.
15. E. S. Oswald, L. M. Brown, J. C. Bulinski and C. T. Hung, *J. Proteome Res.*, 2011, **10**, 3050–3059.
16. M. Hummel, J. H. Cordewener, J. C. de Groot, S. Smeekens, A. H. America and J. Hanson, *Proteomics*, 2012, **12**, 1024–1038.
17. J. Cox and M. Mann, *Nat. Biotechnol.*, 2008, **26**, 1367–1372.
18. J. C. Silva, M. V. Gorenstein, G. Z. Li, J. P. Vissers and S. J. Geromanos, *Mole. Cell. Proteomics*, 2006, **5**, 144–156.
19. Y. Levin, E. Hradetzky and S. Bahn, *Proteomics*, 2011, **11**, 3273–3287.
20. J. A. Taraszka, A. E. Counterman and D. E. Clemmer, *Fresenius J. Anal. Chem.*, 2001, **369**, 234–245.
21. S. Houel, R. Abernathy, K. Renganathan, K. Meyer-Arendt, N. G. Ahn and W. M. Old, *J. Proteome Res.*, 2010, **9**, 4152–4160.
22. M. Bern, G. Finney, M. R. Hoopmann, G. Merrihew, M. J. Toth and M. J. MacCoss, *Anal. Chem.*, 2010, **82**, 833–841.
23. E. Rodriguez-Suarez, L. Gethings, K. Giles, J. Wildgoose, M. Staples, K. E. Fadgen, S. J. Geromanos, J. P. C. Vissers, F. Elortza and J. I. Langridge, *Curr. Anal. Chem.*, 2013, **9**(2), 199–211.
24. Y. Levin, *Proteomics*, 2011, **11**, 2565–2567.
25. T. Geiger, J. Cox and M. Mann, *Mol. Cell. Proteomics*, 2010, **9**, 2252–2261.
26. J. D. Egertson, A. Kuehn, G. E. Merrihew, N. W. Bateman, B. X. Maclean, Y. S. Ting, J. D. Canterbury, D. M. Marsh, M. Kellmann, V. Zabrouskov, C. C. Wu and M. J. Maccoss, *Nat. Methods*, 2013, **10**, 744–746.
27. J. A. Vizcaino, R. G. Cote, A. Csordas, J. A. Dianes, A. Fabregat, J. M. Foster, J. Griss, E. Alpi, M. Birim, J. Contell, G. O'Kelly, A. Schoenegger, D. Ovelleiro, Y. Perez-Riverol, F. Reisinger, D. Rios, R. Wang and H. Hermjakob, *Nucleic Acids Res.*, 2013, **41**, D1063–1069.

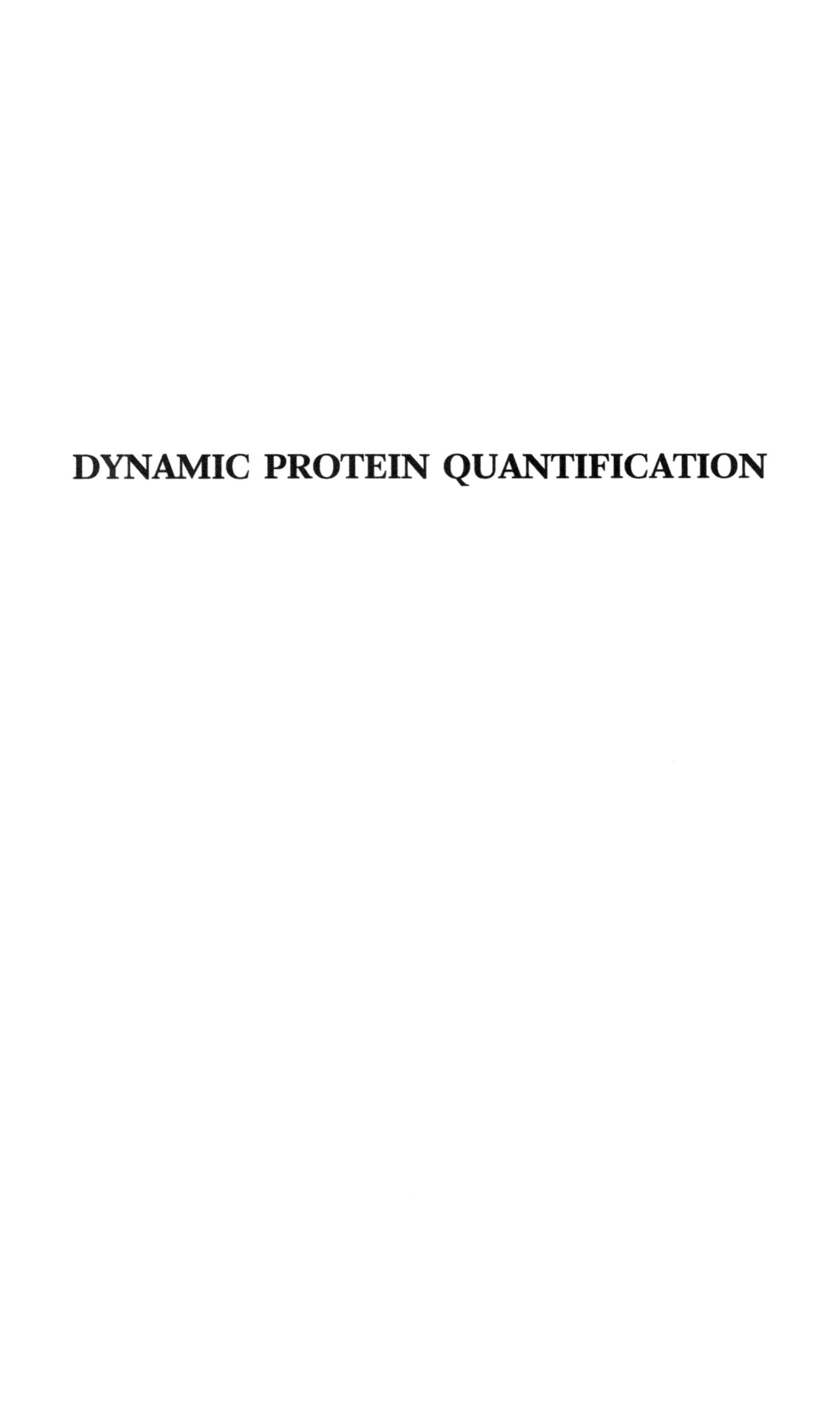

DYNAMIC PROTEIN QUANTIFICATION

CHAPTER 9

Spatial Proteomics: Practical Considerations for Data Acquisition and Analysis in Protein Subcellular Localisation Studies

ANDY CHRISTOFOROU, CLAIRE MULVEY, LISA M. BRECKELS, LAURENT GATTO AND KATHRYN S. LILLEY*

Cambridge Centre for Proteomics, Department of Biochemistry, University of Cambridge, United Kingdom, CB2 1QR
*Email: k.s.lilley@bioc.cam.ac.uk

9.1 Introduction

Localisation and sequestration of proteins within subcellular niches is a fundamental mechanism for the post-translational regulation of protein function. Some such compartments, collectively termed organelles, are physically separated from other niches by lipid bilayers, while others, such as the nucleolus, ribosome, and centrosome, are formed by macromolecular assemblies of proteins and nucleic acids. In simplistic terms this sub-division allows cells to establish a range of distinct microenvironments, each favouring different biochemical reactions and interactions and, therefore, allowing each compartment to fulfil a particular functional role.

New Developments in Mass Spectrometry No. 1
Quantitative Proteomics
Edited by Claire Eyers and Simon J Gaskell

Published by the Royal Society of Chemistry, www.rsc.org

This compartmentalisation of proteins is a dynamic property, and protein trafficking between subcellular niches is a useful mechanism by which to modulate protein function and, therefore, cell behaviour without the requirement for constant cycles of protein degradation and *de novo* synthesis. This is no more evident than in the context of cell–cell signalling, where a stimulus must be transduced from the cell surface to the nucleus or other intracellular niches, often in a short timeframe, in order to elicit a cellular response. Translocation events, such as recruitment of cytosolic proteins to the plasma membrane, and nuclear import and export, are often integral features of signal transduction pathways.

Additionally, the activity of many proteins is highly dependent upon their localisation. For example, β-catenin, a focal node of the canonical Wnt signalling pathway, acts as a component of the adherens junction when peripherally associated with the plasma membrane, forms a 'destruction complex' along with axin and APC in the cytosol, and acts as a transcriptional co-activator in the nucleus. Another example of such multi-functionality is glyceraldehyde-3-phosphate dehydrogenase, which has been reported to re-localise from the cytoplasm to a range of particulate subcellular niches as a stress response, and fulfils distinct functional roles in each of these compartments.[1] Characterising the localisation of proteins and the dynamics of their intracellular trafficking behaviour may, therefore, provide insights into cell states that cannot be readily extrapolated from genomic or transcriptomic data, or by the measurement of protein abundances in a total cell lysate.

The sensitivity, versatility, and high-throughput nature of liquid chromatography-mass spectrometry (LC-MS), coupled with quantification approaches, makes it a well suited technology for such analyses, which has led to the emergence of spatial proteomics, the characterisation of protein localisation by a combination of subcellular fractionation and MS analysis. However, such experiments are not facile. Aspects of experimental design, including subcellular fractionation, MS data acquisition strategy, data processing and analysis, and post-experimental validation, must all be carefully considered in order to construct suitably robust workflows for the characterisation of protein localisation. In this chapter we review several quantitative proteomics methods that have been devised for this purpose, including the application of label-free quantification, differential stable isotope labelling using both *in vivo* metabolic labelling and isobaric tagging, and place them in context with more traditional approaches for studying protein localisation. Furthermore, we consider the strengths and weaknesses of the implementation of different approaches and highlight their complementarity.

9.2 Traditional Approaches for Characterising Protein Localisation

9.2.1 Microscopy-driven Analysis

Protein subcellular localisation is typically determined using microscopy techniques. Protein localisation may be visualised either by immunocytochemistry or by the generation of fluorescent fusion proteins that can be directly observed by confocal microscopy. While these approaches are valuable and well established, there are certain limitations to their applicability.

Immunocytochemical methods are heavily dependent on the availability of antibodies with both high specificity and high affinity for the target epitope. Poor antibody affinity may result in difficulties in obtaining images with sufficient signal quality for analysis, while poor antibody specificity may obfuscate the true location of the target protein, resulting in false classification of subcellular localisation.[2,3]

Fluorescent fusion proteins can be easily followed by confocal microscopy, either in fixed or live cells. However, the generation of fluorescent fusion protein vectors is low throughput, with limited ability to simultaneously characterise multiple proteins. Fusion proteins may also display aberrant function or trafficking due to deleterious effects of the bulky fusion moiety, or as a result of transgene overexpression.[4]

While extensive libraries of fluorescent fusion proteins have been constructed for lower organisms,[5,6] the generation of tools for all tissues of higher organisms presents a daunting challenge. This, however, has not deterred ambitious projects such as the Human Protein Atlas (HPA), an international initiative that aims to systematically characterise the human proteome by use of extensive antibody-based analyses, with over 17 000 antibodies already produced.[4,7,8] While the systematic generation and distribution of such antibody tools is a commendable endeavour, a truly comprehensive antibody atlas requires coverage of all splice variants and post-translational modification states for every protein. As such, the sizeable progress made by the HPA project to date barely scratches the surface of the functional proteome. Interestingly, in a large-scale comparison of immunocytochemistry using HPA antibodies and immunofluorescence microscopy of fusion proteins, only a minority of the assayed proteins were classified with identical localisation by both techniques.[4]

9.2.2 *In Silico* Methods and Analyses

In addition to empirical studies the determination of protein subcellular localisation *in silico* is an established bioinformatics challenge. To date, the majority of methods focus on the prediction of protein localisation from sequence or signalling peptide information. A recent comprehensive review outlines the extensive application of machine learning to sequence data to predict protein–organelle associations, including support vector machines (SVMs), artificial neural networks (ANN), k-nearest-neighbour (kNN) and variants, Markov models, probabilistic Bayesian approaches, and genetic algorithms.[9]

Familiar tools for predicting eukaryotic protein localisation are based on primary amino acid sequences and test for secondary structural motifs, *e.g.*, TargetP, for targeting to the secretory pathway, mitochondria, or plastids.[10]

NLS Mapper predicts nuclear localisation based on α-importin dependent nuclear localisation signals, while WoLF PSORT utilises a wide range of features to predict protein localisation to all organelles.[11,12] The features of proteins, such as the length and amino acid composition of transmembrane domains, are also predictive of probable protein localisation.[13] However, prediction based solely on protein sequence often lack the accuracy and resolution necessary for definitive determination of localisation, especially within the secretory pathway. Such methods also provide no insight into protein translocation events.

9.2.3 Quantitative Mass Spectrometry: A Complementary Technology

MS-based analysis is high throughput, making it an attractive alternative to the aforementioned technologies. Where microscopy-based approaches tend to be targeted on a small number of proteins of interest, with MS it is now quite feasible to routinely characterise thousands of proteins from a single sample within a matter of hours. This throughput also enables protein localisation to be assessed in various cell or treatment states. MS-based workflows are typically not reliant on antibodies or the generation of fluorescent fusion proteins, enabling the analysis of proteins in their native state. MS also has the capability to detect protein isoforms, splice variants, and post-translational modifications. MS is, therefore, both a powerful analytical tool for characterising protein localisation, and one that is highly complementary to the more traditional methods used in such studies. Examples of the application of MS to protein localisation studies are described in detail in section 9.4.

9.3 Subcellular Fractionation

Many proteomics workflows commence with the lysis of cells in chaotropic, detergent-rich buffers in an attempt to capture the abundance and modification state of proteins in their native state. However, such treatments immediately disrupt membranes and protein complexes and, therefore, destroy the spatial organisation within cells. In order to study proteins in the context of the native localisation it is necessary to prepare subcellular fractions in a manner that causes minimal disruption to the integrity of subcellular compartmentalisation within the cell. Typically, such methods involve mechanical or buffer-induced lysis of the plasma membrane with minimal disruption to intracellular organelles, followed by subcellular fractionation.

9.3.1 Separative Centrifugation

Biochemical methods for subcellular fractionation are commonly based upon centrifugation. The simplest such approach is differential centrifugation, in which a cell lysate is subjected to a series of centrifugation steps with incremental increases in centrifugal force and/or time. Dense organelles, such as the nucleus and mitochondrion, will sediment at relatively low centrifugal force, while other organelles and large protein complexes sediment more efficiently at higher forces, thus allowing crude, stepped separation of subcellular compartments.

An alternative and higher resolution approach is density gradient ultracentrifugation, in which a cell lysate is placed into a medium, such as sucrose or iodixanol. When subjected to ultracentrifugal force, organelles and large protein complexes float or sediment to a point of equilibrium depending on their respective buoyant densities. Density gradients can be established to be discontinuous in order to concentrate organelles at specific interfaces, or continuous in order to provide maximum resolution of organelles with similar physical properties, such as components of the secretory pathway. While these techniques are commonly utilised for organelle enrichment, achieving total purification of an organelle by separative centrifugation is challenging, as discussed in section 9.4.

9.3.2 Differential Permeabilisation

An alternative approach that is particularly useful for characterising soluble proteins is to treat cells with a series of buffers that are designed to differentially permeabilise specific subcellular membranes in order to sequentially extract proteins from subcellular niches. For example, the plasma membrane may be permeabilised in order to extract soluble cytosolic proteins, followed by permeabilisation of the nuclear membrane to extract proteins from the nucleoplasm. Many commercial products designed for subcellular fractionation utilise one or both of these two strategies, although the efficiency of such kits is largely unassessed, with commercial materials typically demonstrating enrichment of specific individual protein markers by Western blotting, which, as discussed in section 9.4.1, may give an unrealistic estimation of organelle purity.

9.3.3 Immunocapture and Affinity Purification

Another method for enriching membranes of specific interest is immunocapture, which is a process of precipitating membrane(s) of interest using high specificity antibodies that are cross-linked to agarose or magnetic beads. Such techniques are typically performed as an additional enrichment step after crude separation of membranes using centrifugal methods, and can be used to reduce the level of contamination by other organelles. For example, in a recent paper, immunoprecipitation was used to isolate a SYP61 compartment from *Arabidopsis thaliana* and then MS used to catalogue co-isolated proteins in an attempt to reveal the role of this trans-Golgi network-like compartment in exocytosis.[14] However, such approaches are imperfect and contaminant proteins from other subcellular compartments are often found to persist in the immunoprecipitated fraction.

Biotinylation of cell surface proteins, followed by streptavidin affinity purification from a total cell lysate or crude subcellular fraction, is a common approach for the enrichment of plasma membrane and extracellular matrix proteins.[15] A more sophisticated version of this technique makes use of a recombinant peroxidase that can be targeted to other subcellular niches where it biotinylates proteins in close proximity. The combination of targeted localisation of the enzyme with enrichment of labelled proteins is powerful as it potentially enables sub-organellar levels of resolution that are otherwise difficult to achieve. This approach has been used to study the localisation of proteins within mitochondria.[16]

9.3.4 Zone Electrophoresis

An alternative approach to enrich organelles is the application of free flow electrophoresis (FFE). FFE is a liquid-based separation method that can be used in a variety of operation modes and allows separation of species based on their electrophoretic mobility of the isoelectric point and can used analytically and preparatively.[17] An FFE device typically consists of two narrowly spaced vertical plates, which contain a carrier buffer. A sample is continuously injected in the carrier buffer and components are separated based on the differential effect the perpendicularly applied electric field has on their migration through the carrier buffer. The laterally separated components are then collected into a number of fractions. To date, the most common mode of FFE for separating organelles has been zone electrophoresis (ZE), where separation is achieved based on the charge-to-size ratios of component organelles. The limited success of this technique to 'purify' many subcellular components can be attributed to the very small differences in the electrokinetic properties of the majority of organelles. Recent successes, however, have involved the coupling of the method with pre-enrichment using density centrifugation. Heazlewood and colleagues used this approach to good effect to produce a highly enriched *Arabidopsis* Golgi fraction, which was subsequently been catalogued using LC-MS/MS.[18] A weakness of this method is that contaminant proteins, which remain associated with the organelle during the FFE process, will persist throughout the subsequent analysis, and the method does not allow the discrimination of cargo proteins from the rest of the organellar proteome.

9.3.5 Selecting a Suitable Fractionation Workflow

Regardless of the specific method(s) utilised for fractionation, the analyst must decide upon an appropriate compromise between three competing demands: (i) the degree of organelle enrichment in the collected fractions; (ii) the quantities of protein that are recovered; and (iii) the integrity of membranes and proteins after fractionation.

The appropriate balance of these factors is highly dependent on the particular biological question, and there is, therefore, not a universal procedure for sample preparation. For example, if a study aims to characterise protein localisation within the secretory pathway then a very crude fractionation scheme may not provide sufficient resolution for distinguishing between organelles, while an elaborate and protracted preparation may result in protein yields that are insufficient for analysis. By contrast, if nuclear import and export were of interest, then a simpler fractionation scheme may be sufficient.

9.4 Mass Spectrometric Analysis

In addition to the multiple methods for the preparation of subcellular fractions there are also many different MS data acquisition strategies that can be used to determine protein localisation (Figure 9.1). Some of these approaches are dependent upon organelle purification, while others will tolerate incomplete organelle enrichment. Some determine protein localisation based on their presence or absence in various subcellular fractions, whereas others make use of relative peptide quantification with varying degrees of sophistication in order to determine subcellular distributions of analysed proteins.

9.4.1 Organelle Purification and Cataloguing

The most intuitive approach for identifying organellar proteins is to purify the subcellular compartment of interest, extract and digest the proteins, and perform an extensive LC-MS/MS analysis in order to catalogue as many proteins as possible. All proteins identified within the fraction are then considered to be resident proteins.

The biochemical methods used for organelle enrichment, as described in section 9.3, are, for the most part, crude. The generation of subcellular fractions approaching anything close to 100% purity, and in sufficient quantities for MS/MS analysis, is in practice extremely difficult to achieve. This is especially true for organelles of the secretory pathway due to their similar physicochemical properties.

Since organelle cataloguing is dependent upon this unrealistic total purification, it is therefore highly vulnerable to misclassification of contaminant proteins and proteins in transit as organelle residents. As the sensitivity of MS continues to rapidly grow, and such catalogues become more expansive, the presence of contaminant proteins in such preparations becomes increasingly apparent.

Even organelles that can be readily and efficiently enriched, such as the nucleus, mitochondrion, and chloroplast, are also vulnerable to significant contamination by other organelles, even where considerable effort has been made to optimise and verify the fractionation scheme, and verification techniques, such as Western blotting and electron microscopy, suggest that

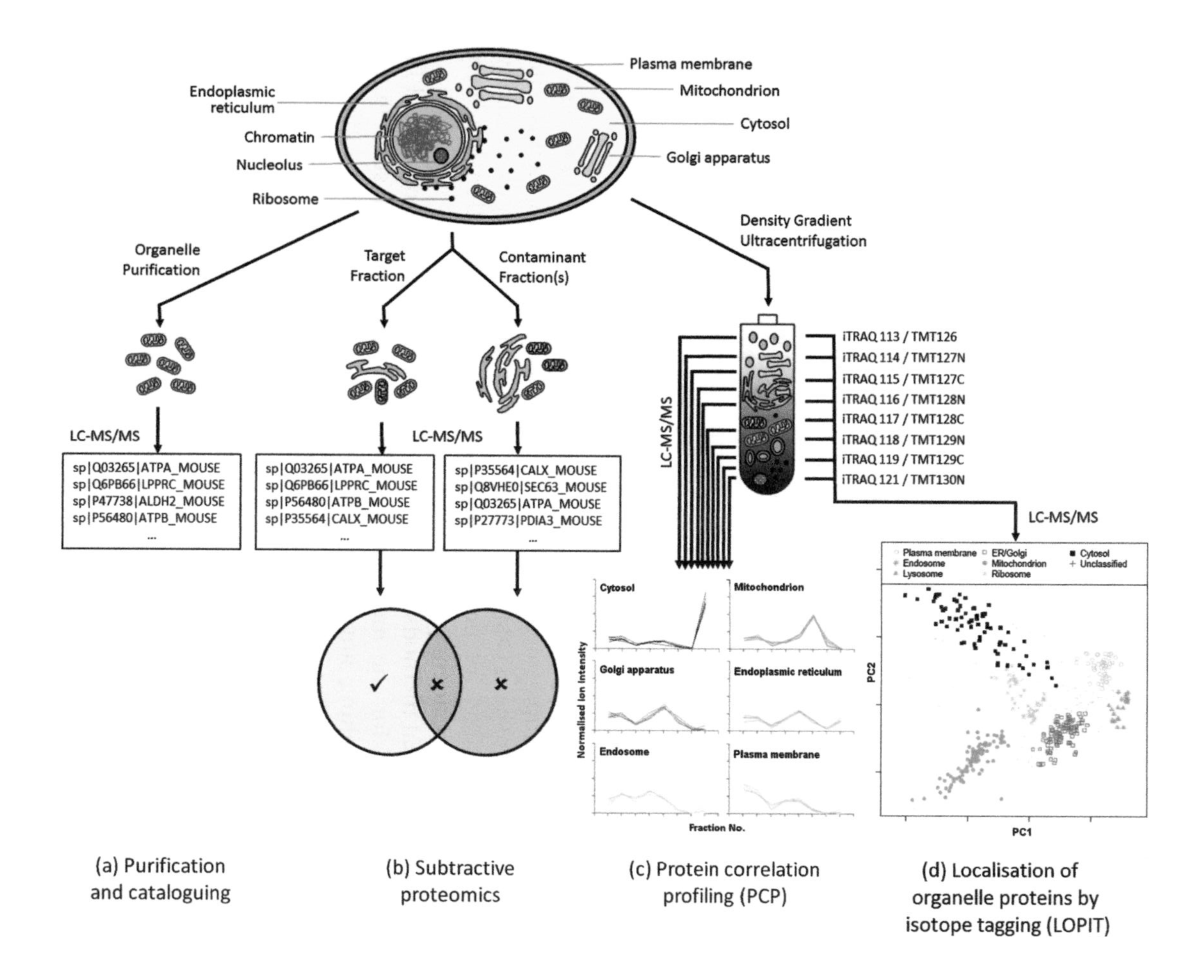

Plasma membrane
Mitochondrion
Cytosol
Golgi apparatus
Endoplasmic reticulum
Chromatin
Nucleolus
Ribosome
Organelle Purification
Target Fraction
Contaminant Fraction(s)
Density Gradient Ultracentrifugation
LC-MS/MS
sp|Q03265|ATPA_MOUSE
sp|Q6PB66|LPPRC_MOUSE
sp|P47738|ALDH2_MOUSE
sp|P56480|ATPB_MOUSE
...
LC-MS/MS
sp|Q03265|ATPA_MOUSE
sp|Q6PB66|LPPRC_MOUSE
sp|P56480|ATPB_MOUSE
sp|P35564|CALX_MOUSE
...
sp|P35564|CALX_MOUSE
sp|Q8VHE0|SEC63_MOUSE
sp|Q03265|ATPA_MOUSE
sp|P27773|PDIA3_MOUSE
...
LC-MS/MS
iTRAQ 113 / TMT126
iTRAQ 114 / TMT127N
iTRAQ 115 / TMT127C
iTRAQ 116 / TMT128N
iTRAQ 117 / TMT128C
iTRAQ 118 / TMT129N
iTRAQ 119 / TMT129C
iTRAQ 121 / TMT130N
LC-MS/MS
Cytosol
Mitochondrion
Golgi apparatus
Endoplasmic reticulum
Endosome
Plasma membrane
Normalised Ion Intensity
Fraction No.
Plasma membrane
Endosome
Lysosome
ER/Golgi
Mitochondrion
Ribosome
Cytosol
Unclassified
PC2
PC1
(a) Purification and cataloguing
(b) Subtractive proteomics
(c) Protein correlation profiling (PCP)
(d) Localisation of organelle proteins by isotope tagging (LOPIT)

the target organelle is pure (Figure 9.2). As a result, high-throughput studies of the nuclear and mitochondrial proteomes are often error-prone. For some experimental designs, this high false classification rate may be tolerable. For example, if high-throughput proteomics data are used for discovery analysis in conjunction with orthogonal techniques, such as large-scale confocal microscopy analysis and electronic inference techniques, then it is possible to filter many of these false discoveries and achieve a more robust false classification rate.[19] However, if high-throughput cataloguing is used in isolation then the reliability of localised identifications will be significantly impaired.

9.4.2 Subtractive Proteomics

One of the more straightforward methodologies that attempts to address the weaknesses of the previously described purification-dependent technique is often referred to as 'subtractive proteomics'.[20] In this approach the subcellular niche of interest is 'purified' in parallel to other organelles that are likely to contaminate this fraction. Proteins in each 'purified' niche are then catalogued by MS. Only proteins that are identified exclusively in the target niche fraction (*i.e.*, that are not detected in any parallel 'contaminant' fractions) are then classified as true organelle residents. For example, Wiederhold and co-workers utilised this approach to distinguish true residents of the yeast vacuole from contaminants, and identified a number of candidate proteins that may demonstrate novel vacuolar functions.[21]

While straightforward, this strategy is sub-optimal in three respects. Firstly, either the niche of interest or the contaminant niche must still be enriched to near 100% purity in order for subtractive proteomics to be effective[22] and, as previously discussed, such high degrees of enrichment are often unfeasible. Many true organelle residents may be excluded as a result of an inability to achieve sufficient levels of organelle enrichment in the target fraction and depletion from contaminant fractions. Secondly, this

Figure 9.1 Schematic of several spatial proteomics workflows. (a) Organelles of interest (in this example, the mitochondrion) may be 'purified' and all proteins identified in this fraction by MS/MS will be classified as organelle residents. (b) In subtractive proteomics workflows subcellular fractions are prepared to enrich both the target organelle and probable contaminants (in this case, the target being mitochondrion and the probable contaminant endoplasmic reticulum). Each fraction is catalogued by MS/MS, and only proteins identified exclusively in the target fraction are considered to be organelle residents. In (c) protein correlation profiling and (d) localisation of organelle proteins by isotope tagging workflows, protein subcellular localisation is determined by the analytical distribution of protein(s) across a fractionation scheme, such as density gradient ultracentrifugation. In the former approach subcellular fractions are analysed by MS/MS in parallel, whereas the latter approach utilises isobaric tagging to enable simultaneous analysis of fractions.

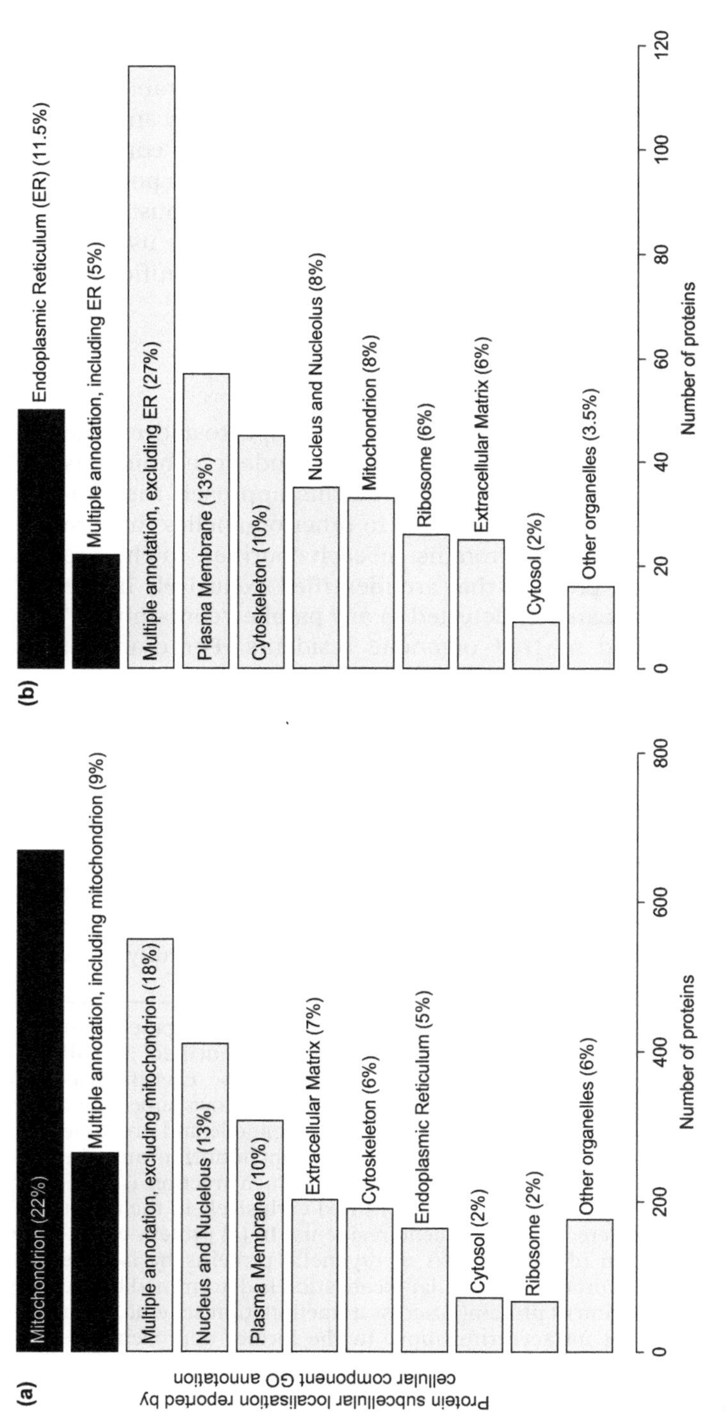

Figure 9.2 Organelle-specific gene ontology (GO) annotation for proteins identified in (a) 'purified' mouse mitochondria[19] (3069 proteins) and (b) enriched *Drosophila melanogaster* endoplasmic reticulum (ER) (759 proteins). Solid black bars represent proteins for which the GO annotation corroborated with the expected mitochondrial or ER localisation, whereas grey bars represent proteins for which the GO annotation is contradictory. For the mouse study, mitochondrial purification was determined by Western blotting with marker protein antibodies and by electron microscopy. For the *Drosophila melanogaster* study, Western blotting analysis indicated that the subcellular fraction was not pure but represented maximal endoplasmic reticulum enrichment from an elaborate separative density gradient. In both cases a majority of proteins identified carry GO annotation that contradicts the expected respective mitochondrial and endoplasmic reticulum localisations. This demonstrates that organelle purification and cataloguing strategies are susceptible to considerable contamination, even where substantial effort is invested in developing and evaluating the purification protocol.

method is vulnerable to the semi-stochastic nature of data-dependent MS/MS acquisition; proteins may be identified in the target fraction but not in the contaminant fraction(s) as a result of run-to-run variability in precursor ion sampling or MS/MS spectral quality rather than true cases of respective presence and absence. This, therefore, increases the false classification rate as contaminants may be deemed to be true organelle residents. Thirdly, this subtractive approach also precludes the possibility of proteins being present in multiple subcellular niches simultaneously, which results in an inability to reliably classify multiply localised proteins, and precludes the use of this strategy for the evaluation of protein trafficking dynamics.

Several of these limitations can be overcome if the subcellular localisation is treated as a quantitative rather than a qualitative problem. By using an MS/MS data acquisition strategy that allows for relative quantification of peptides across the sampled subcellular fractions, it is possible to establish protein localisation even with an imperfect purification scheme. Rather than expecting total organelle purification in any fraction, protein localisation is determined by relative quantification of proteins across all sampled fractions.

Boisvert and colleagues, for example, used triplex SILAC in conjunction with subcellular fractionation in order to measure relative distribution between the cytoplasm, nucleus, and nucleolus of over 2000 proteins in a human colon carcinoma cell line.[23] By enriching each of these compartments from differentially SILAC-labelled cells and carrying out co-analysis using quantitative LC-MS/MS, they were able to show subsets of proteins that co-localised and also dynamically changed in location upon treatment with a topoisomerase inhibitor. Since this quantification method incorporates differential isotopic labels, the three fractions can be pooled and co-analysed by MS/MS, ensuring that the rate and timing of MS sampling and quantification events are equivalent for all subcellular fractions, thus eliminating much of the technical variability of data-dependent MS/MS.

There are also several caveats to this approach. Firstly the SILAC technology requires proteins to be isotopically labelled *in vivo*; therefore, each of the three subcellular fractions for analysis must be derived from a separate cell culture in order to achieve differential labelling. This introduces additional challenges in ensuring that biological and technical variance is controlled during parallel culture and sample preparation. The use of this quantification method also limits such approaches to cell lines and animal models that can be efficiently SILAC labelled. The use of *in vitro* labelling reagents, such as mTRAQ or isotopically enriched formaldehyde (dimethyl labelling), is an alternative option that would allow subcellular fractions to be derived from the same culture as the differential isotopic labelling with such reagents is performed after fractionation.

SILAC labelling and other multiplexed MS^1-based quantification methods also result in increased MS spectral complexity as peptides labelled with each amino acid isotopologue variant will have a distinct mass. In the case of triplex SILAC labelling the resultant spectra require deconvolution prior to quantification, which may introduce quantitative errors. As a result, the multiplexing capacity of SILAC-based analyses is typically limited to three, thus limiting the number of subcellular niches that can be resolved from one another within a single experiment. This approach is, therefore, well suited to the focussed analysis of specific localisation processes, such as nuclear import and export, but is less amenable to simultaneous characterisation of many subcellular niches.

9.4.3 Analytical Fractionation

While total organelle purification is generally unviable, in some cases even achieving a high level of enrichment is challenging; it is nonetheless possible to determine protein subcellular localisation. If a fractionation scheme is devised that produces distinctive and characteristic quantitative distributions for each organelle across the analysed fractions, then it is possible to determine protein localisation based on the overall distribution of that protein despite no single fraction necessarily representing anything close to total organelle purification. By utilising this quantitative approach, dubbed 'analytical fractionation', Christian De Duve and Albert Claude were able to deduce the existence of the previously undocumented lysosome and peroxisome as the quantitative distribution of enzyme activities emanating from these two organelles did not correlate with those of any organelles that were characterised at that time.[24] While De Duve and Claude pioneered such experiments with low-throughput enzymatic assays for quantitative analysis, the principles are transferrable to quantitative MS, enabling such studies to be performed in high throughput.

9.4.4 Protein Correlation Profiling (PCP)

'*Protein correlation profiling*' (PCP) is one technique that applies the principles of analytical fractionation to the MS platform.[25] In this workflow subcellular fractions are individually analysed by LC-MS/MS to determine the relative abundance of peptides across the fractionation scheme, and this information is compiled into protein-level quantitative distributions. A series of χ^2 tests are then performed to determine if the distribution of a protein displays significant positive correlation with those of known organelle markers, and can therefore be classified as an organelle resident.

PCP workflows initially utilised label-free quantification based on peptide ion signal intensities, with peptides from each subcellular fraction analysed in parallel. This approach enables an unlimited number of fractions to be analysed, and therefore is compatible with elaborate subcellular fractionation schema, allowing for high levels of subcellular resolution to be obtained. For example, in a PCP analysis of mouse liver, protein distribution profiles were compiled from 32 subcellular fractions, providing sufficient information to resolve proteins residing in the endoplasmic reticulum, Golgi apparatus, ER/Golgi derived vesicles, early and recycling endosomes, and plasma membrane.[26]

However, since fractions are analysed in parallel, the LC-MS/MS analysis time increases linearly with the number of subcellular fractions generated. Label-free quantitative approaches also introduce additional challenges for post-experimental analysis. Inconsistencies in chromatographic performance, ionisation efficiency, ion transmission, and the irreproducibility of data-dependent MS/MS sampling may all result in missing values and quantitative inaccuracy when attempting to compile distribution profiles.

These complications with PCP data analysis can be overcome by co-analysis of each subcellular fraction with a common, isotopically labelled internal standard. While this approach does nothing to reduce the required LC-MS/MS analysis time, peptide abundance measurements for all subcellular fractions can be normalised against those of the internal standard, improving the accuracy and precision of relative quantification across the subcellular fractions. For example, duplex SILAC quantification has been used to increase the accuracy of PCP studies on protein localisation to the centrosome.[27]

The creation of a suitable internal standard is not necessarily straightforward. 'Super-SILAC mixes' have been developed with the express purpose of improving quantitative performance when comparing multiple cell or tissue types,[28] but a similar mix for the subcellular proteome is not easily constructed. Since protein abundances would be expected to vary substantially between subcellular fractions as a result of organelle separation, developing an internal standard that is within the linear dynamic range for all fractions is challenging. As a result, peptides derived from low abundance proteins may not be detectable in the standard sample and, therefore, cannot be used for normalisation; in cases where the ratio of peptide concentrations between the test fraction and internal standard are large (*i.e.*, efficient enrichment or depletion in the test fraction) the relationship between signal intensity and abundance will be non-linear, and any normalisation would, therefore, introduce inaccuracy to the measurement. In the aforementioned example of PCP-SILAC the internal standard was specifically tailored to ensure maximal sensitivity for centrosomal proteins as this was the focus of this particular study. The PCP-SILAC approach is, therefore, perhaps most effective for experiments in which a small number of subcellular niches are of particular interest rather than for general classification experiments.

9.4.5 Localisation of Organelle Proteins by Isotope Tagging (LOPIT)

'*Localisation of organelle proteins by isotope tagging*' (LOPIT) is another workflow based on analytical fractionation that is, in some respects, similar to PCP. Whereas PCP is a label-free quantitative approach and PCP-SILAC utilises quantification of subcellular fractions relative to a common internal standard, LOPIT utilises higher-dimensional multiplexed quantification *via in vitro* chemical labelling methods. While this approach was initially described utilising ICAT reagents,[29] the LOPIT methodology was soon adapted to take advantage of the superior multiplexing capacity of isobaric tagging reagents, such as iTRAQ 4-plex, TMT 6-plex, and iTRAQ 8-plex.[30]

A typical LOPIT workflow is shown in Figure 9.3. Briefly, a cell or tissue sample is subjected to subcellular fractionation in order to partially separate subcellular niches. Up to eight fractions representing differential enrichment for subcellular niches of interest are then selected, and proteins derived from these fractions are digested to peptides. Peptides from each selected fraction are then chemically labelled with a different isotopologue variant of isobaric tag reagents, such as iTRAQ or TMT. Derivatised peptides from labelled samples are then pooled and analysed by LC-MS/MS. The relative abundance of reporter ions in the MS/MS spectrum of each peptide acts as a surrogate measurement of the relative enrichment of the parent

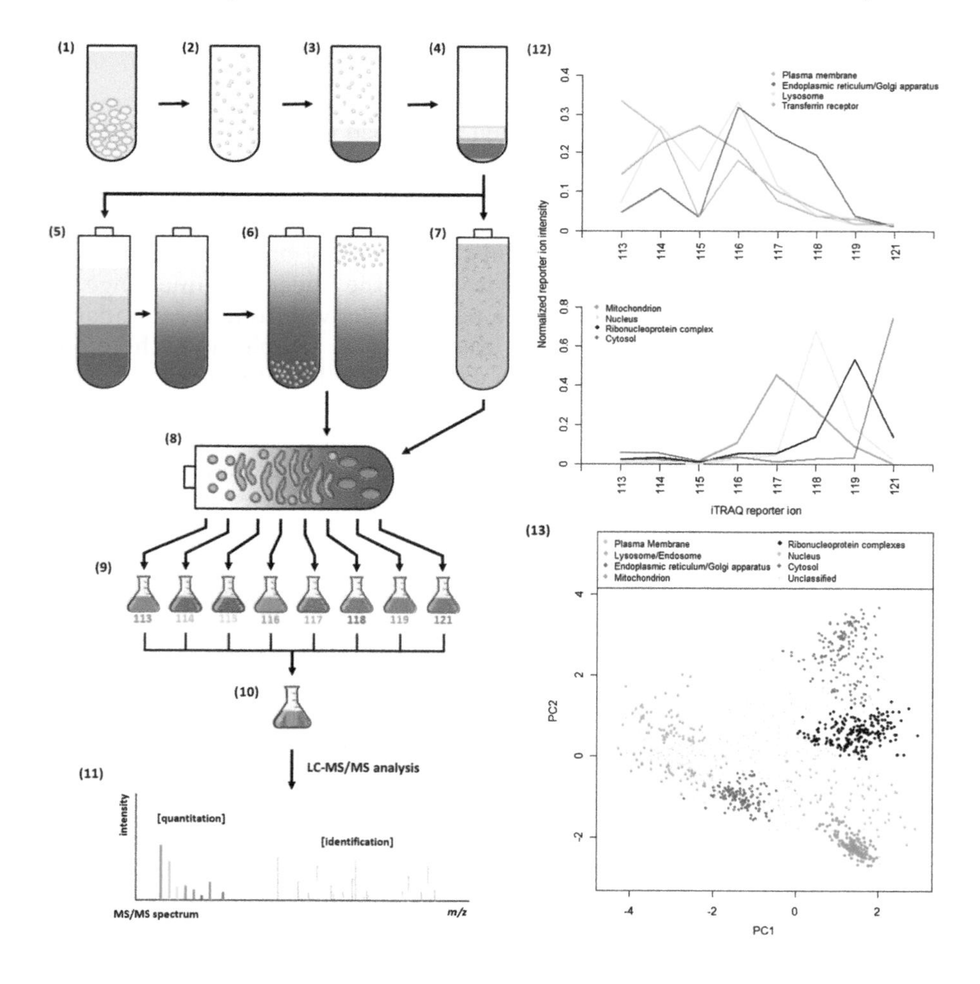

protein in each of the selected subcellular fractions. The reporter ion profile, therefore, recapitulates the quantitative distribution of proteins throughout a fractionation scheme.

This distribution pattern can be used to determine the steady-state subcellular localisation of proteins without the need for total purification of organelles of interest, providing that the fractionation scheme is tailored so that each subcellular niche forms a distinct and characteristic distribution profile. Proteins can then be classified to a subcellular localisation using straightforward multivariate statistical approaches, such as partial least squares-discriminant analysis,[30] or by employing supervised machine learning methods, such as SVMs.[31]

The LOPIT methodology has several significant advantages over the label-free and SILAC-based quantitative strategies previously described. The most attractive feature of this approach is that all subcellular fractions of interest are co-analysed, and the reporter ions in each MS/MS spectrum provide all the quantitative information required to compile a distribution profile for the peptide. Since quantitative information is collected from product ion spectra, this strategy also avoids the weaknesses of multiplexed MS^1-based quantification, such as MS/MS undersampling due to increased spectral complexity, and the division of ion signals.

In vitro labelling approaches, such as isobaric tagging, are also not limited to biological materials that are compatible with SILAC, which has allowed the method to be applied to a wide variety of biological model systems, including

---

Figure 9.3 Schematic overview of the localisation of organelle proteins by isotope tagging (LOPIT) workflow. (1) Cells are harvested and suspended in isosmotic buffer. (2) Cells are lysed with minimal disruption to organelles by passage through a ball bearing homogeniser. (3) The lysate is underlaid with two solutions of iodixanol. (4) Ultracentrifugation concentrates membranes at the interface of the two iodixanol layers, while cytosolic proteins remain in the supernatant. (5) A linear separative density gradient can be pre-formed by layering solutions of increasing iodixanol concentration and allowing them to diffuse for several hours. (6) The crude membrane preparation can then be underlaid or overlaid onto the pre-formed gradient. (7) Alternatively, for self-generating gradients, membranes can be resuspended in a single iodixanol solution. (8) After ultracentrifugation, membranes and large protein complexes are partially separated based on their respective buoyant densities. (9) Fractions representing peak organelle enrichment are selected. Proteins are extracted, digested, and derivatised with isobaric tag reagents. (10) Samples are then pooled for simultaneous LC-MS/MS analysis. (11) Each MS^2 spectrum contains peptide sequence fragment ions for identification and low m/z reporter ions that recapitulate the distribution of the peptide across the fractionation scheme. (12) Despite no single fraction being pure, different subcellular niches will each have a characteristic analytical profile, which can be used for protein classification. (13) The multivariate data can be visualised in two dimensions (2D) by principal component analysis as an intuitive representation of organelle separation.

Arabidopsis callus,[30] whole *Drosophila* embryos,[32] chicken DT40 lymphocytes,[33] and mouse pluripotent embryonic stem cells.[55] Additionally, subcellular fractions for isobaric tag labelling can be derived from the same cell or tissue sample, thus eliminating much of the biological variation that may arise in experimental designs, where each subcellular fraction must be obtained from samples in parallel. The high multiplexing capacity of isobaric tags also allows for many subcellular fractions to be simultaneously analysed.

As with PCP, the ability to construct these detailed analytical profiles enables high levels of subcellular resolution to be obtained. However, whereas PCP allows for an unlimited number of subcellular fractions to be analysed, the number of fractions that can be analysed in a single LOPIT experiment is restricted by the multiplexing capacity of the labelling reagents. The multiplexing capacity of commercially available isobaric tagging reagents currently enables co-analysis of up to 10 samples (with TMT 10-plex), although the development of an expanded range of isotopologue variants coupled with high-resolution MS would enable the capacity of isobaric tag reagents to be expanded to 19-plex or higher.[34] While the distribution profiles are, therefore, compiled from analysis of fewer fractions than is typical for PCP experiments, we find that in practice even 8-plex analysis is sufficient to provide comparable levels of organellar resolution to those observed for 32-plex PCP experiments if the subcellular fractionation scheme is carefully considered.

The most significant weakness of the LOPIT methodology is that the accuracy and precision of isobaric tag quantification for complex mixtures of peptides is often poor. This is due to the contamination of precursor ions by a population of low abundance species that collectively result in a significant but unpredictable distortion of the reporter ions used for quantification in the MS^2 spectrum. For LOPIT experiments, this results in poorer resolution of organelles in the measured distribution profiles than were actually attained by the subcellular fractionation, and therefore reduces the robustness of organelle classification.[35] Several methods have been devised to improve the specificity of precursor ion selection and, therefore, quantitative performance. These approaches include utilising narrowed precursor ion selection windows,[36] acquiring MS^2 spectra only at chromatographic peak apices,[37] employing an additional round of ion selection and fragmentation (*i.e.*, MS^3),[36,38] and quantification *via* peptidotypic 'complement' product ions rather than the low *m*/*z* reporter ions.[39] While none of these approaches currently offer an optimal balance of sensitivity and quantitative accuracy, they represent the best options to date for improving the performance of isobaric tagging studies. Nonetheless, in combination with a suitable fractionation scheme, LOPIT enables robust analysis of subcellular localisation.

9.5 Data Analysis

Given the diversity of data acquisition strategies and MS platforms used for spatial proteomics experiments, there is little consistency in the procedures for performing protein identification, protein quantification, and classification of subcellular localisation from raw spectral data. Each research group will typically have its own data processing pipeline, utilising a range of commercial, freeware, and in-house software packages. This lack of consistency creates a challenge for the comparison of localisation experiments and development of new statistical and bioinformatics tools for data analysis.

Analytical fractionation methods output a vector of relative abundance measurements that approximates protein distribution across the fractionation scheme. A single experiment may yield thousands of peptide and protein gradient profiles. Proteins that belong to the same organelle will co-fractionate and, given sufficient resolution, the distribution pattern of proteins with unknown localisations can be mapped to those of known organelle marker residents using multivariate statistical and machine learning methods.

Traditionally, supervised multivariate statistical and machine learning methods have been employed where a subset of training data, annotated according to known organelle association from public databases, is used to map gradient profiles to subcellular locations. Although supervised methods, such as SVMs, have been shown to map proteins to organelles in experimental proteomics data with high estimated generalisation accuracy,[31] the data often suffers from inherent drawbacks that limit the application of such methods, and the accuracy and relevance of new subcellular associations obtained from it. For example, issues include: (i) incomplete annotation, and therefore lack of training data, for all organelles; (ii) the limited availability of proteins with known organelle membership to use as training data; and (iii) proteins may be present simultaneously in several organelles.

A lack of training data can be particularly limiting when applying a supervised schema if subcellular compartments known to be present in the experiment are not included in data used to create the protein–organelle classifier. Such situations are highly likely to make association errors. The extraction of all organelle-related groupings is a difficult task owing to the time-consuming nature of obtaining reliable protein annotations from databases and, of course, the limited number of proteins with known organelle membership. To address issues (i) and (ii) above, Breckels and coworkers developed a phenotype discovery algorithm, which is used to identify putative organelle clusters in gradient-based MS datasets using a semi-supervised machine learning schema (Figure 9.4).[40] A new pipeline for organelle proteomics data analysis is proposed in which one first uses a novelty detection approach to extract all relevant subcellular compartments, then secondly examines and extracts these profiles to use as labelled training examples for a subsequent round of traditional supervised machine learning to identify protein–organelle localisations.

Robust data analysis using state-of-the-art computational methodologies remains challenging, despite the numerous advances in organelle separation and protein quantification. The pRoloc (http://www.bioconductor.org/packages/devel/bioc/html/pRoloc.html)[41] and MSnbase (http://www.

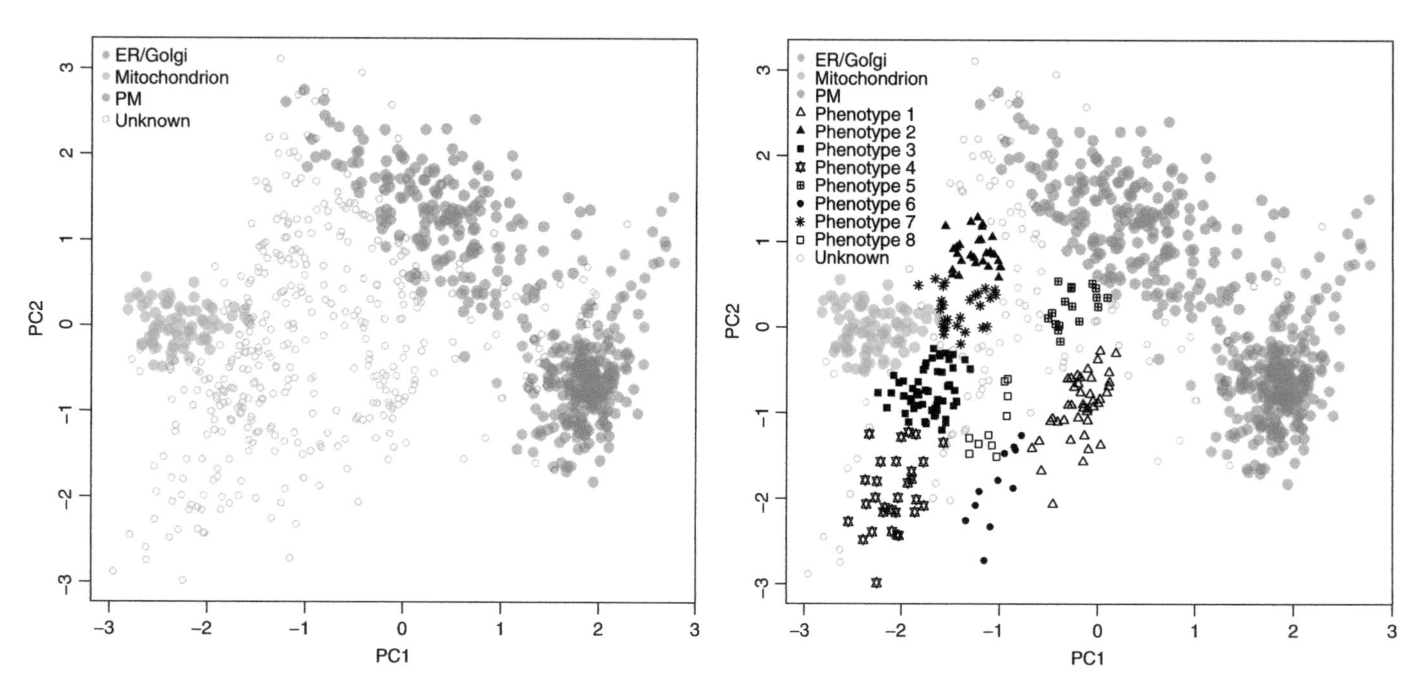

Figure 9.4 Principal components analysis (PCA) plots of LOPIT profiles showing the clustering of proteins according to their density gradient distributions in *Drosophila melanogaster*[32] (left) before and (right) after the application of a phenotype discovery algorithm.[40] Organelle markers used as input for the discovery experiment are shown in PCA plot (left), which include markers from the endoplasmic reticulum (ER), Golgi, mitochondrion, and plasma membrane (PM). New phenotypes identified from the experiment are labelled in PCA plot (right), which include clusters representative of ribosomal subunits (phenotype 1 and 3), proteasome (phenotype 2), nuclear contaminants (phenotype 4), peroxisome (phenotype 6), and proteins with ambiguous or unknown localisation (phenotypes 5, 7, and 8).

bioconductor.org/packages/release/bioc/html/MSnbase.html)[42] R packages provide data manipulation and processing functionalities that can be readily applied to any type of quantitative spatial proteomics data. A range of contemporary supervised approaches and the novelty detection algorithms described above have been implemented in pRoloc as a flexible and robust analysis and visualisation framework.

To date, the existence of robust methods that map proteins to more than one organelle is limited. Hindering the development of such classifiers is the lack of multi-organelle resident proteins, both in terms of size and diversity, which can be used as training data in classifier creation. Approaches that tackle multiple-organelle classification have been published in the literature based on input from amino acid sequence or protein sorting signals,[43–46] some of which have applied probabilistic approaches and ensemble methods.[47–49] However, such approaches are balanced between reflecting a potential biological truth, *i.e.*, distribution of a protein across multiple organelles, and simply reflecting classifier uncertainty when trained with insufficient information. Such approaches have not yet been applied to organelle proteomics experimental output.

Finally, the suitability of specific data analysis approaches used for protein–organelle matching can be influenced by the experimental design employed. For example, multivariate statistical and machine learning methods are intolerant of missing values, which are particularly common for MS workflows utilising label-free quantification, although several multivariate statistical methods have been developed to impute missing values in such experiments.[41,50–52]

9.6 Validation of Localisation Studies

In addition to optimisation and evaluation of the fractionation method, the data acquisition strategy, and data processing workflow, it is also essential that the final outcome of the experiment is also validated so that the classification of protein localisation is suitably robust.

One of the most straightforward methods for assessing the overall classification performance is to compare the experimental data with cellular component gene ontology (GO) annotation and other protein database curation. One way in which such a comparison could be made would be to perform a GO term enrichment analysis. There are multiple tools to facilitate such analyses, for example, BiNGO, a Cytoscape plug-in to assess GO enrichment in biological networks.[53] If the experimental workflow was suitable then one might expect each classification group (*i.e.*, proteins assigned to a particular subcellular niche) to only show significant enrichment of related GO terms.

Alternatively the rate of corroboration between the experimental data and GO annotation (*e.g.*, the number of proteins with supportive GO annotation *vs.* the number of proteins with contradictory GO annotation) could be determined. While GO annotation is imperfect and may not be representative of protein localisation in a particular tissue or treatment state, if the accuracy of GO annotation is assumed to be better than random, then a higher rate of corroboration is generally indicative of higher classification performance and, therefore, a more robust experimental design.

For example, of 100 mouse liver proteins with organelle-specific GO annotation that were classified as plasma membrane residents by PCP,[26] the GO annotation is only supportive of plasma membrane localisation in 25% of cases. Manual examination of these contradictory classifications revealed that there was strong experimental evidence that many of the proteins classified as plasma membrane residents would be expected to reside in a range of other subcellular compartments, including the mitochondrion and nucleus. This low rate of corroboration does not necessarily invalidate the experimental data, but is nonetheless suggestive of poor classification accuracy for this particular subcellular niche. Closer investigation is, therefore, warranted to determine whether the experimental design was sub-optimal. In this case it is possible that the inaccuracy of the label-free quantification method used was resulting in a high rate of false classification, which was the rationale for the development of PCP-SILAC. By contrast, when similar validation of plasma membrane classifications from LOPIT data is performed, the rate of corroboration with GO annotation is typically 80% or higher,[32] and for organelles that are simpler to efficiently enrich, such as the mitochondrion, the rate of corroboration may be above 95%.[56] Thus, the rate of corroboration with GO annotation can be used as a general indicator of classification performance.

Experimental data can also be compared against other high-throughput repositories of data. For example, when we compared protein localisation in HEK293T fibroblasts as determined by LOPIT with the localisation reported by confocal microscopy as part of the HPA project, we observed a high rate of corroboration in localisation where the antibodies used for immunofluorescence were assessed by the HPA project as 'high reliability' (74.3% agreement between microscopy and LOPIT, $n = 109$), but very poor corroboration for antibodies assessed as 'low reliability' (13.8%, $n = 160$). In almost all cases where LOPIT and confocal microscopy data were inconsistent, orthogonal sources of experimental evidence were supportive of the LOPIT classification, suggesting that the MS-based analysis outperformed antibody-based approaches for this particular subset of the proteome.

The above are straightforward approaches for assessing overall performance of a single high-throughput experiment. Another useful validation technique is to perform biological and technical replicate experiments in order to ensure that classification of protein subcellular localisation is consistent.

These methods can be used to 'tune' an experimental workflow to ensure that the high-throughput data is sufficiently reliable for discovery proteomics. For proteins of particular interest, such as proteins where the reported localisation is novel and functionally relevant, or proteins that are found to re-localise in response to a particular perturbation, targeted low-throughput validation by orthogonal approaches, such as microscopy, will further increase the confidence in the observed localisation.

9.7 Conclusions

Subcellular localisation plays a fundamental role in the regulation of protein function, which makes characterisation of protein localisation a valuable line of research. While the use of microscopy to address this biological question is likely to remain ubiquitous, such methods are neither flawless nor universally applicable. For a variety of technical reasons these approaches often yield results that are inconsistent. It is, therefore, important not to become overly reliant on a single technology for determining protein localisation. Spatial proteomics offers an untargeted, high-throughput alternative that is highly complementary to these more focussed techniques.

There are a multitude of biochemical methods for achieving subcellular fractionation, and many MS strategies for addressing the question of protein localisation. Methods based upon organelle purification are relatively simple to perform and the analysis of experimental data is straightforward, but such approaches are also highly error-prone. By contrast, quantitative analysis methods, such as LOPIT and PCP-SILAC, enable characterisation of multiple subcellular niches in a single experiment, enabling more robust classification of the subcellular distribution of proteins. However, these experimental workflows are more protracted and state-of-the-art multivariate analysis methods are necessary in order to attain maximal utility from acquired data. The workflow must, therefore, be tailored to the specific biological question under investigation.

Spatial proteomics will likely become an increasingly powerful tool for characterisation of protein localisation as the speed and sensitivity of MS continues to advance. The ability of mass spectrometers to distinguish between protein isoforms and covalent modification states also makes it a highly attractive approach for the characterisation of the post-translational proteome. In the future, spatial proteomics studies may indeed make use of a variety of strategies. One could envisage first carrying out global analyses using LOPIT and PCP-SILAC, followed by more targeted MS assays using selected reaction monitoring (SRM) or SWATH MS[54] on the same fractions. This would enable interrogation of isoform specific distributions or detection of less abundant proteins below the threshold of detection in the initial analysis. Development of more sophisticated informatics tools will enable improved characterisation of mixed locations and lead to more accurate subcellular annotations within databases and, ultimately, a better understanding of cellular processes.

References

1. C. Tristan, N. Shahani, T. W. Sedlak and A. Sawa, *Cell Signal.*, 2011, **23**(2), 317–323.

2. M. van Noort, F. Weerkamp, H. C. Clevers and F. J. Staal, *Blood*, 2007, **110**(7), 2778–2779.
3. M. T. Maher, A. S. Flozak, A. M. Hartsell, S. Russell, R. Beri, O. N. Peled and C. J. Gottardi, *Biol. Direct*, 2009, **4**, 5.
4. C. Stadler, E. Rexhepaj, V. R. Singan, R. F. Murphy, R. Pepperkok, M. Uhlén, J. C. Simpson and E. Lundberg, *Nat. Methods*, 2013, **10**(4), 315–323.
5. M. Kitagawa, T. Ara, M. Arifuzzaman, T. Ioka-Nakamichi, E. Inamoto, H. Toyonaga and H. Mori, *DNA Res.*, 2005, **12**(5), 291–299.
6. W. K. Huh, J. V. Falvo, L. C. Gerke, A. S. Carroll, R. W. Howson, J. S. Weissman and E. K. O'Shea, *Nature*, 2003, **425**(6959), 686–691.
7. M. Uhlén, E. Björling, C. Agaton, C. A. Szigyarto, B. Amini, E. Andersen, A. C. Andersson, P. Angelidou, A. Asplund, C. Asplund, L. Berglund, K. Bergström, H. Brumer, D. Cerjan, M. Ekström, A. Elobeid, C. Eriksson, L. Fagerberg, R. Falk, J. Fall, M. Forsberg, M. G. Björkland, K. Gumbel, A. Halimi, I. Hallin, C. Hamsten, M. Hansson, M. Hedhammar, G. Hercules, C. Kampf, K. Larsson, M. Lindskog, W. Lodewyckx, J. Lund, J. Lundeberg, K. Magnusson, E. Malm, P. Nilsson, J. Odling, P. Oksvold, I. Olsson, E. Oster, J. Ottosson, L. Paavilainen, A. Persson, R. Rimini, J. Rockberg, M. Runeson, A. Sivertsson, A. Sköllermo, J. Steen, M. Stenvall, F. Sterky, S. Strömberg, M. Sundberg, H. Tegel, S. Tourle, E. Wahlund, A. Waldén, J. Wan, H. Wernérus, J. Westberg, K. Wester, U. Wrethagen, L. L. Xu, S. Hober and F. Pontén, *Mol. Cell Proteomics*, 2005, **4**(12), 1920–1932.
8. M. Uhlén, P. Oksvold, L. Fagerberg, E. Lundberg, K. Jonasson, M. Forsberg, M. Zwahlen, C. Kampf, K. Wester, S. Hober, H. Wernérus, L. Björling and F. Ponten, *Nat. Biotechnol.*, 2010, **28**(12), 1248–1250.
9. P. Du, T. Li and X. Wang, *Expert Rev. Proteomics*, 2011, **8**, 391–404.
10. O. Emanuelsson, S. Brunak, G. von Heijne and H. Nielsen, *Nat. Protoc.*, 2007, **2**(4), 953–971.
11. S. Kosugi, M. Hasebe, M. Tomita and H. Yanagawa, *Proc. Natl Acad. Sci. USA*, 2009, **106**(25), 10171–10176.
12. P. Horton, K. J. Park, T. Obayashi, N. Fujita, H. Harada, C. J. Adams-Collier and K. Nakai, *Nucleic Acids Res.*, 2007, **35**, W585–W587.
13. H. J. Sharpe, T. J. Stevens and S. Munro, *Cell*, 2010, **142**(1), 158–169.
14. G. Drakakaki, W. van de Ven, S. Pan, Y. Miao, J. Wang, N. F. Keinath, B. Weatherly, L. Jiang, K. Schumacher, G. Hicks and N. Raikhel, *Cell Res.*, 2012, **22**(2), 413–424.
15. P. J. Rugg-Gunn, B. J. Cox, F. Lanner, P. Sharma, V. Ignatchenko, A. C. McDonald, J. Garner, A. O. Gramolini, J. Rossant and T. Kislinger, *Dev. Cell.*, 2012, **22**(4), 887–901.
16. H. W. Rhee, P. Zou, N. D. Udeshi, J. D. Martell, V. K. Mootha, S. A. Carr and A. Y. Ting, *Science*, 2013, **339**(6125), 1328–1331.
17. M. Islinger, C. Eckerskorn and A. Völkl, *Electrophoresis*, 2010, **31**(11), 1754–1763.
18. H. T. Parsons, K. Christiansen, B. Knierim, A. Carroll, J. Ito, T. S. Batth, A. M. Smith-Moritz, S. Morrison, P. McInerney, M. Z. Hadi, M. Auer,

A. Mukhopadhyay, C. J. Petzold, H. V. Scheller, D. Logué and J. L. Heazlewood, *Plant Physiol.*, 2012, **159**(1), 12–26.
19. D. J. Pagliarini, S. E. Calvo, B. Chang, S. A. Sheth, S. B. Vafai, S. E. Ong, G. A. Walford, C. Sugiana, A. Boneh, W. K. Chen, D. E. Hill, M. Vidal, J. G. Evans, D. R. Thorburn, S. A. Carr and V. K. Mootha, *Cell*, 2008, **134**(1), 112–123.
20. E. C. Schirmer, L. Florens, T. Guan, J. R. Yates III and L. Gerace, *Science*, 2003, **301**(5638), 1380–1382.
21. E. Wiederhold, T. Gandhi, H. P. Permentier, R. Breitling, B. Poolman and D. J. Slotbloom, *Mol. Cell Proteomics*, 2009, **8**(2), 380–392.
22. F. J. Dieguez-Acuna, S. A. Gerber, S. Kodama, J. E. Elias, S. A. Beausoleil, D. Faustman and S. P. Gygi, *Mol. Cell Proteomics*, 2005, **4**(10), 1459–1470.
23. F. M. Boisvert and A. I. Lamond, *Proteomics*, 2010, **10**(22), 4087–97.
24. C. De Duve and H. Beaufay, *J. Cell Biol.*, 1981, **91**(3), 293s–299s.
25. J. S. Andersen, C. J. Wilkinson, T. Mayor, P. Mortensen, E. A. Nigg and M. Mann, M., *Nature*, 2003, **426**(6966), 570–574.
26. L. J. Foster, C. L. de Hoog, Y. Zhang, Y. Zhang, X. Xie, V. K. Mootha and M. Mann, *Cell*, 2006, **125**(1), 187–199.
27. L. Jakobsen, K. Vanselow, M. Skogs, Y. Toyoda, E. Lundberg, I. Poser, L. G. Falkenby, M. Bennetzen, J. Westendorf, E. A. Nigg, M. Uhlén, A. A. Hyman and J. S. Andersen, *EMBO J.*, 2011, **30**(8), 1520–1535.
28. T. Geiger, J. Cox, P. Ostasiewicz, J. R. Wisniewski and M. Mann, *Nat. Methods*, 2010, **7**(5), 383–385.
29. T. P. Dunkley, R. Watson, J. L. Griffin, P. Dupree and K. S. Lilley, *Mol. Cell Proteomics*, 2004, **3**(11), 1128–1134.
30. T. P. Dunkley, S. Hester, I. P. Shadforth, J. Runions, T. Weimar, S. L. Hanton, J. L. Griffin, C. Bessant, F. Brandizzi, C. Hawes, R. B. Watson, P. Dupree and K. S. Lilley, *Proc. Natl Acad. Sci. USA*, 2006, **103**(17), 6518–6523.
31. M. W. Trotter, P. G. Sadowski, T. P. Dunkley, A. J. Groen and K. S. Lilley, *Proteomics*, 2010, **10**(23), 4213–4219.
32. D. J. Tan, H. Dvinge, A. Christoforou, P. Bertone, A. Martinez Arias and K. S. Lilley, *J. Proteome Res.*, 2009, **8**(6), 2667–2678.
33. S. L. Hall, S. Hester, J. L. Griffin, K. S. Lilley and A. P. Jackson, *Mol. Cell Proteomics*, 2009, **8**(6), 1295–1305.
34. G. C. McAlister, E. L. Huttlin, W. Haas, L. Ting, M. P. Jedrychowski, J. C. Rogers, K. Kuhn, I. Pike, R. A. Grothe, J. D. Blethrow and S. P. Gygi, *Anal. Chem.*, 2012, **84**(17), 7469–7478.
35. A. L. Christoforou and K. S. Lilley, *Anal. Bioanal. Chem.*, 2012, **404**(4), 1029–1037.
36. L. Ting, R. Rad, S. P. Gygi and W. Haas, *Nat. Methods*, 2011, **8**(11), 937–940.
37. S. Y. Ow, M. Salim, J. Noirel, C. Evans and P. C. Wright, *Proteomics*, 2011, **11**(11), 2341–2346.
38. C. D. Wenger, M. V. Lee, A. S. Hebert, G. C. McAlister, D. H. Phanstiel, M. S. Westphall and J. J. Coon, *Nat. Methods*, 2011, **8**(11), 933–935.

39. M. Wühr, W. Haas, G. C. McAlister, L. Peshkin, R. Rad, M. W. Kirschner and S. P. Gygi, *Anal. Chem.*, 2012, **84**(21), 9214–9221.
40. L. M. Breckels, L. Gatto, A. Christoforou, A. J. Groen, K. S. Lilley and M. W. B. Trotter, *J. Proteomics*, 2013, **88**, 129–140.
41. L. Gatto, L. M. Breckels, S. Wieczorek, T. Burger, K. S. Lilley, manuscript in preparation.
42. L. Gatto and K. S. Lilley, *Bioinformatics*, 2012, **28**(2), 288–289.
43. K. C. Chou and H. B. Shen, *J. Proteome Res.*, 2007, **6**(5), 1728–1734.
44. J. Mitschke, J. Fuss, T. Blum, A. Höglund, R. Reski, O. Kohlbacher and S. A. Rensing, *New Phytol.*, 2009, **183**(1), 224–235.
45. Y. Yang and B. L. Lu, *Int. J. Neural. Syst.*, 2010, **20**(1), 13–28.
46. S. Wan, M. W. Mak and S. Y. Kung, *BMC Bioinformatics*, 2012, **13**, 290.
47. S. Briesemeister, J. Rahnenführer and O. Kohlbacher, *Bioinformatics*, 2010, **26**(9), 1232–1238.
48. L. Q. Li, Y. Zhang, L. Y. Zou, Y. Zhou and X. Q. Zheng, *Protein Pept. Lett*, 2012, **19**(4), 375–387.
49. S. Mei, *PLoS One*, 2012, **7**(6), e37716.
50. X. Du, S. J. Callister, N. P. Manes, J. N. Adkins, R. A. Alexandridis, X. Zeng, J. H. Roh, W. E. Smith, T. J. Donohue, S. Kaplan, R. D. Smith and M. S. Lipton, *J. Proteome Res.*, 2008, **7**(7), 2595–2604.
51. Y. Karpievitch, J. Stanley, T. Taverner, J. Huang, J. N. Adkins, C. Ansong, F. Heffron, T. O. Metz, W. J. Qian, H. Yoon, R. D. Smith and A. R. Dabney, *Bioinformatics*, 2009, **25**(16), 2028–2034.
52. N. Nikolovski, D. Rubtsov, M. P. Segura, G. P. Miles, T. J. Stevens, T. P. Dunkley, S. Munro, K. S. Lilley and P. Dupree, *Plant Physiol.*, 2012, **160**(2), 1037–1051.
53. S. Maere, K. Heymans and M. Kuiper, *Bioinformatics*, 2005, **21**(16), 3448–3449.
54. L. C. Gillet, P. Navarro, S. Tate, H. Röst, N. Selevsek, L. Reiter, R. Bonner and R. Aebersold, *Mol. Cell Proteomics*, 2012, **11**(6), O111.016717.
55. A. Christoforou, C. Mulvey, L. M. Breckels, L. Gatto, P. Hayward, A. Martinez Arias and K. S. Lilley, unpublished data.
56. A. Christoforou, C. Mulvey, L. M. Breckels, L. Gatto and K. S. Lilley, unpublished data.

CHAPTER 10

Quantitative Analyses of Phosphotyrosine Cellular Signaling in Disease

HANNAH JOHNSON

Department of Biological Engineering, Massachusetts Institute of Technology, Cambridge, MA 02139, USA
Email: hjohnsn@mit.edu

10.1 Introduction

The reversible and dynamic post-translational modification (PTM) of amino acids through the addition of a negatively charged phosphate group by a protein kinase is known as protein phosphorylation. The most common eukaryotic protein phosphorylation occurs on serine residues (90%), while phosphorylation of threonine residues accounts for 10% and the phosphorylation of tyrosine residues is thought to account for 0.05%.[1] Modification of amino acids in this manor can have a significant impact on the turnover, conformation, activity, protein–protein interactions and subcellular localization of proteins.[1–4] The distinct and dynamic molecular changes that can be brought about by phosphorylation highlight the fundamental role it plays in the regulation of biological processes. Traditionally, the study of protein phosphorylation has been at the level of a single protein, whereby the site of modification(s) was identified through biochemical and analytical chemistry techniques.[5] The structural and functional role of phosphorylation was then ascertained through biochemical

New Developments in Mass Spectrometry No. 1
Quantitative Proteomics
Edited by Claire Eyers and Simon J Gaskell

Published by the Royal Society of Chemistry, www.rsc.org

techniques and structural biology. Throughout recent years the identification and quantification of serine, threonine and tyrosine phosphorylation have significantly increased due to the development of high throughput analytical techniques that have been founded on the introduction of high speed, sensitive mass spectrometry instrumentation. Whilst the analyses of serine and threonine phosphorylation have become more routine, the identification and quantification of tyrosine phosphorylation remain particularly challenging due to the low abundance of this modification in the cell.

The importance of tyrosine phosphorylation in cellular signaling has driven the aspiration to identify and quantify tyrosine phosphorylation signaling to enable the understanding of the role of tyrosine phosphorylation in disease. This chapter describes the use of phosphotyrosine affinity enrichment techniques to enable the phosphorylation site localization and tyrosine kinase signaling status of biological samples. The methods that have been developed to identify tyrosine phosphorylation signaling in biological systems are then highlighted and bring to the forefront the mass spectrometry techniques that are available for the relative and absolute quantification of tyrosine phosphorylation in disease, followed by a summary of recent efforts to facilitate the quantification of phosphotyrosine cellular signaling at the single cell level. Furthermore, this chapter emphasizes functional analyses that can be used to identify the roles of individual and groups of tyrosine phosphorylation sites and highlights specific studies that have used computational modeling to make predictions about biological processes. Lastly, the available bioinformatics tools for data processing are highlighted and the challenges that remain ahead are summarized, with the promise of quantitative proteomics being used for the identification of novel therapeutic targets.

10.1.1 Phosphotyrosine Cellular Signaling

Since the first published instance of tyrosine phosphorylation two decades ago,[5] the specific role of tyrosine phosphorylation as a dynamic process for the intrinsic control of inter- and intra-cellular signaling has become apparent.[1,6] Tyrosine phosphorylation is modulated through the addition of a phosphate group by a tyrosine kinase (TK) or receptor tyrosine kinase (RTK). There are currently 90 known TKs, including 58 RTKs, in the human genome that are responsible for mediating growth factor responses and complex cell–cell interactions.[3] Phosphorylation of tyrosine residues is an effective mechanism of rapid signal propagation that is initiated at the cell membrane throughout the cell. Typically, RTKs receive extracellular stimuli in the form of growth factor stimulation or interaction with a neighboring receptor on an adjacent cell, which in turn induces receptor dimerization or clustering. This inter-receptor interaction leads to auto-phosphorylation of the catalytic domain, which induces activation and potentiation of the kinase activity.[4] These phosphorylation events typically form docking sites for adaptor molecules to bind through Src Homology 2 (SH2) and/or phosphotyrosine binding (PTB) domains and lead to potentiation of signaling downstream.[7] Dynamic intra-cellular signaling cascades are then initiated and transmitted throughout the cytoplasm into the nucleus to transcription factors and result in the mediation of gene expression changes that often lead to a cellular response. Tyrosine phosphorylation signaling is often represented as individual signaling pathways; yet, with the recent understanding of the complexity of these tyrosine phosphorylation signaling systems and the identification of cross talk between the pathways, the idea of signaling networks have evolved. RTK signaling pathways and networks have not been fully annotated and the identification of RTK-specific interactions and connections that occur within specific RTK signal transduction remains a challenge. To complicate matters further, there is also evidence that many of the extra-cellular signaling cues that work through a variety of RTKs converge through a relatively small number of nodes within the signaling network.[8] Such convergence on a particularly small group of interaction mechanisms also makes it challenging to parse apart context-dependent signaling.[8] One can imagine that, in the realms of systems biology, understanding signaling networks downstream of a particular RTK is dependent on the identification and quantification of large numbers of phosphorylation events and their quantification in response to cell perturbations with specific growth factors. Throughout these analyses it is becoming apparent that many different RTKs employ common downstream signaling molecules to facilitate functional changes.

10.1.2 Deregulation of Tyrosine Phosphorylation in Disease

Aberrant phosphorylation of tyrosine residues has been implicated in a number of diseases, such as cancer, inflammation and metabolic disorders.[9–11] The control of tyrosine phosphorylation pins on a delicate balance between TKs and phosphatases that, if deregulated, can have a significant impact on cell growth, angiogenesis, invasion and metastasis, all of which play an important role in cancer progression.[2] To highlight this further, 50% of the known TKs are oncogenes.[12] With respect to cancer, it is well recognized that all cancers arise as a result of gene mutations and epigenetic changes.[13] As such, the availability of whole-genome sequencing has opened up the possibility for personalized medicine for life-threatening human diseases.[14] Yet, while the causes of cancer lie in the genome, the molecular manifestation of this disease, and many others, is at the level of the resulting proteins and biochemical pathways. RTKs are often mutated or overexpressed in disease and aberrant tyrosine phosphorylation signaling has been reported across many different solid tumors and hematological malignancies.[3] RTK overexpression and mutation can lead to constitutive kinase activation by increasing the concentration of dimers and, thus, the frequency of dimerization. For instance, the fusion tyrosine kinase, Bcr-Abl, is present in 95% of all chronic myelogeneous leukaemias (CML) and the tyrosine kinase activity has been found to have the ability to transform normal cells.[15] The epidermal growth factor receptor (EGFR) is often overexpressed in glioblastoma, breast and lung carcinomas.[3] Furthermore, mutation of the EGFR receptor resulting from the loss of the extra-cellular portion of the receptor, results in a constitutively active receptor (EGFRvIII) that induces cellular tumorigenicity across many cancer types.[16] MET signaling is deregulated in non-small cell lung carcinoma (NSCLC) and has been implicated in disease progression.[3] ErbB2 (HER2) is amplified in breast cancer; PDGFR is overexpressed in glioblastoma; and fibroblast growth factor receptors (FGFR1, FGFR2 and FGFR4) are overexpressed across a variety of tumours.[3] Furthermore, the critical role that RTKs play in cancer progression, and their membrane localization, have made them promising therapeutic targets.[17] Thus, despite the low abundance of tyrosine phosphorylation, when compared to serine and threonine phosphorylation, it plays a significant role in the mediation of oncogenic transformation and disease pathogenesis.

It is the importance of TK and RTKs in disease and this remaining challenge in understanding disease progression that makes the application of proteomics to the large-scale phosphotyrosine profiling of cell lines, tumors and tissues particularly important. The identification of signaling that is altered and deregulated in disease through quantitative proteomics analyses is crucial to understand the wiring of the signaling network. Furthermore, by studying the protein and PTM status of a sample, we can gain functional understanding of the pathways that are deregulated in disease. The emergence of functional proteomics approaches has been focused on quantitative protein expression profiling, the quantification of PTMs (particularly relevant to this chapter is tyrosine phosphorylation) and the quantification of kinase activities, which together directly contribute to the functional understanding of a biological system.

10.2 Enrichment Techniques for Phosphotyrosine Profiling

Due to the low abundance of tyrosine phosphorylation in a background of complex, highly abundant biological molecules, it is necessary to reduce sample complexity prior to mass spectrometric analyses by specifically enriching for tyrosine-phosphorylated proteins or peptides.[18]

10.2.1 Phosphotyrosine-specific Antibodies as Enrichment Tools

Antibody-based methods are the current standard for phosphotyrosine protein/peptide enrichment. The larger size of the phosphotyrosine moiety (compared to phosphoserine and phosphothreonine) has led to higher affinity pan-specific antibodies for phosphotyrosine compared to phosphoserine or phosphothreonine. These pan-specific high affinity anti-phosphotyrosine monoclonal antibodies have proved to be a critical tool for the enrichment of tyrosine-phosphorylated polypeptides.[19] Typically, anti-phosphotyrosine antibodies bind to phosphotyrosine proteins or peptides and are immobilized to protein agarose beads prior to washing to remove non-specifically bound proteins or peptides. Phosphotyrosine proteins or peptides are then eluted from the specific antibody in a low pH solution. This strategy has been applied to enrich for phosphotyrosine-containing proteins, where protein elution from the beads is typically followed by proteolysis prior to liquid chromatography tandem mass spectrometry (LC-MS/MS) analysis of the peptides.[20] Phosphotyrosine-specific enrichment at the protein level was applied to the identification of phosphotyrosine proteins in an *in vivo* setting of a zebrafish during embryonic development.[21] The immunoprecipitation (IP) of tyrosine-phosphorylated proteins using a mixture of phosphotyrosine-specific antibodies led to the identification of 800 phosphotyrosine-phosphorylated proteins and interacting partners.[21] Unfortunately, proteolysis of the enriched proteins generates many peptides, the majority of which are not phosphorylated. As a result, detection of the phosphorylated peptides is reduced and the bulk of the instrument time is consumed by identifying non-phosphorylated peptides.[18,22,23] To obtain site-specific information, anti-phosphotyrosine antibodies have been used more recently to enrich for phosphotyrosine peptides prior to mass spectrometry (MS) analysis.[20,24] During this approach peptides are incubated with a cocktail of pan-specific high affinity monoclonal anti-phosphotyrosine antibodies immobilized on protein-G agarose. The use of multiple pan-specific phosphotyrosine antibodies helps to reduce the sequence bias that may occur with a single antibody. Enriching specifically for phosphotyrosine-phosphorylated peptides in this manner is advantageous as it allows the site-specific localization of the PTM. This approach has been demonstrated to facilitate the identification and quantification of hundreds of tyrosine phosphorylation sites in biological samples (see section 10.3).[25]

10.2.2 SH2 Domains as Phosphotyrosine Profiling Tools

One of the ways that tyrosine phosphorylation mediates cellular signaling is through the modulation of protein–protein interactions through Src Homology 2 (SH2) and phosphotyrosine binding (PTB) domains.[7] SH2 domains are the most prevalent phosphotyrosine-binding domain and, as such, they play an integral role in cellular signaling.[4] The specificity of these domains for binding to phosphotyrosine *in vivo* and the important role they play in cellular signaling makes them useful tools for phosphotyrosine profiling and quantification. The binding of SH2 domains to tyrosine-phosphorylated proteins has been utilized to identify the SH2-specific tyrosine phosphorylation status of cell lysates.[26,27] The limitations of this approach has been that there is no direct protein identifications and the resulting data gives only a global "fingerprint" of the SH2 substrate status for each cell lysate.

To overcome this, quantitative MS was utilized to differentially label HeLa cells that were either stimulated with epidermal growth factor (EGF) or unstimulated. Proteins were then enriched for using the Grb2-SH2 domain and differences in the affinity for proteins within the cell lysate were identified through LC-MS/MS. This approach resulted in the identification of 228 proteins, of which 28 proteins were significantly altered with EGF stimulation.[28] While this approach did not result in the identification of tyrosine phosphorylation sites, the endogenous protein–protein interactions were captured in a growth factor-dependent manner.[28]

10.2.3 Tandem Mass Spectrometry Phosphorylation Site Localization

Regardless of the phosphotyrosine enrichment strategy, the current method of choice for the sequence identification and phosphorylation site localization of tyrosine phosphorylation is mass spectrometry. Typically, proteins are proteolyzed (using a suitable proteolytic enzyme) to generate peptides that are then subjected to phosphotyrosine enrichment prior to LC-MS/MS. Trypsin is the most widely used proteolytic enzyme in proteomics workflows due to the high efficiency of cleavage (*C*-terminal to lysine and arginine residues) and the MS suitable size of the resulting peptides generated after proteolytic cleavage. Hundreds of tyrosine phosphorylation sites can be identified from complex biological mixtures following tryptic digestion.[20,24] However, it is clear that a significant portion of the tyrosine phosphorylation sites that are present within the cell are missed due to the generation of tryptic fragments that are not within the m/z range for efficient ionization and/or fragmentation by MS. This has led to the use of multiple proteases, such as Lys-C, Glu-C and Asp-N, within the proteomics workflow to increase the overall coverage of proteins and potential sites of modification.[20]

The most common fragmentation method used for the identification of peptides in proteomics is collision-induced dissociation (CID).[29] Here, collisions with an inert gas result in vibrational excitation leading to fragmentation of the peptide along the peptide backbone to produce product ions diagnostic of the peptides amino acid sequence. However, during CID, serine and threonine phosphorylation often undergo neutral loss of the phosphate group from the peptide backbone, leading to uncertainty as to the site of phosphorylation. Tyrosine phosphorylation is less labile than serine and threonine phosphorylation and the phosphate group typically remains attached to b- and y-series product ions. This often makes the phosphorylation site localization definitive (providing that the b- and y-series ions account for the entire peptide sequence) as the b and y series fragment ions have an addition of 80 Da (HPO_3) where the phosphorylation site resides. Tyrosine-phosphorylated peptides often lose 80 Da from the precursor ion, and infrequently from product ions, which can complicate the MS/MS spectrum and reduce the fragment sequence coverage that is detected for these peptides.[30]

Low molecular weight reporter ions are generated by various types of amino acid modifications and are indicative of a significant amount of information about the peptides sequence, for instance, the presence of the tyrosine-phosphorylated immonium ion is specifically indicated by *m*/*z* 216.043.[31] The presence of this product ion *m*/*z* in a tandem mass spectrum can be used for the selective detection of tyrosine-phosphorylated peptides.[32] Precursor ion scanning has been utilized with a quadrupole time of flight (QToF) instrument, whereby precursor ions that produced *m*/*z* 216.043 following CID were then subjected to additional CID to identify the sequence and site of phosphorylation.[32] This approach works for the analysis of tyrosine phosphorylation in individual proteins; however, in more complex mixtures (such as cell lysates) the dynamic range of the mixture and the low level of phosphorylation present reduce the likelihood that phosphotyrosine peptides are available for detection. In this instance alternate strategies that utilize the enrichment of tyrosine-phosphorylated peptides are often used instead.

As mentioned above, the presence of low molecular mass product ions can generate a wealth of information about the specific sequence details of precursor peptide ions. With the development of the hybrid orbitrap mass spectrometers, CID is typically performed in the linear quadrupole ion trap mass analyzer of the instrument. A significant limitation in the fragmentation of peptides using the ion trap is that the upper limit on the ratio between the precursor *m*/*z* and the lowest trapped fragment ion is 30% due to the decrease in ion stability below this level. For example, fragment ions resulting from the CID of a precursor ion of *m*/*z* 900 will not be detected below *m*/*z* 300, which presents a drawback for the *de novo* sequencing of peptides. This limitation has led to the development of higher energy C-trap dissociation (HCD), which is being increasingly used as the predominant fragmentation method for peptide identifications.[31] In HCD, fragmentation is performed in an octopole collision cell positioned adjacent to the C-trap, circumventing the application of an increased radiofrequency voltage to the C-trap, which decreases trapping efficiency. During fragmentation, the collision cell is sealed gas tight and supplied with a radiofrequency voltage (the direct current applied can be varied) and the cell is filled with an inert gas.[31] The resulting HCD tandem mass spectrum contains ions in the low-mass region and high resolution and high mass accuracy is maintained. This increase in mass accuracy and resolution from the traditional CID in the quadropole ion trap also has an effect on the confidence in peptide identifications.[33] HCD fragmentation also allows detection of the phosphotyrosine immonium ion, which can also be used as a confirmatory metric of the quality of the phosphotyrosine enrichment method employed.

Following fragmentation, MS/MS spectra are searched against a database containing protein sequences from the species of interest. The most common database searching algorithms are MASCOT and SEQUEST.[34,35] These search algorithms are based on the comparison of a computationally derived (*in silico*) product ion spectrum to the experimental MS/MS spectrum.

The comparison between the experimental and theoretical product ion spectra gives rise to the probability that the mass spectrum pertains to the correct peptide sequence in the database. This probability is indicated by score metrics, whereby the higher the score, the higher the likelihood that the mass spectrum is correctly assigned to the peptide sequence it was derived from. The mass accuracy of the precursor ion and product ions is an important factor. As the mass accuracy of the precursor ion masses increase, the width of the search window decreases and the number of possible peptides matching the correct mass decreases. This improves the confidence in the matching of the mass spectra to the correct peptide sequence and reduces potential false positives.

10.3 Quantification of Tyrosine Phosphorylation

To understand the functional role of phosphorylation in disease it is important to move beyond the identification of phosphorylation sites and understand how these phosphorylation sites change both temporally and across different samples. The location and absolute and relative concentrations of phosphorylation at particular sites determine the effect on protein function and regulation. There are a number of techniques to enable the quantification of phosphorylation and, specifically, tyrosine phosphorylation. This section will highlight the most common methods for the relative, absolute and label-free quantification of tyrosine phosphorylation in biological systems and the development of methods for the quantification of phosphotyrosine signaling at the single cell level.

10.3.1 Relative Quantification by Mass Spectrometry

The relative quantification of tyrosine phosphorylation can be carried out as an extension of the enrichment for phosphotyrosine peptides or proteins. Quantification of phosphotyrosine consists of two or more samples being differentially labeled with isotopic variants and combined prior to mass spectrometric analysis. There are a number of methods of stable isotope labeling for quantitative analysis using mass spectrometry. These methods fall broadly into two categories: metabolic labeling, where proteins are differentially labeled using stable isotopically labeled amino acids while cells are grown in culture or labeled using media that exclusively contains a stable isotopically labeled element (most commonly, $^{12}C/^{13}C$, $^{16}O/^{18}O$ or $^{14}N/^{15}N$); or chemical labelling, where proteins or peptides derived from biological samples are differentially labeled with stable isotopes or mass tags during sample processing prior to MS analysis (Figure 10.1).

The most common method of metabolic labeling is stable isotopically labeled amino acids in cell culture (SILAC).[36] During SILAC, cells in culture are grown in the presence of stable isotopically labeled amino acids for a minimum of six doublings during which time the isotopically labeled amino acids fully incorporate into the newly synthesized protein sequences. The

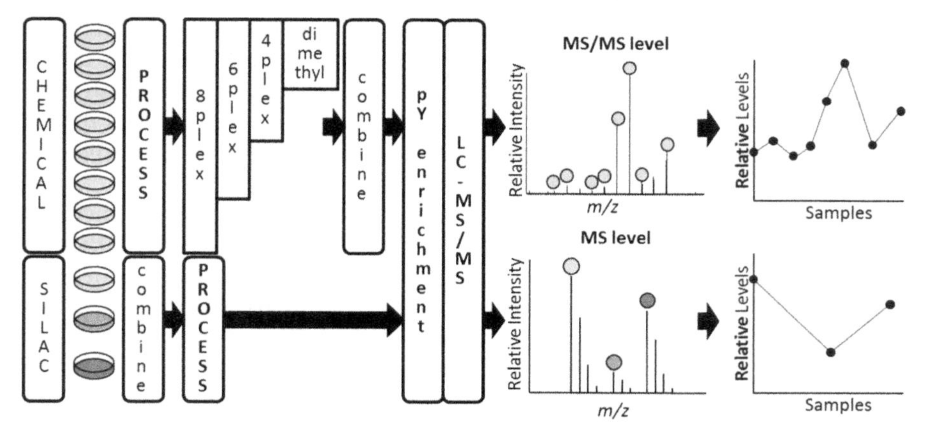

Figure 10.1 Relative quantification of tyrosine phosphorylation by chemical labeling with isobaric tags (iTRAQ 8plex, TMT 6plex and iTRAQ 4plex) or dimethyl stable isotopic labeling. Relative quantification is currently restricted to the quantitative comparison of three samples in a single experiment with SILAC and eight samples with iTRAQ 8plex. "PROCESS" refers to the reduction, alkylation and enzymatic digestion of proteins. Samples are typically combined earlier in the experimental workflow for SILAC analyses than in chemical labeling analyses, as indicated by the "Combine" box. Phosphotyrosine (pY) enrichment is typically carried out at the peptide level and allows phosphorylation site-specific identification. Relative quantification is achieved through the comparison of the relative intensities of reporter ions at the MS/MS level for the isobaric tagging methods. Alternatively, comparison of precursor ion intensities at the MS level for SILAC and dimethyl labeling allows relative quantification.

most commonly used amino acids are $^{13}C_6$-arginine and $^{13}C_6$-lysine as they introduce a consistent mass difference into tryptic peptides; ^{15}N can also be incorporated into these amino acids to allow differentiation by mass, and therefore multiplexing of three samples. To quantify across three different conditions, the "light" samples are unlabeled, "medium" peptides are labeled with $^{13}C_6$-arginine and $^{13}C_6$-lysine and "heavy" peptides are labeled with $^{13}C_6{}^{15}N_4$-arginine and $^{13}C_6{}^{15}N_2$-lysine. The combination of "light", "medium" and "heavy" labeled proteins prior to reduction and alkylation of cysteine residues and proteolytic protein digestion enables the accurate quantification across conditions (Figure 10.1). As the physiochemical properties of peptides with differing stable isotopic content is largely the same, the relative intensities of precursor peptide ions derived from each of the three samples can be directly compared. The use of SILAC reduces biological and technical variation by enabling the combination of labeled and unlabeled samples before commencing sample processing. It also has clear benefits as it enables the labeling of large amounts of protein, which allows the use of a large amount of starting material, which circumvents sensitivity limitations that occur due to the low abundance of tyrosine phosphorylation.

Furthermore, although SILAC is often restricted to cell culture, SILAC labeling of whole organisms has been developed through an exclusive diet of $^{13}C_6$-labelled amino acids across multiple generations.[37,38] Although complete isotopic incorporation using SILAC labeling has been achieved *in vivo*, the high cost of the diet and the extensive time required for complete labeling limits its widespread application.

Chemical labeling is a useful approach for samples that are difficult or impossible to label using metabolic approaches, such as cells with slow doubling times, *in vivo* tissues and tumors. There are a number of chemical labeling strategies that allow the quantitative comparison of at least two different samples or cellular conditions at a given time. Currently, isobaric tags for relative and absolute quantification (iTRAQ®)[39] and tandem mass tags (TMT)[40] facilitate the differential labeling of up to eight independent samples within a single experiment using labeling reagents, such as iTRAQ 4plex, TMT 6plex and iTRAQ 8plex (Figure 10.1; see Chapter 3 for additional information). iTRAQ and TMT labeling consists of the differential labeling of peptides through the reaction of isobaric mass tags with all free α- and ε-amino groups present in the sample, namely the *N*-terminus of a peptide chain and lysine side chains, using *N*-hydroxysuccinimide (NHS) chemistry. iTRAQ- and TMT-labeled peptides are chromatographically indistinguishable and are present at the same mass in the MS spectrum. However, upon fragmentation (CID, HCD) the balance mass is dissociated producing product ions of *m*/*z* 114–117 for iTRAQ 4plex, *m*/*z* 126–131 for TMT 6plex and *m*/*z* 113–121 for iTRAQ 8plex (discounting 120 due to the potential interference of the phenylalanine immonium ion). The relative intensities of these reporter ions in the tandem mass spectrum pertaining to the precursor ion of interest allows the relative determination of the peptide levels and, thus, levels of the site-specific PTM in the samples of origin. iTRAQ quantification has been used extensively for the quantification of tyrosine phosphorylation across a host of different cell lines, tissues and conditions to address multiple biological questions.[24,41,42] An important note is that, due to the low *m*/*z* reporter ion fragments that result after subsequent iTRAQ labeling, the mass spectrometry instrumentation and methodology that is used to identify and quantify phosphorylation has to take this into consideration. For instance, the most common instruments that are used for these experiments are either QToF or orbitrap mass spectrometers. Due to the reduced stability of ions in the low-mass region in ion trap analyzers often utilized for CID in hybrid orbitrap mass spectrometers, HCD is often required to identify low-mass reporter ions required for iTRAQ and TMT quantification. The recent improvement in HCD efficiency has also made this fragmentation method useful for the identification and quantification of peptide ions. The advantage of the QToF instrumentation for iTRAQ experiments is that CID occurs in a collision cell and product ions are filtered through the quadrupole. Thus, product ions within a manually specified *m*/*z* threshold (typically *m*/*z* 50–1800) are identified, encompassing both iTRAQ reporter ions and sequence-specific product ions. Furthermore, there has also been the observation of a reduction in peptide and protein identifications observed for TMT 6plex and iTRAQ 8plex when compared to iTRAQ 4plex.[43] This is thought to occur as a result of the presence of non-peptide sequence product ions arising from the cleavage of the label. It is also thought that there are distinct physicochemical properties of peptides derived from the larger mass tags leading to a difference in the amount of collision energy required for optimal dissociation. Standard conditions could thus lead to more complex uncharacterized fragmentation patterns.[43]

Another method for the chemical labeling of biological samples for relative quantification is the use of dimethyl labeling. Here, formaldehyde (CH_2O) or deuterated formaldehyde (CD_2O) is used to label the *N*-terminus and ε-amino group of Lys residues through reductive amination resulting in the addition of 14 Da or 16 Da, respectively, for each free amine group (Figure 10.1).[44] This labeling strategy is fast, relatively inexpensive and goes to completion without any by-products. As such, this approach has been applied to the quantification of phosphotyrosine signaling in cell culture systems.[45] There can be complications with differentially dimethylated peptides, whereby isotopic variants can have significantly different chromatographic separation profiles leading to potential inconsistencies in relative quantification.

10.3.1.1 Temporal Dynamics

One important question in the study of cell signaling is the temporal effect of growth factor stimulation (or environmental cues) on the phosphorylation network. Relative MS-based quantification experiments have been utilized to address this question. Quantitative iTRAQ 4plex analysis of tyrosine phosphorylation temporal dynamics in HMEC cells at 5, 10 and 30 min after epidermal growth factor (EGF) stimulation led to the identification of 78 tyrosine phosphorylation sites on 58 proteins.[24] From these analyses it was clear that there were a distinct subset of proteins that were significantly increased within 5 min of growth factor stimulation. This quantitative approach was then applied to the quantification of phosphotyrosine following the stimulation of HMEC cells with either EGF or heregulin (HRG, a HER2 receptor growth factor ligand) at distinct time points. This dynamic temporal profile of phosphorylation was then used to decouple phosphotyrosine events that were significantly correlated with cellular migration and proliferation.[42] iTRAQ 4plex quantification was also utilized for the study of the insulin signaling pathway in 3T3-L1 adipocytes following stimulation with insulin for 0, 5, 15 and 45 min, resulting in the quantification of the temporal dynamics of 122 phosphotyrosine sites on 89 proteins.[46] The SILAC quantification of HeLa cells in response to EGF stimulation, followed by the enrichment of phosphorylated peptides, allowed the capture of phosphorylation temporal dynamics in response to EGFR activation across 6600 phosphorylation sites on 2244 proteins. Such studies highlight the critical cross talk between tyrosine phosphorylation at RTKs and serine and threonine phosphorylation in cell signaling.[47]

10.3.1.2 Relative Quantification of Biological Systems

Tyrosine phosphorylation plays an important role in disease and, to understand the effects of this PTM, there have been many studies that have quantified phosphotyrosine across many biological systems. Many of these approaches are currently limited to the profiling of tyrosine phosphorylation across cell lines. SILAC quantification was applied to the determination of EphB2 and ephrin-B1 cell specific signaling networks that control cell sorting. In this approach MS analysis of differentially labeled mixed populations of EphB2 and ephrin-B1 expressing cells allowed the identification of cell-specific tyrosine phosphorylation events. This revealed distinct tyrosine kinases and targets differentially utilized following cell–cell contact.[48] SILAC quantification has also been used to identify substrates of the Src tyrosine kinase: Src was overexpressed in cells and compared to cells with low levels of Src through differential isotopic labeling. Phosphotyrosine proteins were then enriched and the 136 proteins that exhibited significantly increased tyrosine phosphorylation levels (or an increased molecular association) following Src overexpression were categorized as potential substrates.[49] Furthermore, quantification of phosphotyrosine signaling following interleukin-2 signaling highlights the importance of tyrosine phosphorylation in immune signaling.[50] iTRAQ 8plex quantification of phosphotyrosine signaling across immortalized cell lines expressing different ROS fusion tyrosine kinases led to the identification of a phosphorylation mechanism (of E-Syt1) that is responsible for CD74-ROS-driven invasion *in vitro* and *in vivo*.[51] iTRAQ quantification of phosphotyrosine signaling has also been applied to multiple glioblastoma models. iTRAQ 4plex and phosphotyrosine peptide enrichment allowed the identification and quantification of 99 phosphorylation sites on 69 proteins across cell lines expressing different levels of the mutant EGFR receptor (EGFRvIII). This approach enabled the identification of specific signaling that was occurring as a function of the increasing mutant EGFR receptor levels, allowing the identification of specific pathway activation in EGFRvIII cells.[41] More recent advances in the sensitivity of MS instrumentation have enabled the quantitative analysis of human tumors. To this end, iTRAQ 8plex was used to identify the effects of EGFRvIII on tumors *in vivo*. Quantification of 225 tyrosine phosphorylation sites on 168 proteins across eight human tumor xenograft lines allowed the identification of specific signaling present in EGFRvIII tumors. This study also highlighted the significant levels of heterogeneity that exist in the phosphotyrosine signaling network across tumors *in vivo*.[25] Many of these kinases play an important role in cancer progression and the identification of proteins and protein kinases involved in phosphotyrosine cell signaling, in addition to quantification of the sites of modification, could have an important impact on future therapeutics.

10.3.2 Absolute Quantification by Mass Spectrometry

Absolute quantification of tyrosine phosphorylation is more challenging than that of serine and threonine largely due to the decreased overall abundance of tyrosine phosphorylation in the cell. Furthermore, the extensive regulation of tyrosine phosphorylation means that many sites are rapidly turned over with phosphorylation of some sites increasing 100 fold in under a minute and decreasing to baseline levels within 15–30 min. This rapid turnover of tyrosine phosphorylation makes the accurate capturing of phosphorylation stoichiometry in response to growth factor stimulation, or across different systems, difficult. Absolute quantification of phosphorylated proteins has been achieved using the stable isotopic labeling of PTM peptides. In the case of absolute quantification of tyrosine phosphorylation and the determination of phosphorylation stoichiometry, isotopically labeled peptides are typically synthesized, quantified and added to the biological sample in known amounts. Subsequent MS analysis allows calculation of the absolute abundance of phosphorylation as the signal ion intensities of the endogenous peptide and the synthetically introduced peptide can be directly compared.[52,53] Selected reaction monitoring (SRM) depends on the fragmentation of a precursor peptide ion to yield one or more defined product ions that can be used to selectively quantify the peptide of interest within the complex biological matrix. In SRM these specific sequence-dependent product ions at defined m/z values are used to indicate the identification and relative signal intensity of a peptide sequence. This signal intensity can then be compared to the known amount of the identical isotopically labeled standard for accurate quantification. The increased sensitivity for both the identification and quantification achieved through SRM means that it is the method of choice for the absolute quantification of tyrosine-phosphorylated peptides. Furthermore, as the absolute quantity of phosphorylation is comparable across an infinite number of samples, SRM provides greater resolution in temporal measurements and quantification across many different disease states (Figure 10.2).[52,53]

10.3.3 Label-free Quantification by Mass Spectrometry

Label-free quantification requires the reproducible chromatographic separation of peptides prior to MS analysis and has been applied to the quantification of proteins from complex samples with modest success.[54] However, this approach cannot be applied accurately to the quantification of PTMs due to the enrichment methods that are typically required prior to MS analysis. Enrichment steps introduce an additional level of variation, which is often based on a complex number of variables. Furthermore, when identifying tyrosine phosphorylation sites, there is often only one "hit" per peptide. Thus, the application of a label-free technique that is based on the number of times a peptide is identified (*i.e.*, spectral counting) will incorrectly assume that any peptide that is "hit" once is present at the same

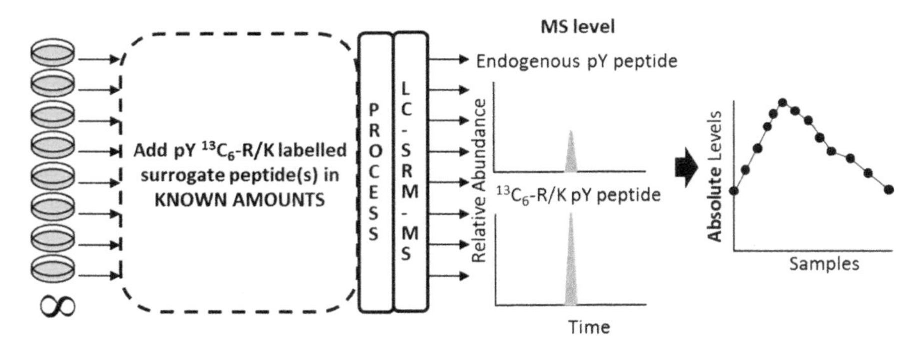

Figure 10.2 Absolute quantification of tyrosine phosphorylation can be carried out across an infinite number of samples. Typically, stable isotopically labeled pY surrogate peptides are added to the cell lysate of interest in known amounts. Proteins are then reduced, alkylated and digested with a suitable protease (indicated by "PROCESS"). Selected reaction monitoring (SRM) MS is then used to identify and quantify endogenous peptides and their stable isotopically labeled counterparts. Comparison of the signal intensities derived from these two peptides allows the absolute quantification of tyrosine phosphorylation.

level (see Chapter 6 for an overview of label-free quantification). In addition, the sample-to-sample variation in data-dependent acquisition (DDA) mode is only ~34%.[42] Thus, it is important to note that the absence of identification does not necessarily mean that the peptide or protein of origin is not present in the sample of interest. Nevertheless, label-free quantification is being applied to the quantification of tyrosine phosphorylation with modest success, with comparisons being performed between signal intensities of the same modified peptides across multiple runs. Phosphotyrosine signaling was profiled across 41 NSCLC cell lines and 150 NSCLC tumors with 4551 phosphotyrosine sites being quantified on 2700 different proteins, significantly increasing the knowledge of tyrosine kinase activation in NSCLC.[55] Label-free phosphotyrosine profiling has also been applied to the identification of signaling networks that are activated in basal breast cancer cells,[56] sarcoma cell lines[57] and prostate cancer xenograft models.[58]

10.3.4 Quantification of Phosphotyrosine Signaling at the Single-cell Level

The quantification of tyrosine phosphorylation signaling is typically carried out on large populations of cells. Despite the recent advances in mass spectrometry instrumentation and techniques, the limit of detection is still not great enough to make the quantification of individual cell signaling possible using the aforementioned techniques. The recent development of a method that couples the use of fluorescence-based flow cytometry antibody tagging and mass spectrometry has made the relative quantification of tyrosine phosphorylation possible at the single-cell level. In this single-cell mass cytometry approach antibodies are labeled with heavy metal isotopes and are subsequently bound to markers of interest present in or on the surface of cells. The levels of each of the antibodies can then be detected using high throughput inductively coupled plasma-mass spectrometry (ICP-MS) to quantify the heavy metal ions, allowing the quantification of dozens of parameters at the single-cell level (Figure 10.3). This approach has been utilized to examine healthy human bone marrow, where 31 proteins or phosphorylation sites, cell viability, DNA content and relative cell size were measured simultaneously.[59] Single-cell mass cytometry has also been applied to the quantification of 14 phosphorylation sites on the tyrosine kinases STAT1, STAT3, STAT5, JAK and ITK. Beyond this, the technique could also simultaneously identify the effects of 27 inhibitors on this system across 14 peripheral blood mononuclear cell (PBMC) types under 96 conditions for mechanistic insight into human disease.[60] These analyses reveal specific single-cell responses and corresponding cell signaling events and allow the identification of signaling heterogeneity.[59,60] There are caveats with this technique: the effects of the binding of a large number of antibodies on signaling in these cells are still unknown but might have an artifactual effect on the quantified signal. This approach is also reliant on the high efficacy of phosphospecific antibodies, which are not readily available for the large majority of identified tyrosine phosphorylation sites. Nevertheless, this technique has great promise in the quantification of tyrosine phosphorylation signaling *in vivo* and can help to address some of the

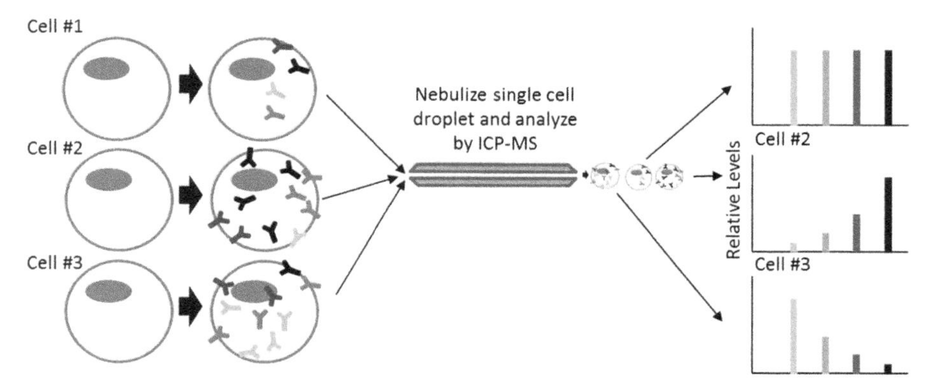

Figure 10.3 Quantification of single-cell phosphorylation signaling. Single-cell mass cytometry consists of the labeling of proteins and phosphorylation sites of interest present in or on the surface of cells using antibodies labeled with heavy metal isotopes. Single cells are nebulized and the levels of each of the different heavy metal isotopes, and thus antibodies, are detected using ICP-MS. Comparison of the signal intensities corresponding to the different heavy metal ions allows the relative quantification of tyrosine phosphorylation at the single-cell level.

current issues facing enhanced understanding of cell signaling, such as sample heterogeneity and the effects that this has on disease progression.

10.4 Functional Analyses

The development of high throughput techniques for the identification and quantification of tyrosine phosphorylation across multiple different cell states, and temporally in cells, tissues and tumors, has led to the generation of a large quantity of data. These phosphorylation sites and quantitative phosphorylation dynamics profiles give us an insight into the way that signaling networks respond at a global level when faced with specific stimulations and perturbations. Yet, the individual site-specific information that is encoded within these data needs further data analyses to understand the specific role that each site may be playing in a disease state. This is often addressed by the time extensive biochemical and biological analysis of these individual phosphorylation sites occurs *in vitro* and *in vivo*. To gain rapid insight from these large sets of data, there are high throughput functional analysis techniques available.

10.4.1 Computational Modeling

One of the ways of understanding quantitative phosphotyrosine signaling data and the role that it plays in exacting changes in biological phenotypes is through computational modeling. One example of this type of analysis is the quantification of phosphotyrosine signaling in HER2 overexpression in HMEC cells in response to EGF and HRG.[42] Partial least squares regression (PLSR) was used to correlate 248 protein metrics to cell migration and proliferation measurements that were carried out in the same cells. Multiple distinct signaling pathways were shown to stimulate migration following EGF stimulation, while HRG stimulation resulted in the activation of a selected subset of proteins involved in the known migration signaling pathway. PLSR analysis revealed quantitative phosphorylation profiles that significantly correlated with cell proliferation and migration.[42] The data were then used to create a reduced model of phosphorylation sites predictive of growth and invasion of HMEC cells in response to EGF or HRG to nine phosphorylation sites on six proteins.[61] PLSR was also applied to EGFRvIII receptor signalling, where four receptor phosphorylation sites were mutated and shown to have a significant impact on the non-mutated phosphorylation levels. The Erk pathway was identified to be negatively correlated with cell growth through PLSR analyses.[62] PLSR has also been utilized for the identification of signaling nodes that are important for driving invasion downstream of two distinct RTKs: EGFR and Met. Quantitative temporal mass spectrometric analyses of tyrosine phosphorylation signaling in response to EGF or HGF at three different concentrations and at two different time points were correlated to invasion measurements at the same growth factor concentrations using PLSR.[63] The top 26 most highly correlated phosphorylation sites were then used to create a reduced model for invasion. While the EGF model was able to accurately predict EGF-driven invasion, it could not predict HGF-driven invasion. Furthermore, the HGF model could predict HGF-driven invasion but not EGF-driven invasion. Distinct signaling nodes were identified to be important for invasion following EGF or HGF. As none of these proteins had been specifically related to EGFR- or Met-mediated invasion prior to these analyses, this highlights the utility of computational modeling in combination with quantitative mass spectrometric analyses.[63]

10.4.2 Bioinformatics Tools

As signaling networks are not fully characterized it is difficult to attribute functional information to the thousands of tyrosine phosphorylation site identifications and quantitative information generated through MS analyses. To understand the role of these phosphorylation sites it is important to identify the functional relevance. A common method for the functional annotation of proteins is gene ontology (GO) analysis, whereby the functional role of the protein and phosphorylation site (if the data is available) can help to identify whether there is an enrichment of a particular functional group of proteins in the dataset that could be responsible for driving a particular function (pantherdb.org/panther/ontologies.jsp).[64] When specifically trying to identify the kinases that may be responsible for the phosphorylation sites that are identified within an MS analysis, Scansite (scansite.mit.edu/) can be used to investigate whether there are kinase motifs present within experimental data. The algorithm queries input sequences and returns signaling motifs along with a score allowing the incorporation of proteins into potentially new signaling pathways.[65] Another algorithm that extracts phosphorylation motifs from MS datasets is the Motif-X algorithm (motif-x.med.harvard.edu/motif-x.html).[66] In order to enable the analysis of quantitative MS PTM data, web-based interfaces have been developed. PTMScout (ptmscout.mit.edu/) enables the storage and data analysis of peptides with PTMs and, in particular, tyrosine phosphorylation. This program applies the Scansite algorithm to quantitative MS data to allow the identification of kinases that may be responsible for the identified tyrosine phosphorylation. Furthermore, this tool serves as an on-line repository and facilitates the comparison of identified PTM sites to those already in the literature.[67]

Beyond functional analyses is the data analysis of quantitative phosphorylation data to better understand the difference between sample conditions. Common methods of compartmentalizing these data are the use of unsupervised clustering methods. The most commonly used clustering methods are self-organizing maps,[24,41] fuzzy c-means clustering,[47] and affinity propagation clustering.[25] These clustering methods group together similar trends within the data to enable the identification of potential biological links between phosphorylation sites.

10.5 Future Perspectives: Quantitative Proteomics to Identify Novel Therapeutic Targets

Developments in the identification and quantification of tyrosine phosphorylation can provide insight into potential drug targets and lead to the identification of phosphorylation events that may be predictive of clinical outcomes. Furthermore, advances in MS sensitivity now provide the ability to carry out quantitative phosphorylation analyses in clinically relevant systems.[25,68] As tyrosine kinases are mutated or altered in a variety of disease states, it is easy to imagine that the identification of the levels of phosphorylation and kinase activities within a system may give information pertaining to potential therapeutic targets for a disease. The use of tyrosine kinase inhibitors (TKIs) in solid tumors has been unsuccessful in stopping tumor growth, with tumor resistance almost invariably occurring. This leads to many remaining questions as to how this resistance is occurring given the specific targeting of the driver TK. For instance, despite the initial effectiveness of the TKI lapatinib against HER2-driven breast cancers, patients relapsed after treatment. To understand these mechanisms of TKI resistance, quantitative mass spectrometry approaches have been utilized. Lapatinib-resistant breast cancer cell lines were compared to their sensitive counterparts and phosphotyrosine immunoaffinity enrichment, followed by MS, uncovered the activation of Src family kinases in resistant cell lines. Further experiments to block Src kinase activity with inhibitors were found to restore sensitivity to lapatinib.[69] Moreover, tyrosine phosphorylation profiling led to the identification of Met receptor phosphorylation in EGFRvIII-driven glioblastoma cell lines. Combinatorial targeting of Met and EGFR led to the effective killing of cancer cells *in vitro*.[41] These data outline the utility of quantitative phosphotyrosine analyses for the identification of novel therapeutic strategies in disease.

The techniques for understanding phosphotyrosine signaling across biological systems outlined in this chapter allow the identification of activated signaling pathways that are critical for disease progression. The application of these approaches to human tumor samples holds the potential to enable the identification of signaling events critical for disease pathogenesis, allowing for the development of personalized therapy approaches based on signaling networks, rather than the genomic status exclusively.

References

1. T. Hunter, *Curr. Opinion Cell Biol.*, 1989, **1**, 1168.
2. T. Hunter, *Cell*, 1995, **80**, 225.
3. P. Blume-Jensen and T. Hunter, *Nature*, 2001, **411**, 355.
4. T. Pawson and P. Nash, *Genes Develop.*, 2000, **14**, 1027.
5. T. Hunter and B. M. Sefton, *Proc. Natl Acad. Sci.*, 1980, 77, 1311.
6. W. Kolch and A. Pitt, *Nature Rev. Cancer*, 2010, **10**, 618.

7. M. B. Yaffe, *Nature Rev. Mole. Cell Biol.*, 2002, **3**, 177.
8. M. Natarajan, K.-M. Lin, R. C. Hsueh, P. C. Sternweis and R. Ranganathan, *Nature Cell Biol.*, 2006, **8**, 571.
9. B. Kaminska, *Biochim. Biophys. Acta*, 2005, **1754**, 253.
10. D. Hanahan and R. A. Weinberg, *Cell*, 2000, **100**, 57.
11. M. F. White, *Canadian J. Physiol. Pharmacol.*, 2006, **84**, 725.
12. P. A. Futreal, A. Kasprzyk, E. Birney, J. C. Mullikin, R. Wooster and M. R. Stratton, *Nature*, 2001, **409**, 850.
13. M. R. Stratton, P. J. Campbell and P. A. Futreal, *Nature*, 2009, **458**, 719.
14. S. Roychowdhury, M. K. Iyer, D. R. Robinson, R. J. Lonigro, Y.-M. Wu, X. Cao, S. Kalyana-Sundaram, L. Sam, O. A. Balbin, M. J. Quist, T. Barrette, J. Everett, J. Siddiqui, L. P. Kunju, N. Navone, J. C. Araujo, P. Troncoso, C. J. Logothetis, J. W. Innis, D. C. Smith, C. D. Lao, S. Y. Kim, J. S. Roberts, S. B. Gruber, K. J. Pienta, M. Talpaz and A. M. Chinnaiyan, *Science Translational Medicine*, 2011, **3**, 111ra121.
15. B. J. Druker, S. Tamura, E. Buchdunger, S. Ohno, G. M. Segal, S. Fanning, J. Zimmermann and N. B. Lydon, *Nature Medicine*, 1996, **2**, 561.
16. C. K. Tang, X.-Q. Gong, D. K. Moscatello, A. J. Wong and M. E. Lippman, *Cancer Research*, 2000, **60**, 3081.
17. T. Pawson, *Eur. J. Cancer*, 2002, **38**, S3.
18. A. Pandey, A. V. Podtelejnikov, B. Blagoev, X. R. Bustelo, M. Mann and H. F. Lodish, *Proc. Natl Acad. Sci.*, 2000, **97**, 179.
19. A. H. Ross, D. Baltimore and H. N. Eisen, *Nature*, 1981, **294**, 654.
20. J. Rush, A. Moritz, K. A. Lee, A. Guo, V. L. Goss, E. J. Spek, H. Zhang, X.-M. Zha, R. D. Polakiewicz and M. J. Comb, *Nature Biotechnol.*, 2005, **23**, 94.
21. S. Lemeer, R. Ruijtenbeek, M. W. H. Pinkse, C. Jopling, A. J. R. Heck, J. den Hertog and M. Slijper, *Mol. Cell. Proteomics*, 2007, **6**, 2088.
22. A. R. Salomon, S. B. Ficarro, L. M. Brill, A. Brinker, Q. T. Phung, C. Ericson, K. Sauer, A. Brock, D. M. Horn, P. G. Schultz and E. C. Peters, *Proc. Natl Acad. Sci.*, 2003, **100**, 443.
23. B. Blagoev, S. E. Ong, I. Kratchmarova and M. Mann, *Nature Biotechnol.*, 2004, **22**, 1139.
24. Y. Zhang, A. Wolf-Yadlin, P. L. Ross, D. J. Pappin, J. Rush, D. A. Lauffenburger and F. M. White, *Mol. Cell. Proteomics*, 2005, **4**, 1240.
25. H. Johnson, A. M. Del Rosario, B. D. Bryson, M. A. Schroeder, J. N. Sarkaria and F. M. White, *Mol. Cell. Proteomics.*, 2012, **11**, 1724.
26. P. Nollau and B. J. Mayer, *Proc. Natl Acad. Sci.*, 2001, **98**, 13531.
27. K. Machida, C. M. Thompson, K. Dierck, K. Jablonowski, S. Karkkainen, B. Liu, H. Zhang, P. D. Nash, D. K. Newman, P. Nollau, T. Pawson, G. H. Renkema, K. Saksela, M. R. Schiller, D.-G. Shin and B. J. Mayer, *Molecular Cell*, 2007, **26**, 899.
28. B. Blagoev, I. Kratchmarova, S.-E. Ong, M. Nielsen, L. J. Foster and M. Mann, *Nature Biotechnol.*, 2003, **21**, 315.

29. D. F. Hunt, J. R. Yates, J. Shabanowitz, S. Winston and C. R. Hauer, *Proc. Natl Acad. Sci.*, 1986, **83**, 6233.
30. M. Salek, A. Alonso, R. Pipkorn and W. D. Lehmann, *Anal. Chem.*, 2003, **75**, 2724.
31. J. V. Olsen, B. Macek, O. Lange, A. Makarov, S. Horning and M. Mann, *Nature Methods*, 2007, **4**, 709.
32. H. Steen, B. Kuster, M. Fernandez, A. Pandey and M. Mann, *Anal. Chem.*, 2001, **73**, 1440.
33. C. K. Frese, A. F. M. Altelaar, M. L. Hennrich, D. Nolting, M. Zeller, J. Griep-Raming, A. J. R. Heck and S. Mohammed, *J. Proteome Res.*, 2011, **10**, 2377.
34. D. N. Perkins, D. J. C. Pappin, D. M. Creasy and J. S. Cottrell, *Electrophoresis*, 1999, **20**, 3551.
35. A. Ducret, I. V. Oostveen, J. K. Eng, J. R. Yates and R. Aebersold, *Protein Science*, 1998, 7, 706.
36. S.-E. Ong, B. Blagoev, I. Kratchmarova, D. B. Kristensen, H. Steen, A. Pandey and M. Mann, *Mol. Cell. Proteomics*, 2002, **1**, 376.
37. M. Kruger, M. Moser, S. Ussar, I. Thievessen, C. A. Luber, F. Forner, S. Schmidt, S. Zanivan, R. Fassler and M. Mann, *Cell*, 2008, **134**, 353.
38. A. Westman-Brinkmalm, A. Abramsson, J. Pannee, C. Gang, M. K. Gustavsson, M. von Otter, K. Blennow, G. Brinkmalm, H. Heumann and H. Zetterberg, *J. Proteomics*, 2011, **75**, 425.
39. P. L. Ross, Y. N. Huang, J. N. Marchese, B. Williamson, K. Parker, S. Hattan, N. Khainovski, S. Pillai, S. Dey, S. Daniels, S. Purkayastha, P. Juhasz, S. Martin, M. Bartlet-Jones, F. He, A. Jacobson and D. J. Pappin, *Mol. Cell. Proteomics*, 2004, **3**, 1154.
40. A. Thompson, J. Schafer, K. Kuhn, S. Kienle, J. Schwarz, G. Schmidt, T. Neumann, R. Johnstone, A. K. Mohammed and C. Hamon, *Anal. Chem.*, 2003, **75**, 1895.
41. P. H. Huang, A. Mukasa, R. Bonavia, R. A. Flynn, Z. E. Brewer, W. K. Cavenee, F. B. Furnari and F. M. White, *Proc. Natl Acad. Sci.*, 2007, **104**, 12867.
42. A. Wolf-Yadlin, N. Kumar, Y. Zhang, S. Hautaniemi, M. Zaman, H.-D. Kim, V. Grantcharova, D. A. Lauffenburger and F. M. White, *Mol. Syst. Biol.*, 2006, **2**, 54.
43. P. Pichler, T. Köcher, J. Holzmann, M. Mazanek, T. Taus, G. Ammerer and K. Mechtler, *Anal. Chem.*, 2010, **82**, 6549.
44. J.-L. Hsu, S.-Y. Huang, N.-H. Chow and S.-H. Chen, *Anal. Chem.*, 2003, **75**, 6843.
45. P. J. Boersema, L. Y. Foong, V. M. Y. Ding, S. Lemeer, B. van Breukelen, R. Philp, J. Boekhorst, B. Snel, J. den Hertog, A. B. H. Choo and A. J. R. Heck, *Mol. Cell. Proteomics*, 2010, **9**, 84.
46. K. Schmelzle, S. Kane, S. Gridley, G. E. Lienhard and F. M. White, *Diabetes*, 2006, **55**, 2171.
47. J. V. Olsen, B. Blagoev, F. Gnad, B. Macek, C. Kumar, P. Mortensen and M. Mann, *Cell*, 2006, **127**, 635.

48. C. Jorgensen, A. Sherman, G. I. Chen, A. Pasculescu, A. Poliakov, M. Hsiung, B. Larsen, D. G. Wilkinson, R. Linding and T. Pawson, *Science*, 2009, **326**, 1502.
49. C. Leroy, C. Fialin, A. Sirvent, V. Simon, S. Urbach, J. Poncet, B. Robert, P. Jouin and S. Roche, *Cancer Research*, 2009, **69**, 2279.
50. N. Osinalde, H. Moss, O. Arrizabalaga, M. J. Omaetxebarria, B. Blagoev, A. M. Zubiaga, A. Fullaondo, J. M. Arizmendi and I. Kratchmarova, *J. Proteomics*, 2011, **75**, 177.
51. H. J. Jun, H. Johnson, R. T. Bronson, S. de Feraudy, F. White and A. Charest, *Cancer Research*, 2012, **72**, 3764.
52. E. Ciccimaro, S. K. Hanks, K. H. Yu and I. A. Blair, *Anal. Chem.*, 2009, **81**, 3304.
53. V. Mayya, K. Rezual, L. Wu, M. B. Fong and D. K. Han, *Mol. Cell. Proteomics*, 2006, **5**, 1146.
54. W. M. Old, K. Meyer-Arendt, L. Aveline-Wolf, K. G. Pierce, A. Mendoza, J. R. Sevinsky, K. A. Resing and N. G. Ahn, *Mol. Cell. Proteomics*, 2005, **4**, 1487.
55. K. Rikova, A. Guo, Q. Zeng, A. Possemato, J. Yu, H. Haack, J. Nardone, K. Lee, C. Reeves, Y. Li, Y. Hu, Z. Tan, M. Stokes, L. Sullivan, J. Mitchell, R. Wetzel, J. MacNeill, J. M. Ren, J. Yuan, C. E. Bakalarski, J. Villen, J. M. Kornhauser, B. Smith, D. Li, X. Zhou, S. P. Gygi, T.-L. Gu, R. D. Polakiewicz, J. Rush and M. J. Comb, *Cell*, 2007, **131**, 1190.
56. F. Hochgrafe, L. Zhang, S. A. O'Toole, B. C. Browne, M. Pinese, A. Porta Cubas, G. M. Lehrbach, D. R. Croucher, D. Rickwood, A. Boulghourjian, R. Shearer, R. Nair, A. Swarbrick, D. Faratian, P. Mullen, D. J. Harrison, A. V. Biankin, R. L. Sutherland, M. J. Raftery and R. J. Daly, *Cancer Research*, 2010, **70**, 9391.
57. Y. Bai, J. Li, B. Fang, A. Edwards, G. Zhang, M. Bui, S. Eschrich, S. Altiok, J. Koomen and E. B. Haura, *Cancer Research*, 2012, **72**, 2501.
58. J. M. Drake, N. A. Graham, T. Stoyanova, A. Sedghi, A. S. Goldstein, H. Cai, D. A. Smith, H. Zhang, E. Komisopoulou, J. Huang, T. G. Graeber and O. N. Witte, *Proc. Natl Acad. Sci.*, 2012, **109**, 1643.
59. S. C. Bendall, E. F. Simonds, P. Qiu, E. D. Amir, P. O. Krutzik, R. Finck, R. V. Bruggner, R. Melamed, A. Trejo, O. I. Ornatsky, R. S. Balderas, S. K. Plevritis, K. Sachs, D. Pe'er, S. D. Tanner and G. P. Nolan, *Science*, 2011, **332**, 687.
60. B. Bodenmiller, E. R. Zunder, R. Finck, T. J. Chen, E. S. Savig, R. V. Bruggner, E. F. Simonds, S. C. Bendall, K. Sachs, P. O. Krutzik and G. P. Nolan, *Nature Biotechnol.*, 2012, **30**, 858.
61. N. Kumar, A. Wolf-Yadlin, F. M. White and D. A. Lauffenburger, *PLoS Computational Biology*, 2007, **3**, e4.
62. P. H. Huang, E. R. Miraldi, A. M. Xu, V. A. Kundukulam, A. M. Del Rosario, R. A. Flynn, W. K. Cavenee, F. B. Furnari and F. M. White, *Mol. Biosystems*, 2010, **6**, 1227.
63. H. Johnson, R. S. Lescarbeau, J. A. Gutierrez and F. M. White, *J. Proteome Res.*, 2013, **12**, 1856.

64. H. Mi and P. Thomas, *Protein Networks and Pathway Analysis*, ed. Y. Nikolsky and J. Bryant, Humana Press, New York, 2009, **563**, 123.
65. M. B. Yaffe, G. G. Leparc, J. Lai, T. Obata, S. Volinia and L. C. Cantley, *Nature Biotechnol.*, 2001, **19**, 348.
66. D. Schwartz and S. P. Gygi, *Nature Biotechnol.*, 2005, **23**, 1391.
67. K. M. Naegle, M. Gymrek, B. A. Joughin, J. P. Wagner, R. E. Welsch, M. B. Yaffe, D. A. Lauffenburger and F. M. White, *Mol.Cell. Proteomics*, 2010, **9**, 2558.
68. N. C. Tedford, A. B. Hall, J. R. Graham, C. E. Murphy, N. F. Gordon and J. A. Radding, *Proteomics*, 2009, **9**, 1469.
69. B. N. Rexer, A. J. Ham, C. Rinehart, S. Hill, M. Granja-Ingram Nde, A. M. Gonzalez-Angulo, G. B. Mills, B. Dave, J. C. Chang, D. C. Liebler and C. L. Arteaga, *Oncogene*, 2011, **30**, 4163.

CHAPTER 11

Next Generation Proteomics: PTMs in Space and Time

DALILA BENSADEK, ARMEL NICOLAS AND ANGUS I. LAMOND*

Centre for Gene Regulation & Expression, College of Life Sciences, University of Dundee, Dow Street, Dundee, DD1 5EH, United Kingdom
*Email: a.i.lamond@dundee.ac.uk

11.1 Introduction

Proteomics has rapidly evolved from the detection and cataloguing of the protein complement of either a cell, sub-cellular organelle or complex,[1] through the quantitative characterisation of proteins and their interactions, to now encompass the comprehensive, large-scale quantitative study of protein dynamics. This includes measuring changes in the complex pattern of post-translational modifications (PTMs) in space and time. We will refer to this new state of the art as "Next Generation" proteomics, reflecting the current depth and detail of analysis that is possible. We anticipate that the application of these new proteomics methods will revolutionise our understanding of cellular function.

Here, we will review methods for the large-scale, spatial and temporal quantitative analysis of some of the most commonly studied PTMs in mammalian cells, including phosphorylation, acetylation and ubiquitinylation. We will focus our discussion on the use of mass spectrometry (MS) combined with stable isotopic labelling for the quantification of proteins and their PTMs, because this has been used so extensively in recent cell

New Developments in Mass Spectrometry No. 1
Quantitative Proteomics
Edited by Claire Eyers and Simon J Gaskell

Published by the Royal Society of Chemistry, www.rsc.org

biology and cell signalling studies, and we will contrast this with label-free methods. We present here an example of an optimised workflow, from sample preparation, through sub-cellular fractionation and liquid chromatography (LC), to data acquisition using high resolution and high mass accuracy Fourier transform mass spectrometry (FT-MS). An integral part of this "Next Generation" workflow includes methods for the efficient storage, analysis, visualisation and sharing of the resulting data. Methods will be described for maximising the protein properties that can be measured in a single experiment by combining appropriate sample preparation, data acquisition and analytical methods. Examples will be shown that illustrate the detection of PTMs and show how these data can be correlated with other measured protein properties, including sub-cellular localisation, in system-wide studies covering a large fraction of the expressed cell proteome.

11.2 Next Generation Proteomics: Addressing the Challenges in Applying Global Proteomics to Cell Biology

In "Next Generation" proteomics we aim to carry out comprehensive proteomic analyses by identifying and characterising a major fraction of the proteins from a specific cell type or tissue, including isoforms and modified states, in a high throughput manner, while maintaining high sensitivity for the detection of low-copy number proteins. However, this task presents significant analytical challenges, including the high degree of protein complexity and the huge dynamic range of proteins expressed in such complex biological mixtures. Proteins in a total cell lysate span a wide range of concentrations, which can exceed six orders of magnitude in cells[2] and eight orders of magnitude in body fluids.[3] While, in principle, MS is effective for detecting proteins at any of these concentrations, the competition for ionisation makes it difficult to detect low-abundance analytes in the presence of more abundant proteins. Mass spectrometric analyses of complex biological samples are, therefore, limited usually by the preferential identification of the most abundant proteins in a sample fraction. Moreover, the complexity of biological samples and the number of peptides derived from them, in shotgun proteomics, can overwhelm even the most sensitive and fastest of current generation mass spectrometers.

It is, therefore, helpful for "Next Generation" proteomics to incorporate pre-fractionation methods to address these limitations by presenting the mass spectrometer with sub-proteomes of reduced complexity. A variety of sample preparation schemes at the level of either intact proteins, or peptides derived from protein digestion, including electrophoresis, chromatography and others, have been incorporated into the proteomics workflow to reduce the complexity of the protein samples. While pre-fractionation at the peptide level addresses the dynamic range issues and enhances the detection of low abundance proteins, thereby increasing the depth of the proteomics analysis, it results in the loss of valuable biological information because it is often not possible to distinguish which protein isoform a given peptide is derived from.[4] Moreover, despite successful reports using a single preparative method with LC-MS analysis, the combination of several approaches may be required for comprehensive proteomic analyses. Since the aim of "Next Generation" proteomics is a comprehensive characterisation of the proteome through space and time, we aim to preserve as much biological information as possible and to maximise the depth of the proteome coverage. Figure 11.1 illustrates the benefits of pre-fractionation in preserving

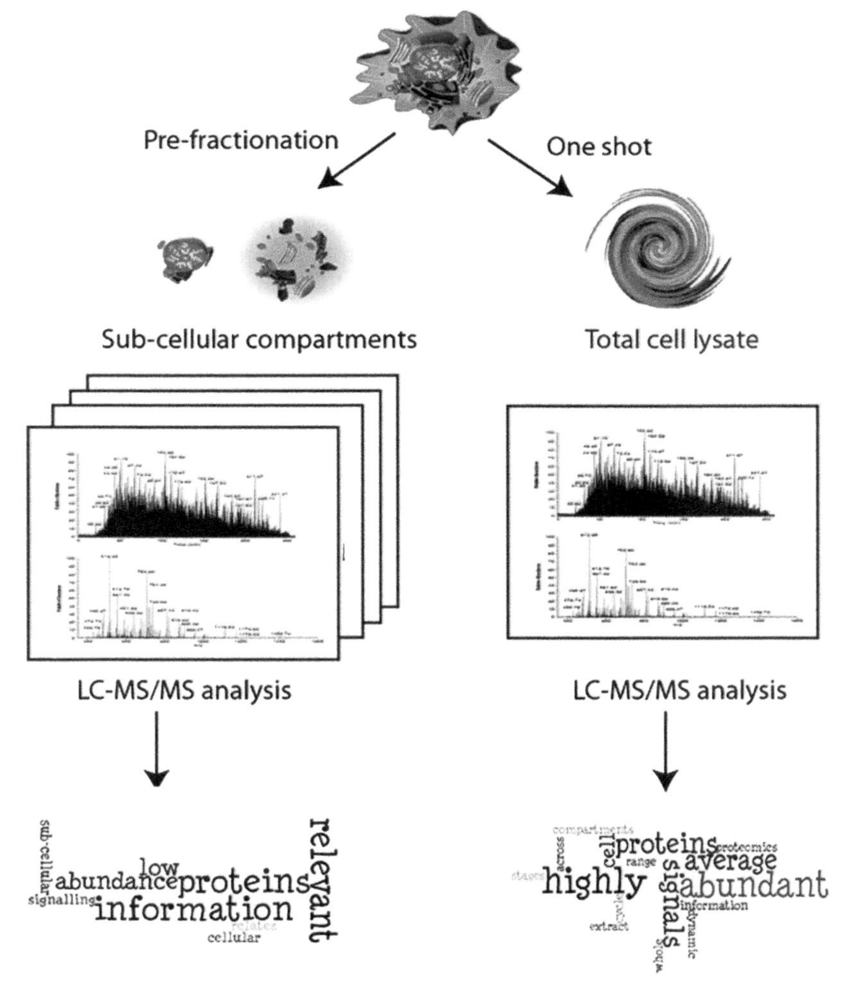

Figure 11.1 Comparison of pre-fractionation *versus* one-shot total lysate analysis highlighting the substantial increase in the depth of the proteome coverage and the relevant biological information obtained from pre-fractionation.

biological information. We have, therefore, favoured the combination of multiple pre-fractionation techniques during sample preparation, namely sub-cellular fractionation and size fractionation of intact proteins. Peptide-level fractionation is applied in specific cases in PTM analysis and will be discussed later.

11.2.1 Sample Preparation: Pre-fractionation

11.2.1.1 Sub-cellular Fractionation

The incorporation of sub-cellular fractionation techniques into the proteomics workflow is a powerful means of addressing the dynamic range issues in global proteomics by reducing sample complexity and thus allowing the detection of changes in low-abundance proteins and a variety of signalling complexes at the organelle level. More importantly, however, it underpins valuable multi-dimensional biological information that would otherwise be averaged across multiple cellular compartments. Finally, analysing the proteome of sub-cellular organelles provides additional information, either on proteins that shuttle between different cellular compartments, or proteins that change localisation in response to a given stimulus. Thus, sub-cellular fractionation is integral to our "Next Generation" proteomics methodology.

Sub-cellular fractionation consists of two major steps: i) disruption of cellular organisation through homogenisation and ii) fractionation of the homogenate to separate the different populations of organelles and cell structures. The most frequently used techniques are differential, rate-zonal and isopycnic gradient centrifugation, which separate organelles based on differences in their density, size and/or shape. Alternatively, proteins from different cellular compartments could be separated/enriched according to their different solubility characteristics by using a range of buffers with increasing protein solubilisation capacity—typically by increasing either detergent concentration or strength. Both approaches may be combined depending on the compartments of interest. The experimental conditions are flexible and adjustable (both in cells and tissues) and can be adapted for several organelles and cell types.

While there are established purification protocols for many organelles,[5,6] further optimisation is often required to maximise recovery and reproducibility of sub-cellular protein extracts from different cell types and to maximise compatibility with later steps in the proteomics workflow, for example, by altering buffer composition, salts and/or detergents. In many of our studies we have focused on isolating the nucleolus for comprehensive nucleolar proteomics.[7,8] However, it is clear that full understanding of the cell proteome requires more systematic fractionation rather than the analysis of any single organelle. Therefore, we have subsequently extended our approaches to include the characterisation of proteomes in the cytoplasm and nucleus, as well as the nucleolus.[9,10] More recently, our protocols have been expanded to allow efficient enrichment also of other sub-cellular compartments, thereby increasing the amount of biologically relevant data we can obtain from analysing the proteome.[11] Figure 11.2 is a schematic representation of our extended sub-cellular fractionation protocol.

It should be noted that achieving complete purity *via* fractionation is, in practice, virtually impossible due to inevitable leakage, combined with disruption and damage of structures when cells are broken open, resulting in cross contamination. Compartmental localisation of proteins in cells is also further complicated because many proteins—including sometimes putative organelle markers—are present in more than one organelle. However, fractions can still be obtained that are reproducibly enriched for specific sub-cellular compartments and structures and these can then be used to identify and characterise separate pools of proteins with distinct functions and properties. It is also possible to use computational tools to correlate the cross-fraction distribution pattern of protein intensities with that of known compartment markers to estimate their localisation, or even to apply

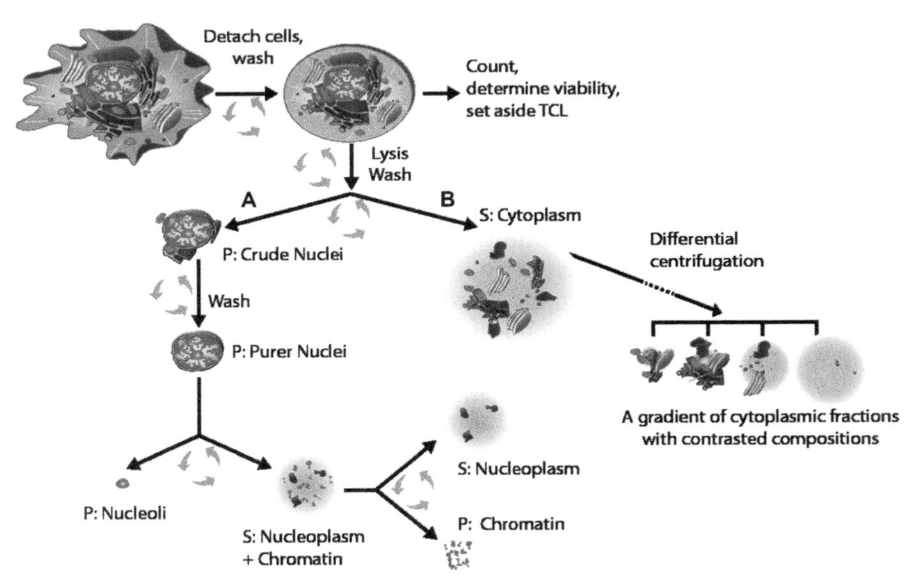

Figure 11.2 Outline of an extensive sub-cellular fractionation method optimised for cells grown in culture. Briefly, adherent cells are detached and washed with ice-cold phosphate-buffered saline (PBS) before mechanical lysis in an isotonic buffer. A) Crude nuclei are pelleted by centrifugation and washed to remove impurities before being broken by sonication. The nucleoli are separated from the nucleoplasm by centrifugation over a sucrose cushion. The nucleoplasm and chromatin are sequentially separated using buffers with increasing solubilisation capacities. B) The cytoplasm is subjected to differential centrifugation to yield several membrane pellets and a cytosolic supernatant. TCL denotes total cell lysate, S denotes supernatant and P denotes pellet.

a non-biased, "organelle discovery" approach to unveil new groupings of co-purifying proteins of potential interest (see Chapter 9 for further information).[12,13] Nonetheless, an issue that, at present, cannot be addressed by mass spectrometric approaches, which measure average values derived from populations of cells, is the fact that protein localisation patterns vary between individual cells, even within a theoretically clonal population. For example, this can arise as a result of protein variation at different cell cycle stages and under different growth conditions, as well as from stochastic effects. Therefore, caution should be used when interpreting the results of proteomic analyses in terms of the sub-cellular protein distribution at a single-cell level, because the proteomics data represent averaged values across cell populations.

11.2.1.2 Size Fractionation

One dimensional (1D-) and two dimensional (2D)-electrophoresis have been the traditional pre-fractionation partners of Liquid chromatography-mass spectrometry (LC-MS) from its early days. In a typical proteomics experiment the proteins from either a total cell lysate, or from sub-cellular fractions, are size-fractionated by 1D sodium dodecyl sulphate (SDS)-polyacrylamide gel electrophoresis (PAGE). The gel is then cut into multiple slices before digesting the proteins using either trypsin or other proteases. The peptide hydrolysates are then extracted from the gel, desalted and analysed by liquid chromatography-tandem MS (LC-MS/MS).[14,15] By combining the information obtained from analysing the raw MS data with the size information from SDS-PAGE, identified peptides can often be assigned to different isoforms and/or functionally distinct pools of the same proteins.[4]

While recognising the benefit of separating proteins based on size, as described above, SDS-PAGE followed by in-gel digestion and peptide extraction prior to LC-MS/MS analysis is laborious and, therefore, is not ideal as a convenient high throughput method for "Next Generation" proteomics.

Moreover, gel-based proteomics suffers from several limitations; namely, reduced sample capacity that can be resolved by SDS-PAGE (although this is dependent on the size of the gel), reproducibility of fractionation between replicates and material losses incurred during peptide extraction. In addition, we note that some proteases, *e.g.*, Glu-C, are not efficient when used for in-gel digestion (our unpublished observations). As an alternative, therefore, size exclusion chromatography (SEC) can be used, which provides higher capacity for size-dependent pre-fractionation. SEC separation can be carried out under denaturing conditions as a pre-fractionation tool; alternatively, native conditions that preserve protein complexes intact can also be used with buffers reflecting near-physiological ionic conditions. Furthermore, the stability of protein complexes and the strength of intra-complex interactions can be assessed by comparing fractionation profiles obtained as a result of using different buffer systems with varying ionic strengths and detergent concentrations.

11.2.1.3 Immunoprecipitation and Affinity Selection

Immunoprecipitation (IP) can be used as an orthogonal technique to enrich for specific proteins, complexes or modified forms of proteins. However, a protein of interest can interact with a wide range of protein partners, which bind non-stoichiometrically to generate a number of complexes with different affinities and specificities. Distinguishing specific, but low-abundance, interacting proteins from the larger number of low-affinity, but abundant, contaminant proteins that are inevitably recovered using commonly used methods, such as pull-down or IP strategies, can therefore be difficult. Another complicating factor in comparing data between IP experiments can be differences in the affinities and specificities of different antibody reagents. Many antibodies do not discriminate efficiently between protein isoforms, while PTMs can also alter antibody interactions with protein substrates.

A key goal in most areas of cell biology is the characterisation of the protein components of multi-protein complexes through the reliable discrimination of specific protein interaction partners from non-specific interactions. One approach for such discrimination is to use stable-isotope labelling by amino acids in cell culture (SILAC) combined with a negative control, which provides an objective measure of background, *i.e.*, contaminant proteins that bind non-specifically. We have previously performed a systematic assessment of proteins in human cell extracts that bind non-specifically to different types of affinity supports termed the "bead proteome."[16] However, this bead proteome related most directly to the specific experimental conditions used for these measurements. In a more general approach to assessing non-specific protein binding we performed an integrative data analysis that incorporated results from a large number of independent SILAC pull-down experiments performed either in whole cell, cytoplasmic or nuclear extracts prepared from different cell lines. This allowed the identification of a wide range of cellular proteins that routinely bind to the affinity matrix and that must, therefore, be regarded as potential non-specific interactors/contaminants whenever they are identified in protein interaction studies.[16] When comparing a large number of data sets, common contaminants are detected much more often than interaction partners that are truly specific for a given bait protein. This led to the development of the Protein Frequency Library (PFL),[17] a probability-based annotation for determining the likelihood of a protein being a real interaction partner in an IP experiment based upon the frequency at which it is detected over multiple experiments. The PFL has been incorporated into the PepTracker® environment (see section 11.5), where it can be applied *via* a user-friendly interface to filter out non-specific interactors for bait proteins tested in pull-down experiments (http://www.peptracker.com/datavisual/).

Similar approaches for distinguishing contaminants from *bona fide* interactors in immunopurifications have incorporated quantitative proteomics using metabolic labelling[18] combined with tagging[19] and/or control of expression levels.[20]

11.3 Quantitative Proteomics

The development of fast and sensitive high resolution mass spectrometers, together with efficient chromatographic separation procedures, has allowed unprecedented detection of large numbers of proteins in complex biological samples. The aim of these studies is often not only to identify thousands of proteins in a given cell type or tissue, but also to quantify dynamic changes in expression levels and/or in PTMs under different conditions. Quantitative information has, thus, become an integral part of proteomics. Several quantitative methods, such as SILAC,[21] isotope coded affinity tags (ICAT),[22] isobaric tags for relative and absolute quantitation (iTRAQ),[23] tandem mass tags (TMT)[24] and dimethyl labelling[25] have been developed to explore the dynamics of whole proteomes and provide relative comparisons of protein abundances under different conditions. Although the methods above may differ in their process, they are all based on the concept of comparing protein levels between samples by virtue of differential incorporation of unique stable isotopes into all the cellular proteins from each individual sample. In recent years technological advances in chromatography systems, as well as in computational and data analysis tools, have added label-free approaches to the arsenal of quantitative proteomics tools. In the following section we will review approaches used for relative quantitative proteomics.

11.3.1 Metabolic Labelling

Metabolic labelling requires the incorporation of the isotopic labels, which are typically stable isotopes of C, N and/or H, in the growth media during cellular metabolism and protein synthesis. Early reports of successful metabolic labelling[26,27] in MS-based proteomics exploited methods that have been long used by structural biologists to express stable-isotope-labelled proteins for NMR. Mann and co-workers subsequently introduced SILAC to the proteomics community.[21,28] This approach, applicable to both eukaryotic and prokaryotic cell cultures and also to model organisms, uses stable-isotope-labelled amino acids (usually arginine and/or lysine) that contained combinations of ^{13}C, ^{15}N and ^{2}H in the culture media. These labelled amino acids eventually become fully incorporated into the proteome of the cell population after prolonged growth.[21,28] With the exception of deuterium-containing amino acids, the stable-isotope-labelled amino acids generally exhibit the same chromatographic and mass spectrometric behaviour as their natural-abundance counterparts, but can be easily distinguished by MS due to their mass differences. The ratio of the ion intensities of each of the isotopic peaks corresponding to the same peptide, which can be determined accurately, provides a direct measure of the relative amounts of the respective proteins present in the separately labelled cell populations. This intensity ratio can thus be measured separately for each peptide identified from a given protein and the accuracy of the final value estimated for the cognate protein is consequently improved by averaging over all the separate peptides assigned to that protein.

In theory any amino acids can be used to label the proteome in the SILAC strategy. However, since trypsin has, to date, been the protease of choice in shotgun proteomics, using stable-isotope derivatives of arginine and lysine is optimal as it ensures that every tryptic peptide contains at least one labelled residue and can thus be used for quantification. This is not to say that other amino acids cannot be used. Indeed there are reports of using stable-isotope-substituted derivatives of leucine,[29] valine[30] and tyrosine[31] (in addition to lysine and arginine) to label proteins. Metabolic labelling using amino acids incorporating heavy isotopes has also been extended to whole model organisms, such as nematodes,[32,33] drosophila,[32] chicken[30,34,35] and mice.[36]

11.3.2 Chemical Labelling

While SILAC provides a powerful experimental strategy in many circumstances, metabolic labelling may not always be a feasible option, especially when working with either clinical samples or tissues isolated from larger organisms. One way of overcoming the necessity of having isotopically labelled reference material, particularly for clinical proteomics, is to create a reference proteome from different SILAC-labelled cultured cells, which represent either the cell types or tissues under study. In this approach, known as "super SILAC", Geiger and co-workers labelled four human breast cancer-derived cell lines and five astrocytoma/glioblastoma-derived cell lines to use as reference standards for quantitative analysis of clinical samples.[37,38] However, it is not always possible to create reference standards that accurately reflect the proteome of the clinical samples.

Chemical labelling has also been utilised as a method of quantifying proteins from either tissues or cell extracts, which are either isolated or cultured under conditions not amenable to metabolic labelling. To do this, protein extracts from the samples to be compared are typically prepared using a proteomics pipeline similar to that previously described, and the peptides (usually tryptic) are differentially labelled post cleavage using natural-abundance and stable isotopically labelled chemical modifiers. In the following section we will briefly review some chemical derivatisation techniques, which have been reported in the literature.

11.3.2.1 MS-level Quantification

Isotope-coded affinity tag (ICAT) reagents[22] consist of three functional elements: a reactive group for derivatising the sulfhydryl group of cysteine residues, an isotopically coded linker and an affinity tag to enrich the labelled peptides. The two different protein populations to be compared are digested and differentially derivatised using ^{12}C- and ^{13}C-containing ICAT reagents; equal amounts of digested protein are then mixed and affinity purified. The peptides are then quantified by measuring, in the MS mode, the relative signal intensities for pairs of peptide ions of identical sequence, which originate from the different samples. A clear disadvantage of the ICAT technology, however, is the fact that it limits quantification to cysteine-containing peptides. Cysteine is one of the least abundant amino acids in the proteome and, while this strategy was originally introduced to minimise sample complexity, quantification using only cysteine-containing peptides not only reduces the accuracy of the measurement per protein, since fewer peptides per protein are analysed, but also reduces the number of proteins that can be quantified from a complex mixture.

Guanidination of peptides was initially introduced to convert lysine residues to homo-arginine to improve their detection in matrix-assisted laser desorption ionisation (MALDI)-MS by increasing their gas phase basicity.[39] However, this strategy was later used also for relative quantification. By carrying out the derivatisation reaction using ^{14}N- and ^{15}N-labelled *O*-methyl isourea, it is possible to differentially label analytes for comparison.[40] As the use of trypsin digestion alone would result in a loss of quantitative information from arginine-containing tryptic peptides and, hence, reduces the number of proteins that can be quantified, guanidination is therefore better coupled with Lys-C digestion.

Dimethyl labelling is probably the least expensive and the most straightforward stable-isotope labelling strategy for quantitative proteomics.[25,41] It uses formaldehyde to label both the *N*-termini of peptides, as well as the ε-amino groups of lysine residues through reductive amination. The reaction takes place in less than five minutes without any detectable by-products.[25] By incorporating ^{13}C and ^{2}H into formaldehyde, peptides from separate pools of proteins can be differentially labelled and quantified in the MS analysis. Dimethyl labelling thus appears to have all the attributes of the ideal chemical labelling method. However, it has not yet been extensively used and validated by the wider proteomics community due, in part, to the slow incorporation of the method into widely available proteomics software used for quantification. It is also possible that some scientists are reluctant to use this method because hydrogen cyanide is released as a by-product of the labelling reaction.

11.3.2.2 Tandem MS-level Quantification

Isobaric tag for relative and absolute quantitation (iTRAQ)[23] and tandem mass tag (TMT)[24] based methods allow the relative quantification of multiple samples at the tandem MS stage (see Chapter 3). Currently, up to eight samples can be quantified using iTRAQ and six using TMT. In this technique the samples to be compared are first digested, then differentially labelled using either iTRAQ or TMT and finally combined for MS analysis. The MS-level analysis cannot distinguish the differentially labelled peptides because the incorporated tags are isobaric. However, when subjected to collisional activation, the peptides dissociate, giving rise to different reporter ions depending on which "flavour" of isobaric tag was used for labelling. The ratios of these different tags are used as surrogate measurements for the relative peptide abundance and, thus, relative quantification of the protein from which they derive. While this approach has its advantages, the low *m*/*z* values of the reporter ions makes iTRAQ and TMT best suited for beam instruments, because the reporter ions will not be observed in ion traps. This is, in part, due to the low mass cut-off of such instruments, also known as "the one-third rule",[42] but also due to the differences in collisional dissociation between ion traps and beam-type instruments. New fragmentation techniques were introduced to overcome this limitation and include: i) pulsed Q dissociation (PQD), which was developed by Thermo Fisher Scientific and implemented exclusively on their linear ion trap mass spectrometers.[43,44] This patented technique generates MS/MS spectra that are similar to those obtained by collision-induced dissociation (CID), while retaining the *m*/*z* fragments that are typically excluded from CID spectra. It was shown to offer similar quantification performance compared with a quadrupole time-of-flight Q-ToF instrument.[45,46] This technique suffers from poor fragmentation efficiency and was later superseded by ii) higher energy C-trap dissociation (HCD), which yields fragment ions similar to those observed in beam instruments and, therefore, can be used in hybrid orbitrap instruments for iTRAQ/TMT-based quantification.

The main disadvantage of isobaric tag-based quantification is that it struggles to produce reliable relative protein abundance estimates and has inherent problems with precision and accuracy especially when applied to complex samples.[47] The precision of the measurements is correlated with reporter-ion signal intensity; the precision of measurements decreases as the signal-to-noise ratio of the reporter ions decreases. Several informatics and statistical approaches have been introduced to address the precision issues.[47,48] Potentially, a larger problem with iTRAQ/TMT is its poor accuracy; equimolar reporter ratios are quantified with reasonable accuracy but, as the expected fold change increases, the reporter ion ratios become increasingly underestimated, resulting in the compression of reporter ion ratios toward one.[46,49–51] This arises, at least in part, from co-eluting peptides with similar *m*/*z* values, which are co-selected during ion isolation and fragmented during CID. As the majority of these contaminants will be at an equimolar ratio across the reporter ion tags, they will contribute an equal background value to the reporter ion signals, thus reducing the calculated ratios. Additional information on iTRAQ/TMT quantification strategies can be found in Chapter 3.

11.3.3 Label-free Quantification

Most relative quantification techniques include isotopic labelling of the samples and, therefore, require additional sample preparation steps, which can result in experimental variability. Label-free LC-MS quantification methods have been introduced to determine the relative abundance of proteins in different samples, either isolated from cells or organisms grown under different conditions, and this, as the name suggests, is carried out without the need for either metabolic or chemical labels. The idea followed from the observation that the electrospray process gives rise to a signal response that is linearly correlated with increasing concentration of the analytes. There are two main ways of doing label-free quantification: i) spectral counting (SC) and ii) peak measurements (see also Chapter 6 for further information). An overview of the different methods for quantification is presented in Figure 11.3.

In SC the total number of peptide spectra matching a protein is taken as an indicator of the protein abundance.[52,53] The aim of SC is thus to quantify protein abundances by counting the number of detected tryptic peptides and their corresponding mass spectra. However, this approach may bias protein abundance in favour of larger proteins if the number of peptide matches is the only parameter considered, because digesting larger proteins will produce more peptides and, therefore, more MS/MS spectra. Normalisation strategies are, therefore, needed to account for protein sequence lengths (number of tryptic peptides) and the ionisation efficiency of different peptides. It should also be noted that the method relies on the stochastic nature of MS/MS and is, therefore, incompatible with applying dynamic exclusion during data acquisition.

We will focus here on methods for measuring the relative amounts of peptides (and thereby proteins) by direct evaluation of peak measurements (peak intensities and peak areas) as this has been the method of choice in our laboratory whenever SILAC is not possible.

In contrast to isotopic labelling in which the differentially labelled lysates are mixed, an obvious issue with label-free approaches is the experimental error arising from the fact that the samples to be compared are processed separately. This adds an additional degree of freedom to the accuracy of measurement because of the chances of variation between samples. The LC-MS analysis of each sample is also carried out separately and only at the final stage are the LC chromatograms overlaid and quantified. Other potential sources of variation, and thus experimental error, include unequal sample loading on the reverse-phase (RP) column and run-to-run variation in the LC-MS/MS process. Sample preparation for label-free quantification is, therefore, of the utmost importance and protocols should be streamlined as much as possible to ensure that the samples are processed identically. Methods for determining peptide loading on the RP column are equally important both for maintaining optimal chromatographic resolution and for the subsequent quantification steps. Finally, since peak measurement quantification relies on overlaying the chromatographic profiles of the samples being compared, robust chromatography with outstanding performance and reproducibility is a pre-requisite for label-free techniques. High resolution nano-LC systems, "splitless" nano-flow pumps, where temperature and pressure can be effectively controlled, high resolution columns with optimal packing material, minimal dead-volume connections

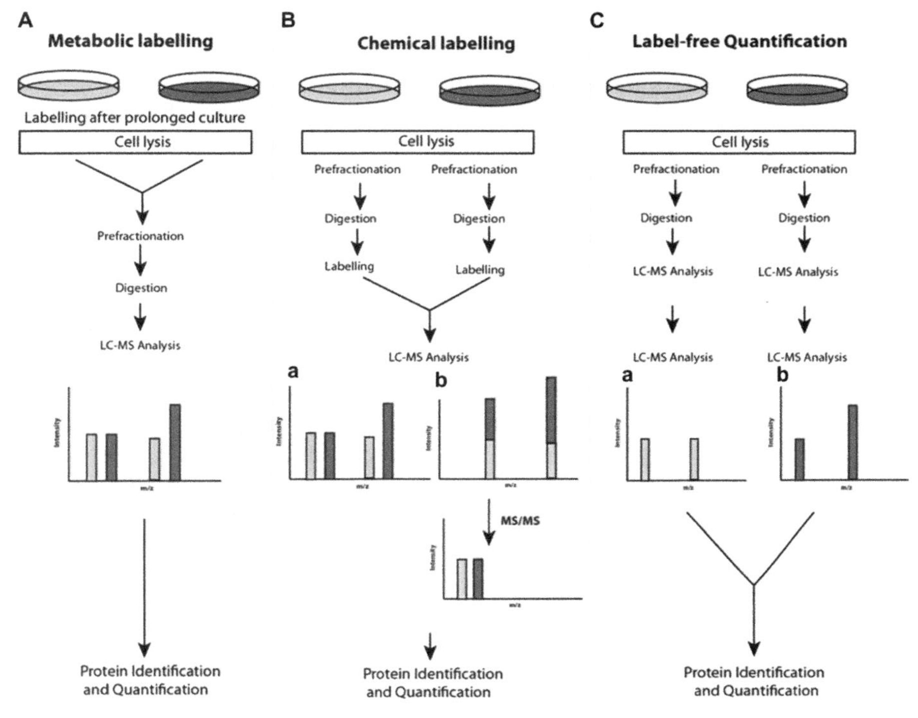

Figure 11.3 Summary of the main methods for quantitative proteomics. A) In metabolic labelling (SILAC) the analytes are differentially labelled by incorporating the different isotopes in the culture media (different shades of grey). After cell lysis, the proteins are pre-fractionated, digested and analysed by LC-MS. The proteins are quantified by measuring the relative ratios of their peptides at the MS level (dark grey *vs* light grey bars). B) In Chemical labelling, protein extracts from the samples to be compared are prepared and digested. The peptides, usually tryptic, are differentially labelled post cleavage using natural-abundance and stable-isotopically labelled chemical tags. There are two types of chemical labelling: a) chemical modifiers that afford quantification at the MS level, such as ICAT and dimethyl labelling, where the peptides originating from the different analytes have different *m*/*z* values and their ratios can be measured in the MS scan and used as a measure for relative protein quantification; b) chemical modifiers that allow quantification at the MS/MS level, *i.e.*, iTRAQ and TMT. The differentially labelled peptides are not separated in the MS1 scan (shown here as vertically stacked light and dark grey bars) and only give rise to separate reporter ions after dissociation in the MS2 scan. The ratio of these reporter ions is used to quantify the peptides and the proteins they originated from. C) In label-free experiments the protein extracts to be compared are processed and digested separately. The different proteolytic digests (different shades of grey) are analysed by LC/MS separately but in an identical manner (a and b). The quantification is performed as the last step at the data-analysis level.

and, of course, optimal sample loading that avoid saturation of the columns and do not adversely affect the chromatography, are probably the most important parameters in a label-free experiment.

After data acquisition, computational methods are required to align the chromatograms to correct for any shifts in retention times and perform background subtraction and intensity normalisation to improve the accuracy of the quantification. Reproducibility is particularly important in label-free methods, since increased variability is introduced by analysing samples in parallel. In principle, normalisation can also correct for differences in peptide loading on the RP-LC column but should not be relied upon. It is important to mention that the quantification in label-free methods is carried out at the MS survey scan level, while the MS/MS is used to determine the sequence of the eluting peptides. It is, therefore, important to maintain a balance between peptide identification and peptide quantification. This is best achieved with high-duty cycle mass spectrometers, such as fast scanning ion traps (LTQ or the Q Exactive instrument from Thermo Scientific). The most obvious advantage of label-free quantification is the theoretically unlimited number of samples that can be compared and quantified.

11.4 Post-translational Modifications (PTMs)

After protein synthesis, amino acid side chains can be subjected to a wide array of PTMs that can alter their enzymatic activity, binding ability, interacting partners and localisation. In order to fully understand cellular function and cellular signalling special attention should be paid to such regulatory PTMs. The systematic identification and profiling of PTMs poses a significant analytical challenge due to their sub-stoichiometric nature and, in some cases, their instability, which may result in their loss during sample preparation.[54] Here, we will consider protein phosphorylation as an example for the PTM characterisation workflow, although the conclusions are generally applicable to other PTMs as well. Phosphorylation is the most well-studied PTM analysed in proteomics, in part due to its importance in cell regulation and signalling, but also due to the availability of sensitive methods for its detection both *ex vivo* and *in vitro*.

The reversible phosphorylation of proteins on serine, threonine and tyrosine residues (*O*-phosphorylation) is known to be involved in regulating major cellular processes through highly dynamic and complex signalling pathways in eukaryotes. Early estimates suggest that the human genome encodes about 400 Ser/Thr protein kinases and that one third of proteins are phosphorylated at some point in their life cycle, representing >100 000 different phosphorylation sites.[55] Current proteomics methods have achieved the identification of around 19 000 phosphorylation sites in replicate studies of embryonic stem cell (ESC) and induced pluripotent stem cell (iPSC) lines using a combination of dissociation techniques,[56] highlighting the need for high throughput and sensitive methods to identify, characterise and monitor new sites of protein phosphorylation.

In addition, phosphorylation has been reported to also occur on arginine, lysine and histidine residues (*N*-phosphorylation).[54,57,58] While it has been suggested that histidine phosphorylation could represent ~6% of total phosphorylation in eukaryotic cells, its detection is rather challenging due to the rapid hydrolysis of the phospho-amide bond under the acidic conditions typically employed for proteomics sample preparation and analysis. In comparison, phosphoester bonds are relatively more stable and are by far the most studied type of phosphorylation and will be the focus of the following section.

High throughput global proteomics analyses have identified large numbers of phosphorylated peptides in different cell types, thus highlighting the extent of phosphorylation taking place in total cell lysates and the diversity of phosphorylated proteins. Phosphorylation was initially estimated to be distributed in the ratio of (90 : 10 : 0.05; Ser : Thr : Tyr),[59] while recent large-scale proteomics studies reported a slightly revised distribution of phosphorylation ratios of (88 : 11 : 1; Ser : Thr : Tyr).[60,61] However, most studies have focused on identifying phosphorylation sites averaged across the whole proteome without relating the phosphoproteomes to parameters such as sub-cellular compartments and cell cycle stages, from which valuable biological information could potentially be extracted. Understanding the role of phosphorylation in signalling and its fine regulation will not only benefit from but actually requires the spatio-temporal resolution that can be achieved by combining sub-cellular fractionation with phosphoproteomics analysis using "Next Generation" proteomics approaches.

The phosphoproteome is highly dynamic. In other words there is no single phosphoproteome, but rather many overlapping phosphoproteomes that are relevant under different circumstances and conditions. For example, a given protein may have multiple phosphorylation sites that can be modified in response to different stimuli. Protein purification protocols for PTM analyses should incorporate the use of appropriate stabilising reagents to preserve the integrity of the modified proteome bearing in mind that such experiments will define a snapshot of the proteome. In the case of the phosphoproteome, phosphatase inhibitors should be used at all steps of the sub-cellular pre-fractionation to prevent the removal of phosphate groups by phosphatases present in the cell extracts. Typical phosphatase inhibitor cocktails contain a mixture of inhibitors, such as sodium orthovanadate (tyrosine phosphatases), imidazole (alkaline phosphatase), sodium tartrate (acid phosphatase), EDTA (alkaline phosphatase, protein phosphatase 2), okadaic acid, β-glycerophosphoric acid and/or sodium pyrophosphate (Ser/Thr phosphatases). Sodium fluoride can also be added as a broad spectrum Ser/Thr phosphatase inhibitor.

11.4.1 PTM Enrichment

MS-based protein phosphorylation analysis on a proteome-wide scale remains a great challenge, hampered by the complexity and dynamic range of protein expression within a cellular system and multi-site phosphorylation at sub-stoichiometric ratios at the individual protein level. For the systematic analysis of protein phosphorylation, either reduction of sample complexity or enrichment of the phosphopeptide pool is a necessary prerequisite. Immobilised metal affinity chromatography (IMAC)[62] and strong cation exchange chromatography (SCX), either alone or in tandem, have emerged as the most widely used chromatographic-based enrichment strategies.[60,63,64]

However, SCX provides little fractionation of the phosphorylated peptides as it separates peptides based on total charge and local charge distribution. Peptide mixtures to be enriched by IMAC are, therefore, more similar in acidic properties, thus not favouring any phosphopeptide (*e.g.*, multiple phosphorylated peptides) in particular. However, acidic peptides tend to compete with phosphorylated peptides reducing the specificity of the enrichment.

Hydrophilic interaction liquid chromatography (HILIC)[65] on the peptide level has been described for phosphopeptide fractionation prior to enrichment, which both enriches the phosphopeptide pool and efficiently fractionates the remaining peptides. HILIC has been shown to improve the selectivity of the metal affinity resin in the IMAC step up to 95% (from 60–70%) by Annan and co-workers.[66] The resulting depletion of non-phosphorylated peptides allows for more efficient use of the MS duty cycle as the instrument spends nearly all of its time analysing phosphorylated peptides. We have adopted their protocols for systematic pre-enrichment of phosphopeptides using HILIC, followed by optimised TiO_2[67,68] enrichment of phosphopeptides. Tryptic peptides are separated on TSKgel Amide-80 columns using a shallow inverse organic gradient. Under these conditions, peptide retention is based on its overall hydrophilicity, so that hydrophilic peptides are retained longer on the column, resulting in a separation that is truly orthogonal to the reverse phase. The hydrophilic fractions, rich in phosphorylated peptides, are subject to phospho-enrichment without the need for chemical derivatisation of the acidic groups. Our phosphopeptide enrichment workflow is summarised in Figure 11.4.

Other PTMs that have similar physico-chemical properties to phosphopeptides, *i.e.*, polarity or hydrophilicity, such as hydroxylation and ADP-ribosylation, are co-enriched under these conditions, which represent a largely unexplored side product of such phosphoenrichment techniques; this is well exemplified in a recent report by Hay and co-workers,[69] in which large phosphoproteomics data sets were reanalysed and interrogated for new PTMs. These studies demonstrated that ADP-ribosylation and phosphate-ribosylation on glutamate, aspartate and arginine were also co-enriched with phosphopeptides and were missed simply because the data were not processed with these modifications in mind.

Another PTM that is co-purified during TiO_2 chromatography is *N*-linked sialylated glycoproteins. Larsen and co-workers[70] have shown that this method can be applied to crude membrane mixtures to effectively enrich

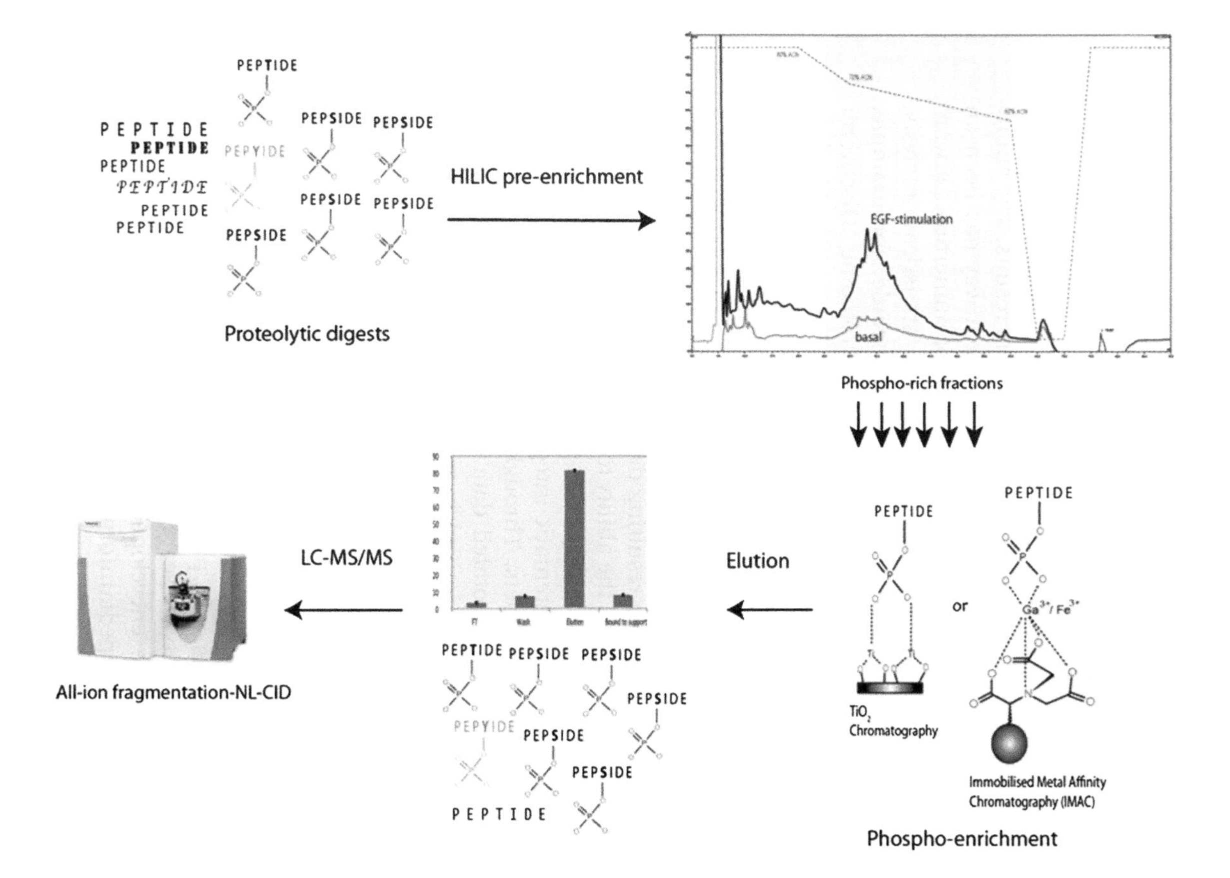

Figure 11.4 An example of a PTM specific enrichment protocol for phosphorylated peptides. After the protein extract has been digested, the proteolytic hydrolysate is pre-fractionated using HILIC as a first step to enriching phosphorylated peptides based on both their increased hydrophilicity, (compared to their non-phosphorylated counterpart) and their longer retention on the column. The phospho-rich fractions are subjected to TiO_2 chromatography as a second step of enrichment, during which phosphorylated peptides are selectively bound by the TiO_2 support while the unmodified peptides are washed away. After several washes, the phosphorylated peptides are eluted using aqueous ammonia at pH 11. The phosphorylated peptides are analysed by LC-MS/MS using phospho-specific methods, such as all ion fragmentation neutral loss-triggered MS/MS (AIF-NL-MS/MS).

what they referred to as the "sialiome". They later showed that the combination of TiO_2 and HILIC, with and without an ion-pairing agent, could be used to enrich for *N*-glycosylated peptides.[71] The number of PTMs detected in a single enrichment experiment can, therefore, be maximised through reanalysis of existing data with new questions.

11.5 Data Analysis and Big Data

The ultimate goal of any quantitative proteomics experiment is to extract as much biological information as possible from numerical results. The ability to achieve this goal in the fastest and the most efficient way has been an active area of research, which led to the development of numerous software packages for protein identification, quantification, statistical analysis, pathway analysis and interaction networks. These include, for example, Mascot[72] and Mascot distiller (Matrix Science), Proteome Discoverer (Thermo Scientific), ProteinPilot™ (AB SCIEX), Trans-Proteomics Pipeline (TPP),[73,74] Progenesis (Nonlinear Dynamics) and Maxquant.[75] As MS instruments become faster, more sensitive and capable of generating increasingly larger amounts of raw data, and our experimental methodologies grow more sophisticated (see Figure 11.5 for a typical "Next Generation" proteomics workflow), software development is even more crucial to handle these technical advances and the resulting data explosion.

A major challenge we foresee is the ability to maximise the knowledge that can be derived from large-scale proteomics data sets. Our strategy in addressing this challenge is the systematic integration and comparison of results from many separate data sets. Through the managed integration of many highly structured and annotated data sets, the amount of new biological information that can be derived is greater than the sum of the individual data sets. A similar strategy can be applied to integrating information from many types of biological experiments over and above proteomics studies, including genomic, transcriptomic and metabolomic data. We have sought to meet this big data challenge by developing PepTracker® (http://www.peptracker.com), a collection of new software tools for management, visualisation and analysis of integrated proteomics data sets with associated complex sets of metadata. Integration of data sets, combined with novel visualisation tools, also provides new opportunities for sharing data with the wider community *via* searchable web-based viewers. To achieve this, we have developed the encyclopedia of proteome dynamics (http://www.peptracker.com/encyclopediaInformation), a searchable, online resource available to the community that includes data from large-scale temporal and spatial proteomics studies and will continue to expand as additional data from further complex multi-dimensional studies continue to be added.[11] (See Figure 11.6 for an output example of the encyclopaedia of proteome dynamics.)

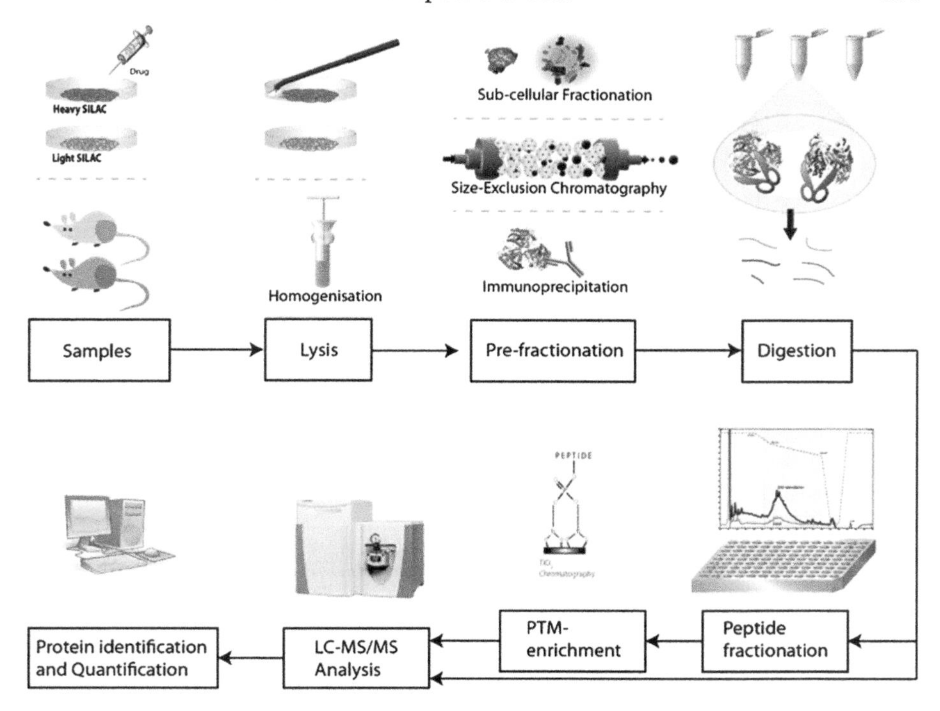

Figure 11.5 Summary of "Next Generation" proteomics workflow for the comprehensive quantitative spatio-temporal characterisation of the proteome. When applicable, the analytes to be compared are differentially labelled in culture (SILAC). After cell lysis, the lysates can be pre-fractionated prior to digestion using a wide range of techniques, such as sub-cellular fractionation, SEC or IP. The digests can be further fractionated, for example, using either HILIC or SCX, before PTM enrichment protocols. Finally, the processed samples are analysed by LC-MS/MS and the data are analysed to identify and quantify proteins. This workflow is highly flexible and can be customised depending on the goal of the study.

11.6 Concluding Remarks

In summary, the past decade has witnessed dramatic advances in the field of proteomics at all levels, starting from sample handling, through data acquisition, to data analysis/mining. Nowadays, MS-based proteomics is becoming more readily available to non-experts and its impact on all fields of molecular and cell biology cannot be underestimated. We anticipate "Next Generation" proteomics to expand our knowledge of the proteome and increase the amount of biologically relevant information that can be extracted from the in-depth characterisation of the dynamic properties of proteins.

We foresee that mass spectrometric techniques will continue to play a central role in proteomics and this, in turn, will push the limits of the

HDAC1_HUMAN

LOCALISATION IN U2OS CELLS

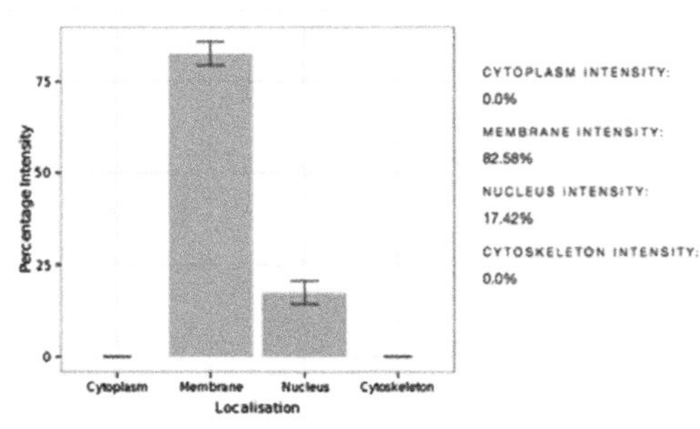

LOCALISATION IN HCT116 CELLS

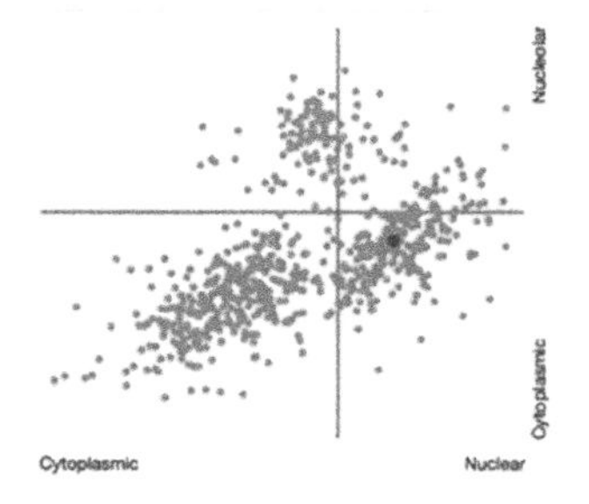

LOCALISATION:
Nuclear
MASS SPECTROMETRY INTENSITY:
26118000.0
FURTHER DETAILS:
Francois-Michel Boisvert, Yun Wah Lam, Douglas Lamont and Angus I. Lamond (2010)
A Quantitative Proteomics Analysis of Subcellular Proteome Localization and Changes Induced by DNA Damage
Molecular & Cellular Proteomics

Figure 11.6 Screen capture of the output of the encyclopaedia of proteome dynamics (EPD), showing the sub-cellular distribution of the HDAC1 protein. Top panel: data showing the relative distribution of a single protein across four cellular compartments of U2OS cells; error bars represent the standard deviation derived from three biological replicates. Bottom panel: distribution of the proteome of HCT 116 cells across three cellular compartments. For comparison, the same protein shown in the top panel is highlighted.

current technologies, with instrumentation becoming faster and more sensitive to allow comprehensive analysis of cells in culture, tissues and, ultimately, single cells.

With such technological advances in instrumentation, the need for faster and more efficient analytics will be greater than ever, paving the path for more systematic and quantitative interrogation of data for protein variants, such as isoforms and post-translationally modified proteins. Generating exponentially bigger data sets will create new challenges to develop innovative analysis methods able to extract relevant biological information. We expect that, as quantitative proteomics becomes faster, more rigorous and more comprehensive, it will ultimately become routinely used in clinical diagnostics.

MS-based proteomics will continue to generate unique data sets that are highly complementary to other types of genomic and post-genomic data (*e.g.*, RNA seq and ChIP seq). This will give rise to new opportunities to integrate proteomics data with genomics, transcriptomics and metabolomics to achieve the goal of a comprehensive system-wide understanding of cellular processes.

References

1. M. Wilkins, J. C. Sanchez, A. A. Gooley, R. D. Appel, I. Humphery-Smith, D. F. Hochstrasser and K. L. Williams, *Biotechnol. Genet. Eng. Rev.*, 1996, **13**, 19–50.
2. P. Picotti, B. Bodenmiller, L. N. Mueller, B. Domon and R. Aebersold, *Cell*, 2009, **138**, 795–806.
3. L. Anderson and C. L. Hunter, *Mol. Cell. Proteomics*, 2006, **5**, 573–588.
4. Y. Ahmad, F. M. Boisvert, E. Lundberg, M. Uhlen and A. I. Lamond, *Mol. Cell. Proteomics*, 2012, **11**, M111.013680.
5. J. D. Castle, *Current Prot. Immunol.*, 2003, **56**, 8.1B.1–8.1B.57.
6. J. D. Castle, *Curr. Prot. Prot. Sci.*, 2004, **37**, 4.2.1–4.2.57.
7. H. M. Moore, B. Bai, F. M. Boisvert, L. Latonen, V. Rantanen, J. C. Simpson, R. Pepperkok, A. I. Lamond and M. Laiho, *Mol. Cell. Proteomics*, 2011, **10**, M111.009241.
8. S. Boulon, B. J. Westman, S. Hutten, F. M. Boisvert and A. I. Lamond, *Mol. Cell*, 2010, **40**, 216–227.
9. F. M. Boisvert, Y. Ahmad, M. Gierlinski, F. Charriere, D. Lamont, M. Scott, G. Barton and A. I. Lamond, *Mol. Cell. Proteomics*, 2012, **11**, M111.011429.
10. F. M. Boisvert, Y. W. Lam, D. Lamont and A. I. Lamond, *Mol. Cell. Proteomics*, 2010, **9**, 457–470.
11. M. Larance, Y. Ahmad, K. J. Kirkwood, T. Ly and A. I. Lamond, *Mol. Cell. Proteomics*, 2013, **12**, 638–650.
12. T. P. J. Dunkley, R. Watson, J. L. Griffin, P. Dupree and K. S. Lilley, *Mol. Cell. Proteomics*, 2004, **3**, 1128–1134.
13. L. M. Breckels, L. Gatto, A. Christoforou, A. J. Groen, K. S. Lilley and M. W. B. Trotter, *J. Proteomics*, 2013, **88**, 129–140.
14. A. Shevchenko, M. Wilm, O. Vorm and M. Mann, *Anal. Chem.*, 1996, **68**, 850–858.
15. A. Shevchenko, H. Tomas, J. Havlis, J. V. Olsen and M. Mann, *Nat. Protocols*, 2007, **1**, 2856–2860.
16. L. Trinkle-Mulcahy, S. Boulon, Y. W. Lam, R. Urcia, F. M. Boisvert, F. Vandermoere, N. A. Morrice, S. Swift, U. Rothbauer, H. Leonhardt and A. I. Lamond, *J. Cell Biol.*, 2008, **183**, 223–239.
17. S. Boulon, Y. Ahmad, L. Trinkle-Mulcahy, C. l. Verheggen, A. Cobley, P. Gregor, E. Bertrand, M. Whitehorn and A. I. Lamond, *Mol. Cell. Proteomics*, 2009.
18. A. J. Tackett, J. A. DeGrasse, M. D. Sekedat, M. Oeffinger, M. P. Rout and B. T. Chait, *J. Proteome Res.*, 2005, **4**, 1752–1756.
19. F. E. Paul, F. Hosp and M. Selbach, *Methods*, 2011, **54**, 387–395.

20. M. Selbach and M. Mann, *Nat. Methods*, 2006, **3**, 981–983.
21. S. E. Ong, B. Blagoev, I. Kratchmarova, D. B. Kristensen, H. Steen, A. Pandey and M. Mann, *Mol. Cell. Proteomics*, 2002, **1**, 376–386.
22. S. P. Gygi, B. Rist, S. A. Gerber, F. Turecek, M. H. Gelb and R. Aebersold, *Nat. Biotech.*, 1999, **17**, 994–999.
23. P. L. Ross, Y. N. Huang, J. N. Marchese, B. Williamson, K. Parker, S. Hattan, N. Khainovski, S. Pillai, S. Dey, S. Daniels, S. Purkayastha, P. Juhasz, S. Martin, M. Bartlet-Jones, F. He, A. Jacobson and D. J. Pappin, *Mol. Cell. Proteomics*, 2004, **3**, 1154–1169.
24. A. Thompson, J. Schaefer, K. Kuhn, S. Kienle, J. Schwarz, G. Schmidt, T. Neumann and C. Hamon, *Anal. Chem.*, 2003, **75**, 1895–1904.
25. J. L. Hsu, S. Y. Huang, N. H. Chow and S. H. Chen, *Anal. Chem.*, 2003, **75**, 6843–6852.
26. Y. Oda, K. Huang, F. R. Cross, D. Cowburn and B. T. Chait, *Proc. Natl. Acad. Sci. USA*, 1999, **96**, 6591–6596.
27. H. W. Lahm and H. Langen, *Electrophoresis*, 2000, **21**, 2105–2114.
28. S. E. Ong, I. Kratchmarova and M. Mann, *J. Proteome Res.*, 2002, **2**, 173–181.
29. H. Jiang and A. M. English, *J. Proteome Res.*, 2002, **1**, 345–350.
30. M. K. Doherty, C. Whitehead, H. McCormack, S. J. Gaskell and R. J. Beynon, *Proteomics*, 2005, **5**, 522–533.
31. R. Amanchy, D. E. Kalume, A. Iwahori, J. Zhong and A. Pandey, *J. Proteome Res.*, 2005, **4**, 1661–1671.
32. J. Krijgsveld, R. F. Ketting, T. Mahmoudi, J. Johansen, M. Artal-Sanz, C. P. Verrijzer, R. H. A. Plasterk and A. J. R. Heck, *Nat. Biotech*, 2003, **21**, 927–931.
33. M. Larance, A. P. Bailly, E. Pourkarimi, R. T. Hay, G. Buchanan, S. Coulthurst, D. P. Xirodimas, A. Gartner and A. I. Lamond, *Nat. Methods*, 2011, **8**, 849–851.
34. M. K. Doherty, L. McClean, I. Edwards, H. McCormack, L. McTeir, C. Whitehead, S. J. Gaskell and R. J. Beynon, *British Poultry Science*, 2004, **45**, S27–S28.
35. M. K. Doherty, L. McLean, J. R. Hayter, J. M. Pratt, D. H. L. Robertson, A. El-Shafei, S. J. Gaskell and R. J. Beynon, *Proteomics*, 2004, **4**, 2082–2093.
36. M. Kruger, M. Moser, S. Ussar, I. Thievessen, C. A. Luber, F. Forner, S. Schmidt, S. Zanivan, R. Fassler and M. Mann, *Cell*, 2008, **134**, 353–364.
37. T. Geiger, J. Cox, P. Ostasiewicz, J. R. Wisniewski and M. Mann, *Nat. Methods*, 2010, 7, 383–385.
38. T. Geiger, J. R. Wisniewski, J. Cox, S. Zanivan, M. Kruger, Y. Ishihama and M. Mann, *Nat. Protocols*, 2011, **6**, 147–157.
39. F. L. Brancia, S. G. Oliver and S. J. Gaskell, *Rapid Comm. Mass Spectrom*, 2000, **14**, 2070–2073.
40. S. Warwood, S. Mohammed, I. M. Cristea, C. Evans, A. D. Whetton and S. J. Gaskell, *Rapid Comm. Mass Spectrom.*, 2006, **20**, 3245–3256.
41. P. J. Boersema, R. Raijmakers, S. Lemeer, S. Mohammed and A. J. R. Heck, *Nat. Protocols*, 2009, **4**, 484–494.

42. E. J. Want, B. F. Cravatt and G. Siuzdak, *ChemBioChem*, 2005, **6**, 1941–1951.
43. J. C. Schwartz, USA Pat., US6949743 B1, 2005.
44. J. C. Schwartz, J. P. Syka and S. T. Quarmby, *53rd ASMS Conference on Mass Spectrometry*, San Antonio, Texas, 2005.
45. T. J. Griffin, H. Xie, S. Bandhakavi, J. Popko, A. Mohan, J. V. Carlis and L. Higgins, *J. Proteome Res.*, 2007, **6**, 4200–4209.
46. M. Bantscheff, M. Boesche, D. Eberhard, T. Matthieson, G. Sweetman and B. Kuster, *Mol. Cell. Proteomics*, 2008, **7**, 1702–1713.
47. N. A. Karp, W. Huber, P. G. Sadowski, P. D. Charles, S. V. Hester and K. S. Lilley, *Mol. Cell. Proteomics*, 2010, **9**, 1885–1897.
48. S. Y. Ow, M. Salim, J. Noirel, C. Evans, I. Rehman and P. C. Wright, *J. Proteome Res.*, 2009, **8**, 5347–5355.
49. L. V. DeSouza, A. D. Romaschin, T. J. Colgan and K. W. M. Siu, *Anal. Chem.*, 2009, **81**, 3462–3470.
50. V. G. Keshamouni, G. Michailidis, C. S. Grasso, S. Anthwal, J. R. Strahler, A. Walker, D. A. Arenberg, R. C. Reddy, S. Akulapalli, V. J. Thannickal, T. J. Standiford, P. C. Andrews and G. S. Omenn, *J. Proteome Res.*, 2006, **5**, 1143–1154.
51. M. A. Kuzyk, L. B. Ohlund, M. H. Elliott, D. Smith, H. Qian, A. Delaney, C. L. Hunter and C. H. Borchers, *Proteomics*, 2009, **9**, 3328–3340.
52. H. Liu, R. G. Sadygov and J. R. Yates, *Anal. Chem.*, 2004, **76**, 4193–4201.
53. J. D. Venable, M. Q. Dong, J. Wohlschlegel, A. Dillin and J. R. Yates, *Nat. Methods*, 2004, **1**, 39–45.
54. S. Klumpp and J. Krieglstein, *Biochimica et Biophysica Acta (BBA) - Proteins and Proteomics*, 2005, **1754**, 291–295.
55. H. Zhang, X. Zha, Y. Tan, P. V. Hornbeck, A. J. Mastrangelo, D. R. Alessi, R. D. Polakiewicz and M. J. Comb, *J. Biol. Chem.*, 2002, **277**, 39379–39387.
56. D. H. Phanstiel, J. Brumbaugh, C. D. Wenger, S. Tian, M. D. Probasco, D. J. Bailey, D. L. Swaney, M. A. Tervo, J. M. Bolin, V. Ruotti, R. Stewart, J. A. Thomson and J. J. Coon, *Nat. Methods*, 2011, **8**, 821–827.
57. H. R. Matthews, *Pharmacology & Therapeutics*, 1995, **67**, 323–350.
58. J. Puttick, E. N. Baker and L. T. J. Delbaere, *Bioch. Bioph. Acta (BBA) - Proteins and Proteomics*, 2008, **1784**, 100–105.
59. T. Hunter and B. M. Sefton, *Proc. Natl. Acad. Sci. USA*, 1980, 77, 1311–1315.
60. J. Villen, S. A. Beausoleil, S. A. Gerber and S. P. Gygi, *Proc. Natl. Acad. Sci. USA*, 2007, **104**, 1488–1493.
61. J. V. Olsen, B. Blagoev, F. Gnad, B. Macek, C. Kumar, P. Mortensen and M. Mann, *Cell*, 2006, **127**, 635–648.
62. L. Nuwaysir and J. Stults, *J. Am. Soc. Mass Spectrom.*, 1993, **4**, 662–669.
63. B. Macek, I. Mijakovic, J. V. Olsen, F. Gnad, C. Kumar, P. R. Jensen and M. Mann, *Mol. Cell. Proteomics*, 2007, **6**, 697–707.
64. M. Hilger, T. Bonaldi, F. Gnad and M. Mann, *Mol. Cell. Proteomics*, 2009, **8**, 1908–1920.

65. A. J. Alpert, *J. Chromatography A*, 1990, **499**, 177–196.
66. D. E. McNulty and R. S. Annan, *Mol. Cell. Proteomics*, 2008, 7, 971–980.
67. M. W. H. Pinkse, P. M. Uitto, M. J. Hilhorst, B. Ooms and A. J. R. Heck, *Anal. Chem.*, 2004, **76**, 3935–3943.
68. M. R. Larsen, T. E. Thingholm, O. N. Jensen, P. Roepstorff and T. J. D. Jorgensen, *Mol. Cell. Proteomics*, 2005, **4**, 873–886.
69. I. Matic, I. Ahel and R. T. Hay, *Nat. Methods*, 2012, **9**, 771–772.
70. M. R. Larsen, S. S. Jensen, L. A. Jakobsen and N. H. H. Heegaard, *Mol. Cell. Proteomics*, 2007, **6**, 1778–1787.
71. B. L. Parker, G. Palmisano, A. V. G. Edwards, M. Y. White, K. Engholm-Keller, A. Lee, N. E. Scott, D. Kolarich, B. D. Hambly, N. H. Packer, M. R. Larsen and S. J. Cordwell, *Mol. Cell. Proteomics*, 2011, **10**, M110.006833.
72. D. N. Perkins, D. J. C. Pappin, D. M. Creasy and J. S. Cottrell, *Electrophoresis*, 1999, **20**, 3551–3567.
73. X. j. Li, P. G. A. Pedrioli, J. Eng, D. Martin, E. C. Yi, H. Lee and R. Aebersold, *Anal. Chem.*, 2004, **76**, 3856–3860.
74. P. A. Pedrioli, in *Proteome Bioinformatics*, ed. S. J. Hubbard and A. R. Jones, Humana Press, London, 2010, vol. 604, pp. 213–238.
75. J. Cox and M. Mann, *Nat. Biotech.*, 2008, **26**, 1367–1372.

CHAPTER 12

Experimental and Analytical Approaches to the Quantification of Protein Turnover on a Proteome-wide Scale

AMY J. CLAYDON,† DEAN E. HAMMOND† AND ROBERT J. BEYNON*

Protein Function Group, Institute of Integrative Biology, University of Liverpool, Liverpool L69 7ZB, UK
*Email: R.Beynon@liverpool.ac.uk

12.1 The Significance of Proteome Dynamics

The process of proteostasis, whereby the level of each protein in a cell or biological fluid is maintained, is not fully understood. What is clear, however, is that it is achieved by balancing the opposing processes of protein synthesis, folding and trafficking (inputs to a protein pool) with intracellular degradation or secretion (outputs from the pool). The proteome of any organism is in a flux and, when a cell is in steady-state, proteins are synthesised and degraded in an equilibrium that ensures constant protein levels. It follows that a change in protein abundance is mediated by

†These authors made an equal contribution to the chapter.

New Developments in Mass Spectrometry No. 1
Quantitative Proteomics
Edited by Claire Eyers and Simon J Gaskell

Published by the Royal Society of Chemistry, www.rsc.org

adjustment of the rates of synthesis and/or degradation—or 'protein turnover'. A high rate of turnover enables a rapid and responsive change in protein abundance at the cost of the metabolic energy that is required. As such, high turnover may be restricted to those biological processes that demand responsive adjustment of protein abundance. The appreciation of the importance of protein turnover has now impacted on proteomics and there is emerging methodology that aims to measure the turnover rate of individual proteins within a proteome (see refs. 1–4 and 5 for a list of turnover studies to date). Because protein turnover can continue in the absence of any change in protein abundance, turnover studies must measure the flux of amino acids through the protein pool; in proteomics studies this inevitably means the flux of stable-isotope precursors. The precursors can be amino acids, simple molecules, such as isotopically labelled water or ammonium chloride, or metabolic precursors, such as glucose (reviewed in refs. 5 and 6). The nature of the precursor, and the 'metabolic distance' from the amino acid pool, influences the pattern of incorporation of label into proteins and the analytical methods that are used to analyse this incorporation. However, in this chapter, we mainly address the use of stable-isotope-labelled amino acids for the elucidation of protein turnover rates and discuss the experimental and analytical strategies that can be deployed to recover turnover parameters in proteome studies.

12.2 Labelling Strategies for Turnover Studies

Proteome dynamics can be assessed either by measuring the incorporation of labelled precursors into newly synthesised protein ('labelling' strategies), or by following the loss of label from previously labelled protein ('unlabelling' strategies). Kinetically and analytically, the two approaches are formally equivalent; the choice of strategy is influenced by the organism under study, the analytical approach to be adopted and the budget (selectively labelled stable-isotope amino acids are high cost chemicals). To illustrate, cells grown in culture lend themselves well to an unlabelling strategy as they are often undergoing exponential growth and consequent expansion of the biomass. It is straightforward, therefore, to obtain fully labelled cells after the passage of seven cell doublings. By contrast, with intact organisms, such as the mouse, a labelling strategy is far more cost effective, as the generation of fully labelled animals (over several generations) simply to monitor their return to the unlabelled state would be very resource intensive.

An optimal proteome turnover experiment has to meet multiple criteria. First, and ideally, the labelled precursor should not be metabolisable into another amino acid that would be incorporated into the protein differently. Secondly, the organism under study should not be competent in the synthesis *de novo* of the selected amino acid as this would dilute the degree of labelling of the amino acid in the precursor pool. In bacteria or yeast, strains auxotrophic for the chosen amino acid (*i.e.*, incapable of *de novo* synthesis) may be available. However, this requirement is not essential, as witnessed by the many expression systems that achieve complete labelling of recombinant proteins. Provided the precursor-labelled amino acid is provided in the medium in adequate amounts, the parsimonious nature of microbial metabolism will repress induction of the biosynthetic pathway and the exogenous precursor will be preferentially used. Thirdly, in a 'perfect' experiment, the labelled amino acid, when released from the protein pool, would not be re-incorporated into the protein; this reutilisation can give the impression of slower turnover of the target protein. However, a moment's consideration will reveal the illogicality of this requirement: if an amino acid is incorporated into a protein (as it must be to permit monitoring), then reutilisation has to take place. In practice the effects of reutilisation can be ameliorated by ensuring that the precursor pool is rapidly diluted, whether by excess unlabelled amino acid, as a 'chase', or by dilution through the mass of pre-existing protein undergoing turnover (*e.g.*, in animals).

12.2.1 Label Choice

There have been many excellent examples of turnover studies that involve elemental labelling, usually by uniform incorporation of ^{15}N or ^{2}H;[7–10] such elemental uniform labels were used in about 40% of the studies reviewed by ref. 5. In such studies the pattern of label incorporation can be complex, reflecting incorporation at many sites. Many other studies have used stable-isotope-labelled amino acids as the precursor. The criteria for selection of a suitable amino acid precursor include abundance in the proteome (so that many peptides contain informative label incorporation data), metabolic fate and potential for precursor pool dilution through *de novo* synthesis.

Amino acids commonly used in stable-isotope labelling with amino acids in cell culture (SILAC) experiments are arginine and lysine,[11] primarily because these precursors favour single sites of incorporation in tryptic peptides. Furthermore, both arginine and lysine are available in +6 Da isotopic variants, simplifying downstream analysis. However, peptides that have multiple sites of labelling (which is needed to check the relative isotope abundance of the precursor pool, as discussed later) will be relatively infrequent obligatory or partial miscleavage products, and may be large and evade detection. Arginine is also metabolised to proline in eukaryotic cells, leading to a disruption of labelling patterns; proline-containing peptides might not, therefore, respond as expected due to an extra mass shift introduced by [$^{13}C_5$]proline. Although short-term experiments may not be compromised by such factors,[12] this confounding factor can be minimised[13] or corrected computationally,[14] if necessary.

In studies with fully deuterated amino acids we have observed transamination-mediated loss of the α-carbon deuteron, either fully[4] or partially.[2] For example, in a study in the mouse using [$^{2}H_8$]valine as a precursor, we observed partial conversion to [$^{2}H_7$]valine *in vivo*; a loss of 1 Da was consistent with the exchange of a metabolically labile hydrogen atom.

Transamination seems to be rapid, and subsequent protein labelling mass spectra reflect the incorporation of two isotopic variants of the same amino acid. This can be advantageous for manual data analysis due to the characteristic and recognisable labelling pattern, but most software, including protein identification search engines, will struggle to recognise the labelled peptide precluding high throughput automated analysis.

In cell culture, recycling of labelled amino acids into proteins can be minimised by ensuring a rapid change in the extent of precursor labelling, most readily affected by a 'chase' with unlabelled amino acid. With more complex systems, such as animals, this is not practical and, as amino acids are incorporated into newly synthesised proteins, unlabelled amino acids from the degradation of existing proteins will enter the protein pool and alter the precursor relative-isotope abundance (RIA). This is easily resolved if all amino acids have the same probability of incorporation into new proteins; the RIA of the precursor at the point of synthesis will be calculated from the mass spectral data; however, if there is bias in the reutilisation, for example, if amino acids from degraded proteins are incorporated preferentially, then the turnover rate calculated will be inaccurate. Alternatively, if labelled amino acids are incorporated into new protein from the precursor pool in preference to other natural amino acids in the pool, this will also have an effect. For example, in our studies where we have supplemented an existing protein containing animal diet with free, labelled amino acids, we acquired evidence of a diurnal variation in labelling of the precursor pool.[2]

12.2.2 Control of Precursor Enrichment and the Significance of Precursor RIA

There is a general assumption that the precursor pool should transition from fully labelled to fully unlabelled, or *vice versa*, such that the RIA of the precursor should be shifted very quickly from one to zero, permitting subsequent monitoring of the loss of label from the protein pool, or it must transition quickly from zero to one, after which proteins are monitored for the incorporation of label. In practice the shift between the extremes of zero and one is never achieved. Even in cell systems in which the precursor pool can be readily manipulated, the isotopic purity of the amino acids is often not complete: label incorporation is typically 98 to 99 atom percent excess. In animal systems the pre-existing biomass will contribute a substantial pool of unlabelled amino acids that will dilute the degree of labelling of exogenously provided precursors, even if the diet is fully labelled. In practice we would argue that the degree of labelling is irrelevant, provided that the separation between the extremes is known and that it is large enough to permit high quality determination of protein turnover parameters. In whole animal systems complete labelling prior to the commencement of the turnover study would be extremely slow and rather costly. When a 'labelling' approach is taken, the amino acid precursor pool will, therefore, contain both labelled and unlabelled amino acids. Unlabelled amino acids are derived from pre-existing unlabelled amino acids in the pool, degradation of unlabelled proteins and, often, unlabelled amino acids from the labelling source itself—usually through the food or water. To avoid feeding a fully synthetic diet, which could be unpalatable and have unknown effects on the metabolism of the animal (which would, in turn, affect the rate of turnover of some proteins), we have adopted a protocol whereby stable-isotope-labelled amino acids are added to standard laboratory chow to produce a diet with a final precursor RIA of 0.5.[1,2,15] Similarly, for studies using [$^2H_2$]O, the label will be distributed into, and diluted by, the unlabelled water content of the animal. Partial isotopically enriched drinking water is provided to achieve a steady labelled state enrichment at around 5%.[8,16]

12.2.3 Sampling Times and Frequency

Although, in principle, it is possible to calculate the turnover rate from a single time point,[4] the lack of knowledge of the shape of the labelling/unlabelling profile could compromise recovery of an accurate turnover parameter. Turnover experiments should determine the degree of labelling of proteins over several half-lives for that particular protein. This creates problems in proteome studies, where individual proteins can differ in k_{deg} by several orders of magnitude. Moreover, in labelling or unlabelling studies on cells growing exponentially in culture or in rapidly growing young animals, biomass expansion creates a limited window of opportunity as the protein pool is gaining or losing label through growth. To extract maximal information on the labelling trajectory of individual proteins, sampling times must be able to provide sufficient data for both high turnover and low turnover proteins. High frequency sampling will acquire the turnover rates of high turnover proteins, but will generate large numbers of unnecessary samples for low turnover proteins. At the other extreme, low sampling rates run the risk of not being able to generate sufficient points to define the trajectory of high turnover proteins. We favour a sampling strategy that captures several early time points, closely spaced (to define high turnover components) and gradually slowing to infrequent sampling to build the profile of low turnover proteins. In animals the labelling times are defined by the intrinsic metabolic rate of the species, and also by the tissue under investigation. In the mouse most proteins in the liver reach maximal labelling about seven days after label administration (high turnover proteins reach this point much earlier, of course), whereas skeletal muscle proteins are still largely unlabelled to any sufficient degree after fourteen days, despite the precursor RIA in both tissues having reached the same plateau value within similar time frames.[2] A further complication derives from cells with life-time kinetics, such as erythrocytes or some proteins from spermatozoa.[1] To track the turnover of these proteins requires more extended labelling periods, dependent on the lifespan of the cells themselves. Delays in equilibration of the precursor pool can also preclude very rapid sampling, and extremely high turnover proteins may have completed their progression to final RIA before the first sample can be taken.

12.3 Solutions for the Analysis of Proteome Dynamics Data

12.3.1 Measurement of Precursor RIA

The precursor RIA is a critical parameter. It is specified in design, but experimental and metabolic factors can both conspire to distort the actual

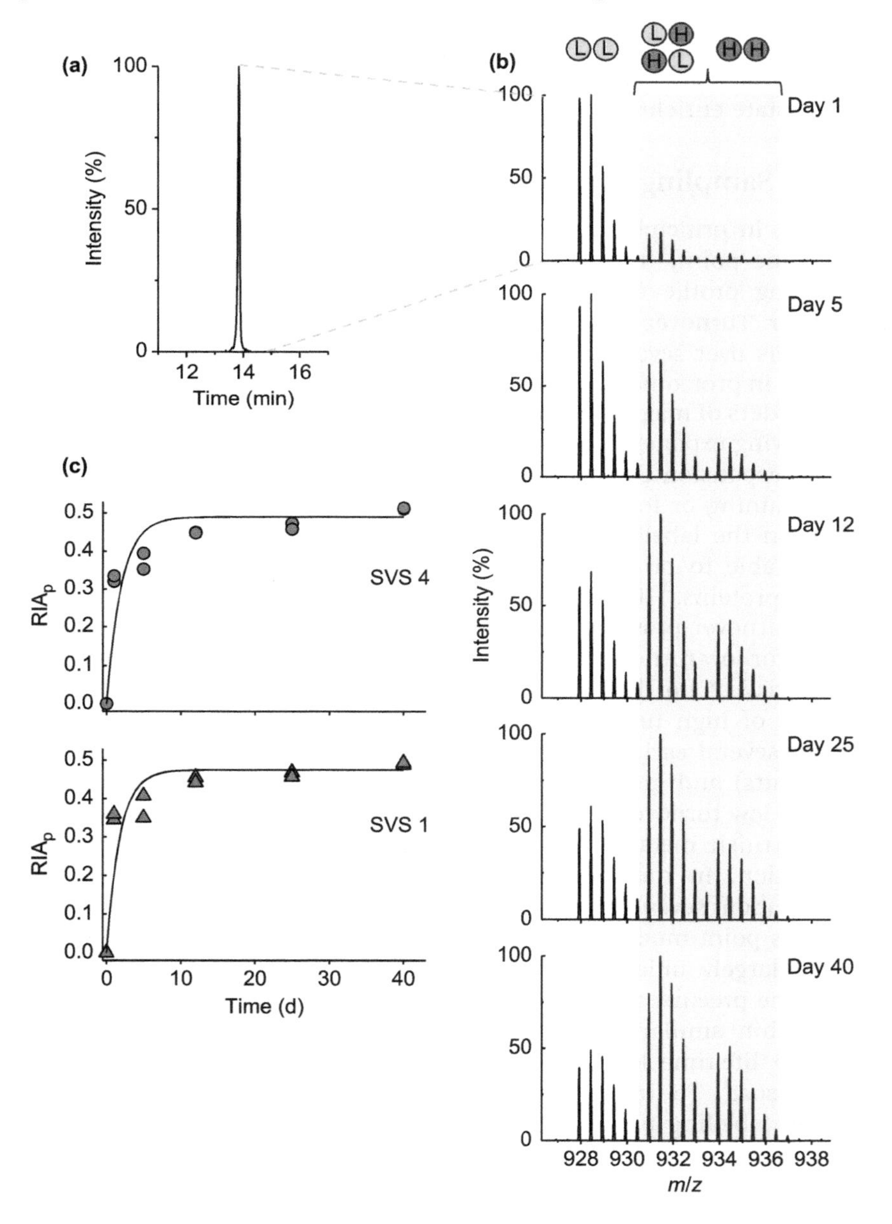

parameter from the expected value, and there is much to be said for including the experimental acquisition of this parameter as part of the workflow. This requires no adjustment of the protocol, as the pattern of peptide labelling will give a direct assessment of the precursor RIA at the point of synthesis of the protein, most simply acquired by mass spectrometry of peptides. The degree of labelling of the precursor pools (RIA_p) determines the likelihood of a labelled amino acid being incorporated into proteins. In experiments where the precursor RIA is (by design) less than one, peptides with a single potential labelling site derived from newly synthesised proteins can either contain a labelled ('heavy', 'H') or unlabelled ('light', 'L') instance of the amino acid. Moreover, the unlabelled peptide could have been recently synthesised or derived from existing biomass. It is essential that the precursor RIA be known and, in practice, this is readily achieved by inspection of peptides that contain more than one instance of the labelled amino acid.

If such a peptide contained two incorporation sites, then three peptide ions could be observed: fully light (LL), partially labelled (positionally, HL or LH) and fully heavy (HH) (Figure 12.1). Both the HL/LH and HH peptides must have been derived from newly synthesised protein and, thus, can provide a direct measure of the precursor RIA_p at the point of protein synthesis (eqn 12.1).

$$RIA_p = 2R/(1 + 2R), \tag{12.1}$$

where $R = I_{HH}/I_{HL}$ and I_{HH} and I_{HL} are the intensities of the HH and HL peaks, respectively.

Similar solutions exist for peptides containing three instances of the amino acid.[15,17] In cultured cell systems the intent is often to create boundary RIA values of one or zero, and so the parameter of importance becomes the growth rate of the cells, which determines the rate of dilution (k_{dil}) of label into daughter cells as the parent cells divide. In adult animals, growth is low, but the RIA of the precursor pools cannot easily be fixed at one; it would require an extended period of labelling, possibly over multiple

Figure 12.1 Generic approach to the calculation of precursor RIA. To recover the value of the precursor RIA at any time point, the mass spectrum of a peptide containing more than one instance of the labelled amino acid is required. The example here is of a doubly labelled peptide with RIA_∞ of approximately 0.5, which has been identified and quality-controlled by the extracted ion chromatogram (panel a). The summed intensities of the doubly labelled (HH) and singly labelled (HL) peptide isotope profiles (panel b) can then be used to calculate the RIA of the precursor pool at the point of protein synthesis. In panel c the time-dependent changes in the precursor RIA are calculated using peptides from two different mouse proteins (SVS4, P18419 and SVS1, Q6WIZ7) in the same tissue. In this example (an intact animal) there is about a one day delay in the equilibration of the precursor RIA with the dietary input pool.

generations, either using a fully synthetic diet or a very complex diet based on selective labelling of proteins from single cell organisms.

Turnover studies with whole animal models predominantly start with a precursor RIA of zero, label is then introduced and precursor incorporation is monitored over time. Even if a fully synthetic diet was designed (RIA = 1), the labelled precursor pool would still be diluted by unlabelled amino acids from the pre-existing biomass and the true precursor RIA would still need to be calculated. Given such complexities, we have based our studies on a natural diet supplemented with one stable-isotope-labelled crystalline amino acid to an RIA of 0.5. In such studies, and indeed in any labelling system, a quick check for peptides containing multiple instances of the amino acid being labelled is an effective method to confirm that the experimental design is not compromised, whether from incompletely labelled precursors or by dilution of the precursor pool by degradation-derived amino acids or *de novo* biosynthesis.

12.3.2 Analytical Approaches to the Measurement of Protein Turnover Rates

It is not fully appreciated that the analysis of protein turnover kinetics brings with it some specific demands in terms of data analysis. A requirement to track the flux of label into, or out of, the protein pool, even when the protein concentration does not change, immediately precludes label-free methods. Furthermore, and unlike SILAC-type experiments, the abundance ratio of unlabelled to labelled peptides will change, by definition, over time and biological replicates that are sampled at different times cannot be aggregated as they will define the trajectory of the turnover process. We have used the term 'dynamic SILAC'[6,12] to reflect that the fact that the SILAC ratio changes throughout an experiment, which is a situation rather different to the more common two-way or three-way SILAC experiments.

In all turnover studies the key measured parameter is the RIA of each peptide at time t (RIA_t). A generic first order curve can be fitted to a set of (t, RIA_t) data using eqn 12.2.

$$\mathrm{RIA}_t = \mathrm{RIA}_0 + (\mathrm{RIA}_\infty - \mathrm{RIA}_0)(1 - \mathrm{e}^{-kt}), \tag{12.2}$$

where RIA_0 is the starting value of the precursor pool, RIA_∞ defines the final value and k is the first order rate constant for degradation. This equation applies equally well to both labelling and unlabelling experiments. When $t = 0$, the equation reduces to $\mathrm{RIA}_t = \mathrm{RIA}_0$, and when $t = \infty$, $\mathrm{RIA}_t = \mathrm{RIA}_\infty$. The changes in RIA as a function of time are readily modelled using different beginning and ending values of RIA and eqn 12.2 applies equally to either experimental methodology. Moreover, this equation covers those situations where complete labelling is not attained during the labelling period (Figure 12.2).

In principle eqn 12.2 has three unknowns (RIA_0, RIA_∞ and k). It is possible to fit the turnover curve and obtain best-fit values for all three parameters, but it is rare to have sufficient data points to recover all three parameters with

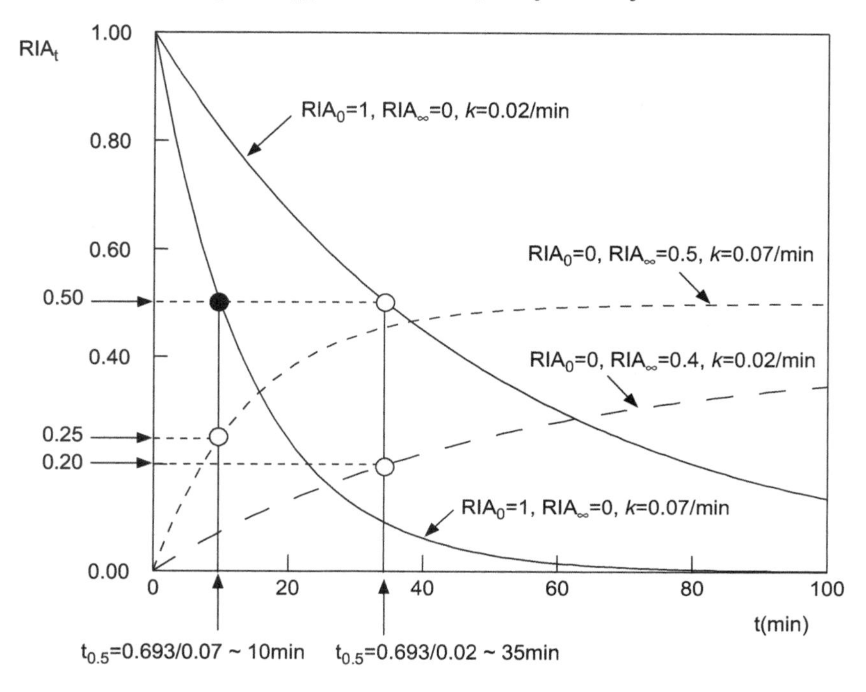

Figure 12.2 The scope of turnover studies. All turnover studies are 'labelling' (dashed lines) or 'unlabelling' (solid lines) experiments, reflecting the incorporation or loss of a stable-isotope tracer within the protein pool. It is common for 'unlabelling' studies to aim for pre-labelled proteins at RIA = 1, and monitor the decline in labelling to the complete replacement of label (RIA = 0). When the transition has reached halfway, this defines the half-life of the protein, but this is best expressed as the degradation first order rate constant, k. Both the span of RIA change and the rate constant dictate the trajectory of the (t, RIA$_t$) curve.

good confidence. It is preferable to derive system values of RIA_0 and RIA_∞ by additional experiments or analyses, external to the curve fitting stage (see above). If this is possible, then the remaining unknown in the equation is k. Because the equation is nonlinear in k, any one of a range of iterative, nonlinear optimisation algorithms will yield statistically reliable values.

12.3.3 Acquisition of Precursor Peptide Intensities for Labelled and Unlabelled Variants

The development of software to aid in the analysis of data containing heavy–light peptide pairs has been primarily focused on SILAC-type experiments to determine the heavy : light ratios of peptides in corresponding samples exposed to different treatments/conditions. Previously, there were limited software tools allowing the time-dependence of SILAC ratios to be extracted from data sets or the targeted identification of peptides with SILAC ratios altered from the mean. Subsequently, software offering comparative analysis of multiple SILAC-labelled samples in an automated fashion was developed; if the multiple samples are differentiated only by sampling time, then protein turnover rates can be calculated. However, the crucial parameters of a turnover experiment, as discussed above, are the precursor pool RIA and non-linear curve fitting to extract the degradation rate constant (k_{deg}). The ability of software to accurately determine these parameters can be used as a measure of its success, assuming the ability to extract the peptide ratio values is adequate.

12.3.4 Manual Analysis

For a limited exploration or a study focussed on specific proteins, it is arguably more efficient to process the data manually. Peptides from these proteins can be extracted from complex LC-MS profiles, and the area under the curve for the extracted ion chromatograms of labelled and unlabelled variants will yield the H and L parameters to allow calculation of RIA_t. Alternatively, if the mass spectra of the H and L variants can be isolated in the summed mass spectra across the chromatographic peak, the intensities of the peptides can be recovered as the intensities of the H and L peptide ions (Figure 12.3). This methodology is essential when detailed studies on two-dimensional (2D) gel resolved protein spots must be analysed.[2]

12.3.5 Software Solutions for the Calculation of Protein Turnover Rates on a Proteome-wide Scale

The increasing number of studies on proteome turnover[5] have led to the emergence of a range of different software solutions for the analysis of LC-MS/MS data sets in which the peptide mass spectra reveal a time-dependent progression in RIA_t.[3,18–21] Because a well-constructed experiment is able to cover the full range of RIA values between RIA_0 and RIA_∞, standard software packages that are designed to analyse SILAC data are not particularly suitable. Specific packages that recover (t, RIA_t) from datasets and then calculate the first order rate constant have been generated to address this need. Broadly speaking, these packages have been written to accommodate specific experimental systems and labelling protocols, and there are no truly generic solutions at this time. Table 12.1 summarises some of the key features of several of these packages aimed at the analysis of data from turnover experiments using labelled amino acids.

SILACtor[20] is best used (as the name would suggest) for analysis of cell culture turnover experiments where the labelling transition is from all-heavy to all-light or *vice versa*, although loss of label is the more common route. In addition to protein information this software contains features to identify potential peptides of interest in the absence of identification and corresponding sequence data, achieved by generating accurate mass and retention time (AMRT) inclusion lists targeting specific peptides. SILACtor uses chromatographic peak areas of the peptide isotope distributions to

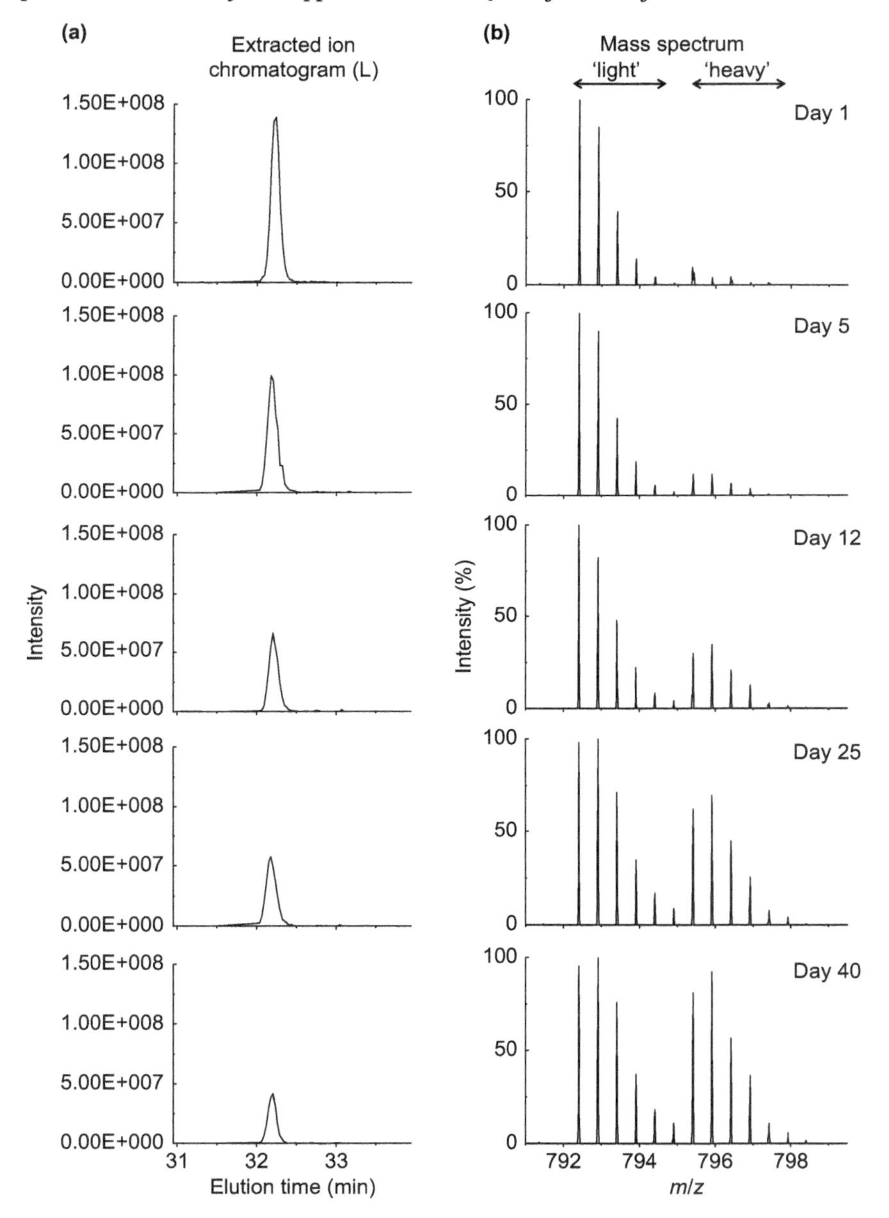

Figure 12.3 Manual analysis of turnover data. To determine the turnover rate of single proteins, it is feasible to isolate a high quality peptide with a monotonic extracted ion chromatogram (panel a, reflecting the abundance of the unlabelled 'light' peptide in a labelling study with $RIA_{\infty} \sim 0.5$). The cumulative mass spectrum spanning the elution of the peptide can then be recovered and the intensity of the unlabelled and labelled components of the spectrum (panel b) are used to derive the RIA_t values prior to non-linear curve fitting in an external software package.

Table 12.1 Summary of some features of software for the analysis of amino-acid-labelled turnover studies.

	Package			
Property	*SILACtor*[20]	*Topograph*[21]	*SILACanalyser*[26]	*Progenesis & PPP/R*[28–32]
Operating system(s) and control environment	Windows and Linux Command line interface	Windows Graphical user interface	Mac, Linux and Windows Graphical user interface and/or command line driven	*Progenesis and Progenesis Post Processor (PPP)*: Windows GUI *R*: Mac, Linux and Windows Command line driven
Connectivity to protein identity	Currently only Mascot search engine outputs supported	Can import search data in many formats (.sqt, .pep.xml, .pepXML, .idpXML, .dat, .ssl, .mzid, .perc.xml, .final_fragment.csv) Also supports BiblioSpec spectral library import	Part of OpenMS; database search results from X!Tandem, Mascot, OMSSA, MyriMatch supported	Progenesis can import search data in many formats from different search engines, including Mascot, ProteinLynx Global Server and PEAKS
Calculation of precursor RIA	Not possible; although maximum % labelling can be entered into software	Uses mass isotopomer analysis of multiply labelled peptides	Not possible—this package is not designed for turnover	Scriptable in R, based on the output from Progenesis/PPP
Commentary on features	Simple .txt/.csv output files	Visual inspection of data after nonlinear curve fitting Data can be exported in .csv/.tsv format	Multiple .txt/.csv output files	Highly flexible analytical workflow using R with .txt/.csv output files from PPP

determine heavy : light ratios. Peptide pairs with sequence identification are grouped by protein to determine the average protein RIA at each time point. A monoexponential curve is fitted to the (t, RIA_t) to elicit a k_{loss} value for each protein. The freely available tools Hardklör[22] and MakeMS2 (http://proteome.gs.washington.edu/software/makems2) are required for peak-picking and preparative analyses prior to execution of SILACtor. An appropriate peptide/protein database search engine (*e.g.*, Mascot[23], Sequest[24]) is also required, the input for which is usually a simplified raw data file (MS peaks extracted) re-written in a specific format (*e.g.*, mzML, mgf). The ProteoWizard library and toolpack[25] are a set of modular and extensible open-source, cross-platform tools and software libraries that facilitate proteomics data analysis, such as the file conversions required for data submission to peptide search engines (http://proteowizard.sourceforge.net).

There are both Windows and Linux versions of SILACtor, executed *via* the command-line, which output a series of tab-delimited .txt files—one for each MS raw data file in the time series—that contain peptide information (identity, sequence, k_{loss}, RIA, R^2 goodness-of-fit value). If replicate analyses have been carried out, a grouped data file for each time point is also produced. Aggregate protein-level data is written to a final .txt file containing k_{loss}, RIA ($\pm$ SEM) and R^2 data for each protein. The output files are compatible with downstream data-mining/extraction applications. SILACtor does not resolve peptides with multiple labelling centres (such as those produced by tryptic mis-cleavage). Lastly, due to the assumption that the samples follow an all-heavy to all-light (or *vice versa*) trajectory, SILACtor does not complete automatic calculation of the precursor pool RIA.

Topograph[21] is a Windows application designed for protein turnover experiments. It deconvolutes peptide isotope envelopes from the time-related series of LC-MS chromatograms and measures abundances from peak areas, from which it calculates turnover rates at the peptide level. Topograph initially calculates the RIA of the amino acid precursor pool from peptides that contain two or more instances of the stable-isotope-labelled precursor amino acid by adopting the approach first used by Doherty *et al.*[15] Topograph constructs the expected distribution of unlabelled (LL), partially labelled (LH and/or HL positional pairs) and fully labelled (HH) peptides at all values of the amino acid precursor pool RIA in 1% increments and finds the closest match of RIA_p to the experimental data. The measured precursor pool RIA is then incorporated into the analysis of peptides containing a single instance of the label, fitting a monoexponential function to recover the degradation rate constant. Topograph accommodates peptides in parallel samples, even if the peptide ion was not selected for fragmentation. The graphical user interface of Topograph is appealing, allowing visual inspection of the RIA_t data and corresponding fitted exponential curve. Output is in the form of text files or as publication-quality graphics. A considerable amount of post-processing and manual filtering of the data provided by Topograph is required to achieve an accurate representation of experimental data, a requirement that is both time-consuming and that introduces a degree of subjectivity into the analysis.

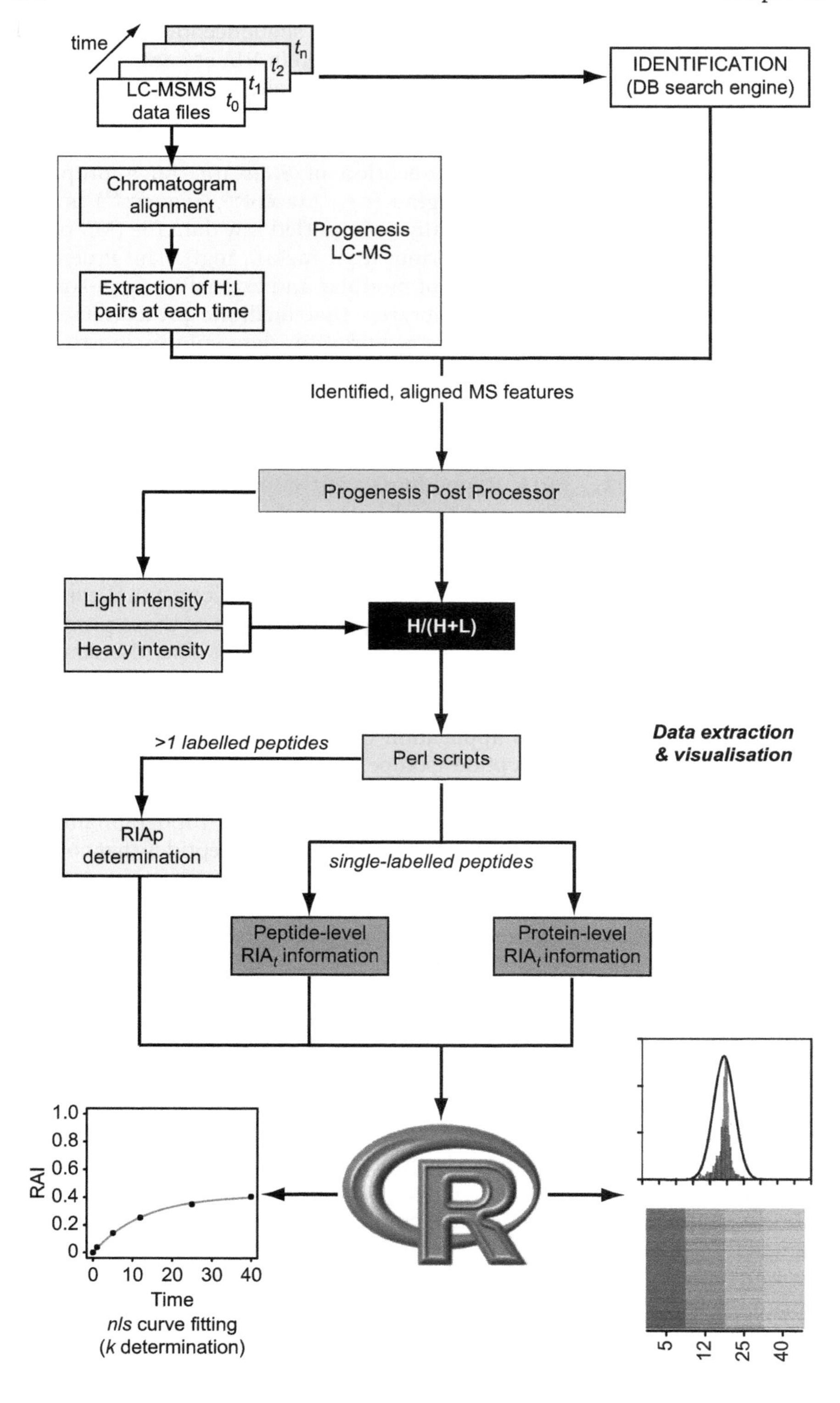
time
t_n
t_2
t_1
LC-MSMS data files t_0
IDENTIFICATION (DB search engine)
Chromatogram alignment
Progenesis LC-MS
Extraction of H:L pairs at each time
Identified, aligned MS features
Progenesis Post Processor
Light intensity
Heavy intensity
H/(H+L)
>1 labelled peptides
Perl scripts
Data extraction & visualisation
RIAp determination
single-labelled peptides
Peptide-level RIA_t information
Protein-level RIA_t information
1.0
0.8
0.6
0.4
0.2
0
RAI
0 10 20 30 40
Time
nls curve fitting
(*k* determination)
5
12
25
40

SILACAnalyzer,[26] integrated into the OpenMS pipeline,[27] is not specifically designed for analysis of protein turnover data. It identifies pairs of isotopic envelopes at fixed *m*/*z* separation with no requirement for prior sequence identification of the peptides. Initially, peptide ions within a range of *m*/*z* values are filtered to form data-point clusters in the time–*m*/*z* plane, which are assigned using hierarchical clustering. Each data point in a cluster corresponds to three intensity pairs, starting with the monoisotopic mass of the light peptide. For a turnover analysis, ratiometric information must be extracted from the output files (gathered from independent runs over the time points of a turnover experiment) and manipulated to obtain the degradation rate constant at the peptide level that has then to be amalgamated to report protein-level information. However, the OpenMS environment creates the opportunity for this additional analysis to be layered on top of the RIA_t acquisition step.

Progenesis LC-MS (Non-linear Dynamics) is a label-free relative-quantification software package that works by aligning successive LC-MS runs, inferring peptide identity from one run to another and retaining intensity data in the absence of identification of the peptide in that particular run. The protein information output contains only relative protein abundances, but the peptide output file provides the identity and intensity of all peptides, so if a variable modification of labelled amino acid is included in the database search, the relative intensities of labelled and unlabelled peptides can be derived. This would demand substantial manual data interrogation, although a Progenesis post-processing tool[28] that includes algorithms to compute the intensity of labelled and unlabelled peptide pairs ($H/(H+L)$) across multiple LC-MS data sets removes a degree of the manipulation necessary.

12.3.6 Open Source Solutions for Turnover Analysis

Based on our own experiences with protein turnover studies we are exploring a flexible approach to the analysis of turnover data, centred on the use of scripting languages and statistical programming tools (*e.g.*, Perl and *R*,[29,30] respectively), outlined in the workflow in Figure 12.4. Central to this workflow/pipeline is a data file containing protein and peptide identification

Figure 12.4 Overall workflow for analysis of proteome-wide turnover data. In our workflow we extract the intensities of labelled and unlabelled peptides using proprietary software coupled with a post-processing tool (although there are open-source solutions). The intensity values of the labelled and unlabelled peptides are then resolved into singly labelled peptides (that are used to generate turnover data) and multiply labelled peptides that are used to provide the RIA_p values at each time point in the labelling curve. Once RIA_0 and RIA_∞ are pre-calculated, the first order rate constant for the turnover of each protein can be calculated by non-linear curve fitting packages, such as those found in the *R* statistical environment, which also provide valuable visualisation tools. In the future we intend to develop this workflow such that a greater part of it resides within this environment.

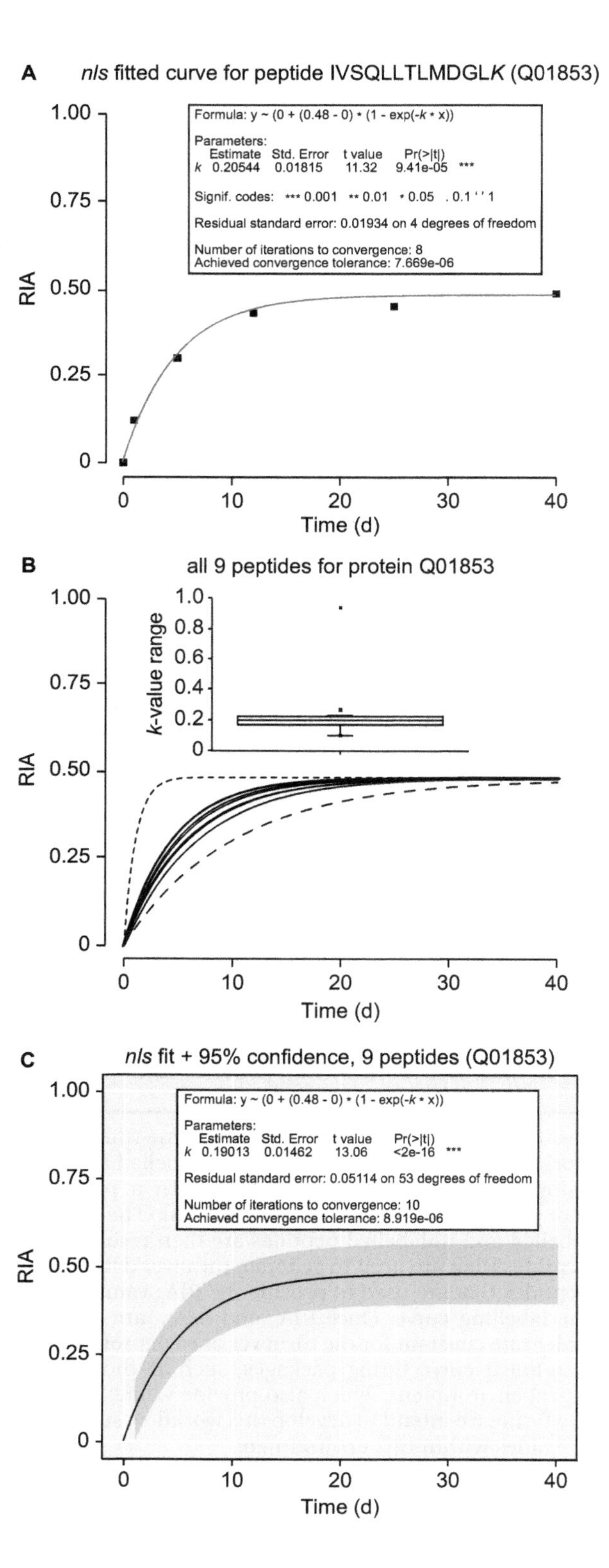
A
nls fitted curve for peptide IVSQLLTLMDGLK (Q01853)
Formula: y ~ (0 + (0.48 - 0) * (1 - exp(-k * x))
Parameters:
Estimate Std. Error t value Pr(>|t|)
k 0.20544 0.01815 11.32 9.41e-05 ***
Signif. codes: *** 0.001 ** 0.01 * 0.05 . 0.1 ' ' 1
Residual standard error: 0.01934 on 4 degrees of freedom
Number of iterations to convergence: 8
Achieved convergence tolerance: 7.669e-06
RIA
Time (d)
B
all 9 peptides for protein Q01853
k-value range
RIA
Time (d)
C
nls fit + 95% confidence, 9 peptides (Q01853)
Formula: y ~ (0 + (0.48 - 0) * (1 - exp(-k * x))
Parameters:
Estimate Std. Error t value Pr(>|t|)
k 0.19013 0.01462 13.06 <2e-16 ***
Residual standard error: 0.05114 on 53 degrees of freedom
Number of iterations to convergence: 10
Achieved convergence tolerance: 8.919e-06
RIA
Time (d)

information together with time-series RIA_t data, or a file containing raw MS1 spectral intensities or area-under-curves (AUCs), from which H/(H + L) calculations can be recovered. For example, this file is generated by the Progenesis Post-Processor package[28] but the equivalent information can also be extracted from other SILAC software tools (see above). These data can then be analysed in *R* to allow for non-linear curve fitting (using, for example, the *nls* [stats] package[31]) and plotting (see Figure 12.5), together with additional data analysis and visualisation. Peptides containing multiple label instances allow recovery of the system-wide RIA_p value. Non-linear curve fitting can be implemented at the peptide or protein level. Although these environments are complex, the accessibility of the software, the accessibility of the individual scripts and the ability to generate high quality graphical outputs using pre-existing and robust packages (such as *R*forProteomics[32]) are strong advocates for this approach.

12.3.7 Optimal Software for the Analysis of Proteome Turnover Data

We suggest that, at the time of writing, there is no optimal solution to the analysis of turnover data. Most solutions are directed to specific experimental designs and contain assumptions that preclude broader applicability to different experimental workflows. A (hypothetical) optimal package for the analysis of proteome-wide protein turnover data would be challenging to produce, but might create a robust, generic solution. The features that might be expected in such a package include:

1. *Compatibility with different labelling protocols*: although experiments that either measure the gain or loss of stable-isotope labels are formally equivalent, the progression of the labelling curve is different. It would be desirable to specify the labelling protocol and anticipate the direction of the transition. Moreover, the selection of different amino

Figure 12.5 Analysis of turnover data in *R*. Panel A: an example of *nls* [stats] fitting carried out on single-peptide-level turnover data, obtained from transitional endoplasmic reticulum ATPase (Q01853), with the model summary shown in the box. Panel B: individual *nls* [stats] fitted curves for all nine peptides identified in the study from protein Q01853, together with a box plot of the range of *k* values in the inset. Panel C: *nls* [stats] fitting carried out on the combined dataset for all nine peptides to obtain the 95% confidence limits and plotted using ggplot2.[33,34] The resultant grouped model summary is displayed in the box. It is obvious that two of the peptides (dotted line) give outlier values for the progression; examination of the data corresponding to these peptides revealed that there are contaminating peptide ions contributing to the recorded intensities. It might be possible to assign other metrics of confidence in individual peptides to refine the k_{deg} estimate.

acids as labelling centres should not impact on the data analysis, and should be specified in the configuration of the analysis.

2. *Calculation of the peptide RIA values at each time point, even if the peptide was not formally identified in each LC-MS run*: in most turnover experiments the protein abundances are constant throughout the labelling period, the change being in the proportional occupation of the 'heavy' and 'light' pools. Thus, one or the other might not be triggered for fragmentation as the abundance declines, even though it should be present in the original sample. Optimal software would extract the heavy and light ions at each time point. Nomination of peptide ions based on the need for identification at each time runs the risk of missing values.
3. *Analysis of precursor RIA*: even if the RIA transition is designed to change between zero and one, or *vice versa*, there is merit in the analysis of the data set to acquire the values of RIA_0 and RIA_∞. This not only provides a health check on the experimental design but also allows determination of the two parameters external to the fitting of the (t, RIA_t) data, reducing the number of parameters and the degrees of freedom in the exponential curve fitting stage.
4. *Use of non-linear optimisation methods for calculation of* k_{deg}: the optimal approach to the calculation of the degradation rate constant is to retain the (t, RIA_t) values in true parameter space and not transform the data (*e.g.*, logarithmically, followed by linear regression). The fitting function should be specified by the user to allow for the incorporation of such variants as biphasic decay or delays in the labelling/unlabelling curves. The function(s) should be externally specified for optimal flexibility.
5. *Handling of missing values and aberrant data*: it would be advantageous to be able to select and highlight peptides that cannot be readily used to determine turnover, such as mis-cleaved peptides or false discoveries/decoy sequences.
6. *Generation of final results table with* RIA_t *values by peptide and by protein*: although the output should not be reported in protein group form (*i.e.*, peptides from the same protein amalgamated to give one value of k_{deg}), it is also valuable to access simple text files defining turnover behaviour on a peptide-by-peptide basis.

The rapidly emerging interest in the measurement of protein turnover on a proteome-wide scale has brought with it a variety of labelling protocols, experimental systems and analytical strategies. At present, there are no studies that are sufficiently well replicated to allow evaluation of the optimal experimental and analytical approach. Yet, the overall experimental workflows and data analyses share many common features, and a convergent platform to collate and cross-compare turnover studies is within reach. The ideas we propose above are readily deliverable and could be made available in an open-source environment with relatively limited resources. When such tools are available, it will be feasible to conduct higher level syntheses of the turnover domain, permit comparison of structural determinants of intracellular stability and relate turnover to the synthetic and degradative capacity of the cell.

Acknowledgments

This work was supported by grants from the Biotechnology and Biological Sciences Research Council (BB/G009112/1) and the Natural Environment Research Council (NE/H009558/1). The authors would like to thank Mike Hoopmann and Nick Shulman for their valuable input whilst testing SILACtor and Topograph (respectively).

References

1. A. J. Claydon, A. Pennington, S. A. Ramm, J. H. Hurst, P. Stockley and R. J. Beynon, *Mol. Cell. Proteomics*, 2012, **11**, 014993.
2. A. J. Claydon, M. D. Thom, J. L. Hurst and R. J. Beynon, *Proteomics*, 2012, **12**, 1194.
3. S. Guan, J. C. Price, S. B. Prusiner, S. Ghaemmaghami and A. L. Burlingame, *Mol. Cell. Proteomics*, 2011, **10**, 010728.
4. J. M. Pratt, J. Petty, I. Riba-Garcia, D. H. Robertson, S. J. Gaskell, S. G. Oliver and R. J. Beynon, *Mol. Cell. Proteomics*, 2002, **1**, 579.
5. A. J. Claydon and R. J. Beynon, *Mol. Cell. Proteomics*, 2012, **11**, 1551.
6. A. J. Claydon and R. J. Beynon, *Methods Mol. Biol.*, 2011, **759**, 179.
7. R. Busch, Y. K. Kim, R. A. Neese, V. Schade-Serin, M. Collins, M. Awada, J. L. Gardner, C. Beysen, M. E. Marino, L. M. Misell and M. K. Hellerstein, *Biochim. Biophys. Acta*, 2006, **1760**, 730.
8. T. Y. Kim, D. Wang, A. K. Kim, E. Lau, A. J. Lin, D. A. Liem, J. Zhang, N. C. Zong, M. P. Lam and P. Ping, *Mol. Cell. Proteomics*, 2012, **11**, 1586.
9. Y. Zhang, S. Reckow, C. Webhofer, M. Boehme, P. Gormanns, W. M. Egge-Jacobsen and C. W. Turck, *Anal. Chem.*, 2011, **83**, 1665.
10. J. C. Price, S. Guan, A. Burlingame, S. B. Prusiner and S. Ghaemmaghami, *Proc. Natl. Acad. Sci. USA*, 2010, **107**, 14508.
11. S. Gruhler and I. Kratchmarova, *Methods Mol. Biol.*, 2008, **424**, 101.
12. M. K. Doherty, D. E. Hammond, M. J. Clague, S. J. Gaskell and R. J. Beynon, *J. Proteome Res.*, 2009, **8**, 104.
13. D. Van Hoof, M. W. Pinkse, D. W. Oostwaard, C. L. Mummery, A. J. Heck and J. Krijgsveld, *Nat. Methods*, 2007, **4**, 677.
14. S. K. Park, L. Liao, J. Y. Kim and J. R. Yates III, *Nat. Methods*, 2009, **6**, 184.
15. M. K. Doherty, C. Whitehead, H. McCormack, S. J. Gaskell and R. J. Beynon, *Proteomics*, 2005, **5**, 522.
16. J. C. Price, W. E. Holmes, K. W. Li, N. A. Floreani, R. A. Neese, S. M. Turner and M. K. Hellerstein, *Anal. Biochem.*, 2012, **420**, 73.
17. M. K. Doherty and R. J. Beynon, *Expert Rev. Proteomics*, 2006, **3**, 97.

18. C. Trotschel, S. P. Albaum, D. Wolff, S. Schroder, A. Goesmann, T. W. Nattkemper and A. Poetsch, *Mol. Cell. Proteomics*, 2012, **11**, 512.
19. F. M. Boisvert, Y. Ahmad, M. Gierlinski, F. Charriere, D. Lamont, M. Scott, G. Barton and A. I. Lamond, *Mol. Cell. Proteomics*, 2012, **11**, 011429.
20. M. R. Hoopmann, J. D. Chavez and J. E. Bruce, *Anal. Chem.*, 2011, **83**, 8403.
21. E. J. Hsieh, N. J. Shulman, D. F. Dai, E. S. Vincow, P. P. Karunadharma, L. Pallanck, P. S. Rabinovitch and M. J. MacCoss, *Mol. Cell. Proteomics*, 2012, **11**, 1468.
22. M. R. Hoopmann, M. J. MacCoss and R. L. Moritz, *Curr. Protoc. Bioinformatics*, 2002, **37**, 1.
23. D. N. Perkins, D. J. Pappin, D. M. Creasy and J. S. Cottrell, *Electrophoresis*, 1999, **20**, 3551.
24. J. Eng, A. McCormack and J. Yates, *J. Am. Soc. Mass Spectrom.*, 1994, **5**, 976.
25. D. Kessner, M. Chambers, R. Burke, D. Agus and P. Mallick, *Bioinformatics*, 2008, **24**, 2534.
26. L. Nilse, M. Sturm, D. Trudgian, M. Salek, P. G. Sims, K. Carroll and S. Hubbard, SILACAnalyzer – A Tool for Differential Quantitation of Stable Isotope Derived Data, in F. Masulli, L. E. Peterson, R. Tagliaferri (eds.), *Computational Intelligence Methods for Bioinformatics and Biostatistics: 6th International Meeting, CIBB 2009, Genoa, Italy, Revised Selected Papers*, Springer, 2009, 45–55.
27. M. Sturm, A. Bertsch, C. Gropl, A. Hildebrandt, R. Hussong, E. Lange, N. Pfeifer, O. Schulz-Trieglaff, A. Zerck, K. Reinert and O. Kohlbacher, *BMC Bioinformatics*, 2008, **9**, 163.
28. D. Qi, P. Brownridge, D. Xia, K. Mackay, F. F. Gonzalez-Galarza, J. Kenyani, V. Harman, R. J. Beynon and A. R. Jones, *Omics*, 2012, **16**, 489.
29. R. Ihaka and R. Gentleman, *J. Comput. Graph. Stat.*, 1996, **5**, 01299.
30. R Core Team, *R Foundation for Statistical Computing*, 2013, http://www.R-project.org/, Accessed 10th October 2013.
31. D. G. Rossiter, *Technical note: Curve fitting with the R Environment for Statistical Computing*, International Institute for Geoinformation Science & Earth Observation (ITC), Enschede (NL), 2009, 1.
32. L. Gatto and A. Christoforou, *Biochim Biophys Acta*, 2013, DOI: 10.1016/j.bbapap.2013.04.032.
33. C. Ginestet, *J. R. Stat. Soc.: A*, 2011, **174**, 245.
34. H. Wickham, *ggplot2: Elegant Graphics for Data Analysis*, Springer, New York, 2009.

APPLICATIONS OF QUANTITATIVE PROTEOMICS

CHAPTER 13

Protein Quantification by MRM for Biomarker Validation

L. STAUNTON,[a] T. CLANCY,[b] C. TONRY,[a] B. HERNÁNDEZ,[a] S. ADEMOWO,[a] M. DHARSEE,[c] K. EVANS,[c] A. C. PARNELL,[d] R. W. WATSON,[a] K. A. TASKEN[b] AND S. R. PENNINGTON*[a]

[a] UCD Conway Institute, School of Medicine and Medical Science, University College Dublin, Dublin 4, Ireland; [b] Department of Tumor Biology, Institute for Cancer Research, Oslo University Hospital, Norway; [c] Ontario Cancer Biomarker Network, Ontario Cancer Biomarker Network, Toronto, Ontario, M5A 2K3, Canada; [d] School of Mathematical Sciences, University College Dublin, Dublin 4, Ireland
*Email: Stephen.Pennington@ucd.ie

13.1 Introduction

A biomarker may be defined as "a characteristic that is objectively measured and evaluated as an indicator of normal biological processes, pathogenic processes or pharmacological responses to therapeutic intervention".[1] While there are a multitude of similar definitions of a biomarker, this is one that works well to highlight some of the key features of a biomarker. The words "measured" and "evaluated" imply that both an analytical process to determine the level of the biomarker and subsequent interpretation of the data are important. Additionally, it has been suggested that a biomarker may be an indicator or a surrogate, serving the purpose of diagnosis or to monitor the prognosis of a disease.[2] This, in turn, implies that a biomarker aids, and indeed may be essential to, a clinical decision making process—and perhaps

New Developments in Mass Spectrometry No. 1
Quantitative Proteomics
Edited by Claire Eyers and Simon J Gaskell

Published by the Royal Society of Chemistry, www.rsc.org

this is the fundamental objective of a biomarker—that it should have clinical utility. For this reason, it has been suggested that potential or candidate biomarkers should not be afforded the title of "biomarker" until they have been demonstrated to have clinical utility—that is until it they have been "validated" and are in use. The biomarker discovery and development process has been compared to and shown to have similarities to drug development and so, by analogy, it could be argued that biomarkers that have been discovered but not yet developed should be called "**new biomarker entities**" and that once they have been selected for development they might be called "**candidate biomarkers**" or "**lead candidates**", and only when they have been validated as having appropriate performance characteristics should they be called "**biomarkers**".

Existing clinical biomarkers, including those used in oncology, rheumatology and many other diseases, guide clinicians in choosing the best course of action for the treatment of a patient and their disease. As noted, these biomarkers are used for early diagnosis, disease prognosis, and the selection of appropriate treatment strategies and subsequent monitoring of treatment response. They are also used to support the identification of aggressive forms of a disease or disease subtypes, as well as the prediction of treatment failure or disease recurrence. Of particular value are **companion biomarkers** that are used to guide the more effective use of a particular therapy or pharmaceutical intervention for an individual patient.[3,4] Collectively, biomarkers underpin "individualised" or "personalised" medicine.[5–7]

In addition to excellent "performance" (high specificity and sensitivity) an ideal biomarker for predicting and monitoring patient response to therapy would be one that could be measured in a relatively non-invasive manner using a readily obtained sample that might be obtained repeatedly from the same patient. For this reason, much interest has focused on use of biofluids, such as blood, urine and cerebrospinal fluid, as the samples of choice for the measurement of biomarkers. There are also instances in which the presence of a biomarker in the main tissue(s) associated with the disease may be more useful and there are many well-established and very valuable tissue biomarkers. Notably, biomarkers that establish the presence of the protein target of therapeutic intervention or alterations in protein pathways associated with the disease can play a highly significant role in the treatment of patients. For example, the presence of HER2 is used to establish over-expression of the epidermal growth factor (EGF) receptor in breast cancer tissue, indicating an aggressive form of the disease that can be treated effectively with agents (such as trastuzumab and lapatinib, which target (inhibit) the EGF receptor).

Biomarkers in current use include a wide range of clinical measurements, ranging from blood cell counts, erythrocyte sedimentation rate, serum analytes and antigens/antibodies and more "molecular" entities, such as single-nucleotide polymorphisms (SNPs), mRNAs and proteins (peptides). For example, biological markers used in rheumatoid arthritis and psoriatic arthritis to establish the presence of inflammation and disease activity include measurement of acute-phase reactants as surrogate markers of the disease process, as well as levels of C-reactive protein (CRP) and fibrinogen in combination with measures of the erythrocyte sedimentation rate, elevated liver enzymes (aminotransferase, serum aspartate and alkaline phosphatase) and serum albumin levels.[8,9] While these, and a huge range of other molecular biomarkers, are in routine use, it is recognised that many have significant limitations and do not provide the desired levels of sensitivity and specificity.[9–11] For example, prostate specific antigen (PSA), a protein that is used to diagnose the presence and recurrence (post-treatment) of prostate cancer (PCa), is widely acknowledged as lacking appropriate specificity for screening purposes and its use has led to widespread over-treatment of the disease.[11–14] Notwithstanding these well known limitations, a large number of protein biomarkers are used to great effect (see Table 13.1).

13.1.1 Biomarker Panels

The measurement of individual protein biomarkers is often used in conjunction with other measurements and/or clinical data to arrive at a "score", which is then used to support a clinical decision; for example, in prostate cancer, nomograms and Partin tables are used[11,15] and, in rheumatology, a series of measures are combined and used to establish disease severity.[16,17] A big effort is made in establishing and testing these scoring methods, which must then be approved for use by relevant disease-specific organisations.[18]

In recent years it has become increasingly apparent that a single protein biomarker, even when incorporated into an overall scoring classification, may not be sufficiently informative. It has been argued that, for a biomarker to be of any significant clinical use, it is vital that it captures the disease course pathophysiology to provide a true reflection of disease outcome.[19] And, it seems likely that a pathological function cannot be solely attributed to a single molecule.[20] It is now accepted that combining and measuring several molecular biomarkers simultaneously, *i.e.*, a biomarker panel or **biomarker signature**, will ultimately provide greater performance. Examples of biomarker panels or biomarker signatures are emerging. An optimal biomarker panel would contain markers whereby each represent a distinct pathophysiological pathway, target presence, disease activity, *etc.* and, when measured as a whole, would offer complementary information and allow improved disease diagnosis, more definitive risk assessment and more successful individualised patient therapy.[21]

Numerous recent publications reflect the efforts to construct and validate biomarker panels. For example, in cardiovascular disease (CVD), multiple biomarkers from the main pathophysiological pathways of CVD—myocardial cell damage, left ventricular dysfunction, renal failure and inflammation—were measured in combination to assess if this would add substantial prognostic information with regard to risk of death from CVD.[22]

Table 13.1 Illustration of some approved cancer biomarkers with the biological sample in which they are measured and their usage (adapted from Pan *et al.* 2011).[46]

Protein target	*Glycosylation*	*Detection*	*Source*	*Disease*	*Biomarker Use*
α-Fetoprotein	Yes	Glycoprotein	Serum	Liver cancer, Hepatocellular carcinoma, Nonseminomatous testicular cancer	Diagnosis
Human chorionic gonadotropin-β	Yes	Glycoprotein	Serum	Testicular cancer, Gestational trophoblastic disease	Diagnosis
CA19-19	Yes	Carbohydrate	Serum	Pancreatic cancer	Monitoring
CA125	Yes	Glycoprotein	Serum	Ovarian cancer, Heart failure	Monitoring
Carcinoembryonic antigen	Yes	Protein	Serum	Colon cancer	Monitoring
Epidermal growth factor receptor	Yes	Protein	Tissue	Colon cancer, Lung cancer	Therapy selection
KIT	Yes	Protein (IHC)	Tissue	Gastrointestinal (GIST) cancer	Diagnosis/therapy selection
Thyroglobulin	Yes	Protein	Serum	Thyroid cancer	Monitoring
Prostate specific antigen	Yes	Protein	Serum	Prostate cancer	Screening/ monitoring/ diagnosis
CA15-3	Yes	Glycoprotein	Serum	Breast cancer, Benign ovarian cysts, Benign lung disease	Monitoring
CA27-29	Yes	Glycoprotein	Serum	Breast cancer	Monitoring
HER2/NEU	Yes	Protein	Tissue	Breast cancer	Prognosis/therapy selection/ monitoring
Fibrin/FDP-fibrin degradation protein	Yes	Protein	Tissue	Bladder cancer, Acute coronary syndrome	Monitoring
Bladder tumour-associated antigen (complement factor H-related protein)	Yes	Protein	Urine	Bladder cancer	Monitoring
CEA and mucin (high molecular weight)	Yes	Protein (IF)	Urine	Bladder cancer, Colon cancer	Monitoring

Proto-oncogene c-Kit: KIT; immunohistochemistry: IHC; immuno florescence: IF.

In this study troponin I, NT-proBNP, cystatin C and hsCRP levels were incorporated into a model with established risk factors and this was shown to improve risk stratification for death from CVD.[22] Similarly, Ahluwalia *et al.* evaluated the comparative prognostic values of four biomarkers, MR-proANP, NT-proBNP, CRP and homocysteine, in addition to conventional risk factors in patients with stable CVD.[23] Findings from this particular study suggested an improvement in risk stratification for CVD, not only for heart failure, but also other major and overall CVD events when combinations of biomarkers were measured.[23] In another recent study O'Hartaigh and co-workers evaluated a panel of nine biomarkers, which represented different bio-physiological pathways thought to be involved in the pathogenesis of CVD.[24] Here, it was demonstrated that the simultaneous addition of nine biomarkers provided further incremental value in terms of immediate adverse CVD prognosis and during a ten-year follow-up, which is in contrast to the use of conventional risk factors alone. The studies described above, and many others like them, use a retrospective approach for the assessment of the potential benefits of biomarker panels. However, in order to bring biomarkers from the discovery stage to clinical validation, prospective studies (clinical trials) are required. This is an important consideration and one that requires a very different approach from the work leading up to this stage in biomarker development.

To summarise, it is widely recognised that new and better biomarkers are needed for a range of diseases and associated therapeutic strategies. Furthermore, as new treatment strategies and new therapeutics are developed, it is clear that there is a huge demand for more effective biomarkers capable of more accurate and precise diagnostic and prognostic capabilities. It can be reasonably expected that such biomarkers — as part of a biomarker panel — will reduce patient morbidity and mortality significantly.[6] Bluntly, existing biomarkers have limitations and new biomarkers are needed.

13.1.2 Protein Biomarker Discovery and Development

It has been widely assumed that new and improved performance biomarkers will arise from "-omic" studies but, to date, relatively few have; this begs the obvious question as to why? There are multiple reasons for this lack of success and many of them have been described and discussed at length.[25,26] Many existing biomarkers are proteins (see Table 13.1) and so it stands to reason that proteomics has been expected to deliver new biomarkers. Proteomics, namely the analysis of proteins on a genome-wide scale, has now evolved and has been elaborated to include many elements that range from attempts to use mass spectrometry to measure whole proteomes[27,28] and chemical proteomic strategies to analyse protein activities and their inhibition[29,30] to sophisticated methods for the measurement of protein folding and folded structures to link protein structure with activity.[31,32] Together, these strategies afford the opportunity to understand cellular protein-driven processes in a more holistic manner and form part of a new systems approach to understanding biological processes, including in human health and disease, so contributing to a greater understanding of complex disease pathways. An understanding of perturbations in these pathways during disease initiation and progression may, in turn, lead to the identification new biomarkers. Not surprisingly, proteomic methods have become a key element in protein biomarker discovery and validation.[33]

The process of protein biomarker discovery involves a number of discrete steps (as illustrated in Figure 13.1); however, for many reasons, the process has, to date, enjoyed limited success.[34,35]

The challenges inherent in this long and multi-step process from discovering a new candidate protein biomarker to having it used as a clinically valuable tool are technical, logistical and regulatory/legal. The limitations of existing strategies have been reviewed and discussed in detail.[14,27,36,37] It has been suggested that a major stumbling block in taking a potential biomarker from discovery to clinical utility originates at the outset of the process, whereby misguided attempts have been made to discover potential biomarkers using clinical samples that are readily available without a clear definition of what and how the biomarkers might usefully contribute to clinical decision-making. We and others have argued that biomarker discovery and development should be end-user driven, address a specific clinical question or need, and begin by using the most appropriate samples in the discovery phase.[37] All too often biomarker discovery has used

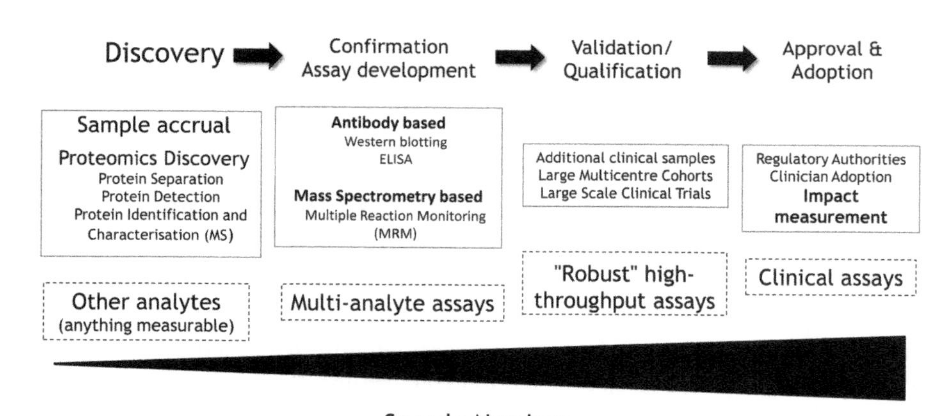

Figure 13.1 Protein biomarker development pipeline. The consecutive analytical stages a candidate biomarker has to go through in order to progress from discovery experiments in a mass spectrometry laboratory and confirmation assay development right through to validation using large sample cohorts and multi-centre trials. Multiple reaction monitoring (MRM) assays have the potential to be used throughout this process.

sophisticated "-omic" (including proteomic) discovery technologies but has undertaken them using inappropriate patient samples.[26]

We suggest that one further reason for this lack of success lies in the confusion of the different discrete end uses that a biomarker may have. For example, for a biomarker that directly measures the likely efficacy of a particular pharmaceutical treatment by virtue of measuring the target protein, it is clear that knowledge of the protein biomarker and its role in the disease pathogenesis may be essential. However, for a biomarker signature that measures directly whether a patient will respond or has responded to a particular therapeutic regime, it may be sufficient for there to be a proven (validated) correlation between the biomarker signature score and patient response. Detailed knowledge of the role of the individual biomarkers within the signature in the disease process in this instance may be superfluous. We have argued that, in this case, the biomarker will be discovered only if relevant patient samples are used, *i.e.*, patient samples that reflect the conditions one wishes to be able to discriminate.[37]

Historically, confirmation of protein changes in proteomics experiments has been undertaken on a limited number of candidates using methods that rely on antibody-based strategies.[38,39] Despite recent impressive and substantial efforts to generate antibodies to human proteins, including the leading "Protein Atlas" project,[40] the development of antibody-based protein assays remains time consuming and extremely challenging. More importantly, the multiplexing of such measurements has had relatively limited success beyond measuring soluble cytokines and other selected proteins for which there are good antibody pairs available. In spite of these challenges there is continued interest in, and active development of, antibody-based multiplexed methods (including in-solution and solid-phase approaches) and these have been successfully applied for the discovery and initial validation of panels of candidate biomarkers.[40]

The relatively recent introduction of multiple reaction monitoring (MRM) to the measurement of peptides, with selected reaction monitoring (SRM) originally being used for small molecule measurements, has afforded the important opportunity and pressing need to develop, in a multiplexed manner, rapid, flexible and relatively cheap assays for measuring multiple candidate biomarkers and to progress them through the biomarker development pipeline (see Figure 13.1).

The ability to measure tens/hundreds and potentially thousands of biomarkers simultaneously has made a huge impact on the process of protein biomarker development, even to the extent that mass spectrometry-based measurements for validation of these candidates on thousands of samples and subsequent conversion to clinical assays is now an area of active academic and commercial interest. The opportunity to have all this data available has, in turn, placed emphasis on the development of appropriate algorithms and statistical tools for evaluating the biomarker panel/signature performance.

The potential success of any candidate biomarker resides in the considerable care and attention to detail applied to all stages of the biomarker pipeline, from identifying the unmet clinical need to designing the discovery studies and, finally, to the rigorous validation of potential candidate biomarkers. Studies that follow recommendations, such as those in the reporting guidelines for the Standards for Reporting of Diagnostic Accuracy, which clearly outline how validation studies should be designed and executed,[41,42] are relatively scarce.

In the following sections we discuss the processes of assembly of new candidate biomarkers into a panel and the prioritisation of individual candidate biomarkers for progression to MRM assay development. In addition, considerations for the analysis of MRM data and the statistical tools that can be applied to candidate biomarkers to support the development of robust candidate biomarker signatures that may be suited for clinical validation will be evaluated. We draw our observations and comments from recent publications and experience of developing MRM assays for candidate biomarkers that have the potential to address key issues and clinical questions in pre-clinical toxicity, inflammatory arthritis (including psoriatic arthritis) and prostate cancer.

13.1.3 Clinical Samples

For any potential biomarker to make a successful transition from the research environment to routine clinical practice, it is imperative that appropriate and well-characterised clinical specimens are used for both discovery and validation. Blood, in the form of serum or plasma, has been of particular interest as a sample for biomarker discovery (and subsequent clinical use) because of its easy accessibility and potential to support repeated sampling and longitudinal (time series) measurements. It is also believed to be a rich source of potential biomarkers of cancers and other diseases.[43] However, the complexity of this biofluid and the wide dynamic range of the concentration of individual proteins within it (a range which spans at least ten orders of magnitude) makes proteomics-based discovery of novel biomarkers challenging.[44] Furthermore, a relatively small number of proteins represent a large proportion of total blood protein, with albumin alone constituting about half of the protein content in blood. It is for these reasons that many researchers choose to remove or "deplete" serum/plasma of higher abundant proteins and/or fractionate the sample prior to undertaking proteomics-based biomarker discovery of the fluid. Such approaches support the analysis of lower abundance proteins that would otherwise be masked by the abundant ones. The extent to which plasma or serum needs to be depleted and/or fractionated to enable the discovery of sufficiently sensitive and specific biomarkers remains the subject of some debate.[45] Other approaches to the analysis of this fluid include the initial enrichment of proteins, such as glycoproteins — a strategy that takes advantage of the realisation that most protein biomarkers in current clinical use are glycoproteins.[46]

Some suggest that tissue samples from the site of disease are better for the protein-based discovery of novel biomarkers. In rheumatology the synovium is a primary site of inflammation and it is increasingly of interest to study the synovial tissue (ST) of patients. Van Kuijk and colleagues suggested that investigation of the protein profile of ST samples of patients with inflammatory arthritis is a potential rich source of biomarkers and may reveal more information on the inflammatory mechanisms of the disease.[47] However, due to the invasive nature of obtaining this sample, the possibility of converting any biomarkers discovered in it to use in a less invasive matrix, such as serum or synovial fluid, represents a significant challenge. There are also limitations in tissue-based biomarker discovery, one of the most obvious being tissue heterogeneity: the disease may be present in highly variable amounts within the different patients' samples. In an attempt to overcome this some have used enrichment techniques, such as laser capture microdissection (LCM), to isolate cells of interest and reduce contamination with surrounding cells.[48] This approach has even been extended to formalin-fixed paraffin-embedded (FFPE) material, which potentially enables discovery experiments to be undertaken using clinically archived FFPE tissue, thus opening up the opportunity to use a vast bank of material deposited in hospitals/biobanks worldwide.[49,50] The application of quantitative proteomics to these workflows has been relatively limited as LCM of fresh or archival FFPE material is an arduous, time-consuming and exacting task that yields very low amounts of protein for subsequent analysis.

13.2 From Candidate Biomarker Discovery to Verification

Moving from discovery to verification is a significant bottleneck in the biomarker pipeline. One of the primary reasons for the lack of new protein biomarkers reaching clinical utility is the struggle to evaluate lengthy lists of candidate biomarker proteins in a statistically relevant number of patient samples with sufficient sensitivity and specificity. It is at the verification stage that hundreds of potential biomarkers need to be screened against hundreds to thousands of patient samples for the evaluation of their true clinical utility.[36] Verification has a singular goal: to determine if there is enough evidence for potential clinical utility. It is achieved by performing pilot studies with the maximum number of candidate biomarkers, the highest possible throughput and the lowest cost to ensure at least a few successes move forward. Verification platforms are optimised for the high-throughput measurement of pre-selected candidate proteins and they are generally more superior in sensitivity, precision and specificity when compared to biomarker discovery platforms.[44,51] There are two common types of assays developed to verify and validate potential protein biomarkers: antibody-based and MS-based assays. Currently, clinical validation relies primarily on immunoassays due to their specificity for the target analyte, sensitivity and high throughput.[52] Enzyme-linked immunoabsorbant assay

(ELISA) is the current gold standard for protein measurement in patient samples. Today, the emergence of multiplexed ELISAs performed in 96-well plates or bead-based formats permit higher throughput and the potential to evaluate biomarker panels. However, ELISAs are not without significant limitations. For example, a wide variety of variables are known to affect the performance characteristics of an ELISA; at present, immunoassays exist for less than a few thousand proteins, multiplexing these measurements is extremely difficult and the generation of appropriate quality antibody reagents to "novel" proteins is time consuming and costly.

MRM assays have emerged as an alternative to affinity-based measurements for the targeted quantification of proteins[53,54] (see Chapter 4 for more detailed information). The emergence of MRM technology as a powerful method for multiplexed biomarker verification is largely due to advances in instrumentation and associated software for assay design and data analysis. MRM assays have three major advantages over antibody affinity assays: (i) high specificity for the protein of interest or its isoform without requiring the use of antibodies; (ii) relatively short timelines for assay development; (iii) high multiplexing capabilities. MRM assays readily support the simultaneous measurement of tens to hundreds of candidate biomarkers in a single liquid chromatography-mass spectrometry (LC-MS) run[55,56] and this has contributed to its increasingly popular use in the verification and validation of candidate biomarkers.[57,58] So far, many research groups have focused their attention on establishing the performance capabilities of instrumentation and methods basing their objectives on the predicted requirements of potential assays of clinical utility.[59,60] Others have focused on using the approach to undertake verification and validation of panels of candidate biomarkers that have potential for clinical utility.[61,62]

13.2.1 Assembly of Candidate Protein Biomarker Panels

The importance of new biomarker discoveries is recognised by the large sums of both private and public funding that has been used to support biomarker discovery and validation studies worldwide. But, despite numerous publications each year (a Pubmed search identifies 1200 biomarker discovery papers alone in 2012 with 530 from January to July 2013), very few candidate biomarkers reach clinical utility. It is now well accepted that biomarker panels or signatures are more likely to have appropriate sensitivity and specificity than the traditional individual biomarker. For example, in breast cancer where the traditional biomarkers, such as ki67, HER2, estrogen and progesterone receptors, are used as predictive and/or prognostic markers, the Oncotype DX test, which measures expression levels of 16 outcome genes and 5 reference genes combined as a recurrence score, is showing promise for improved biomarker performance.[63] In short, large numbers of potential biomarkers and the availability of multiplexed verification by MRM presents a new opportunity to establish biomarker panels of improved utility.

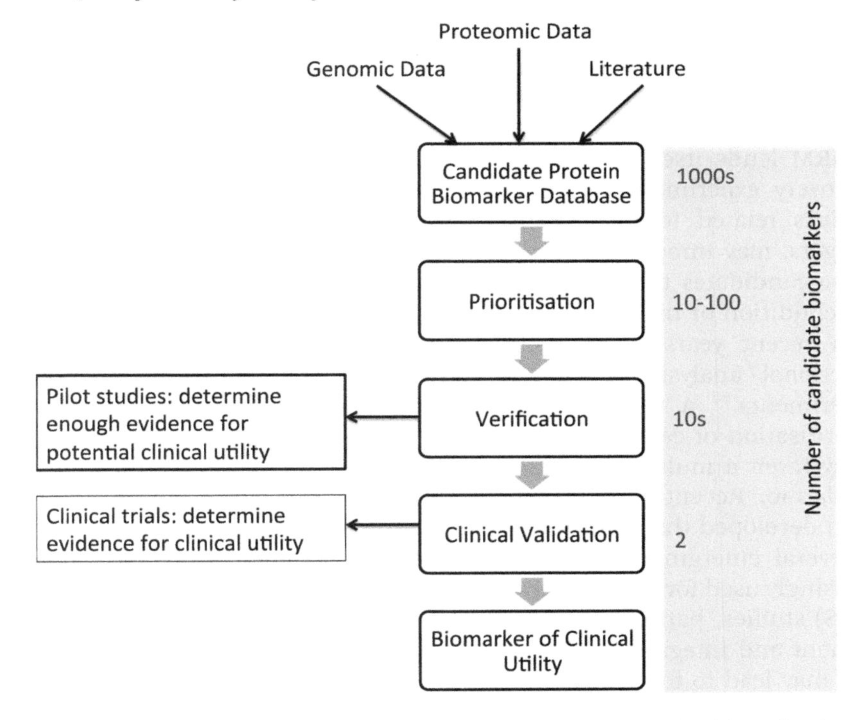

Figure 13.2 Assembly protein biomaker panels. The steps involved in reduction by prioritisation of candidate biomarkers from 1000s in the initial biomarker database to the final one or two potential candidates that demonstrate enough sensitivity and specificity to justify their continuation along the validation path.

It has also become clear that candidate protein biomarker panels can be assembled from multiple sources that include protein discovery experiments, analysis of the literature and of the publicly available databases of gene (transcript) expression (Figure 13.2). Notably, it has been demonstrated that candidates discovered in transcriptomics experiments can subsequently be measured and verified at the protein (peptide) level by MRM[64] so opening the way to developing protein biomarkers based on transcriptomic discovery experiments.

13.3 Integrative Bioinformatics for Biomarker Prioritisation

While MS-based biomarker discovery studies can generate large sets of differentially expressed proteins, such studies are often characterised by small sample sizes and the use of MS technology of modest quantitative sensitivity. To compensate for these limitations, statistical criteria used to determine differential expression in the discovery phase tend to be less stringent than those used in subsequent phases. This enables investigators to cast a wider net on potential biomarkers, thus effectively reducing the false negative rate, but inevitably admitting a disproportionate number of false positive discoveries.

MRM lends itself well to the confirmation of candidates derived from discovery experiments, particularly MS-based approaches. However, constraints related to instrument capacity, manpower and available study budgets, may impose a selection process to identify, or at least to predict, those candidates that present the greatest promise and relevance towards the condition or biological question under investigation.

In recent years many bioinformatics toolkits have emerged for the functional analysis of large protein lists generated from proteomics experiments.[65] A related challenge to that of functional analysis is the prioritisation of candidate biomarkers for selection prior to developing an assay (even a multiplexed assay) for candidate biomarker verification and validation. Recently, more sophisticated bioinformatics approaches have been developed that help address this issue

Several emerging prioritisation strategies are being developed and increasingly used for disease-gene prioritisation in next generation sequencing (NGS) studies, particularly in cancer.[66] These methods, with further development and integration, are possibly adaptable to quantitative proteomics and may lead to improved discovery.

13.3.1 Gene Ontology as a Tool for Biomarker Discovery

Gene Ontology (GO)[67] is an indispensable resource for the characterisation and annotation of candidate proteins. Each GO term is a description of a biological function, which in turn is mapped to a computer-readable identifier. Within GO, each protein is described by a "structured vocabulary", represented as a hierarchical tree with different levels of granularity of functional descriptions for each protein. With the advent of high-throughput assays many bioinformatics web services and tools have been developed to analyse large gene and protein lists using statistical approaches in enrichment[68] (or over-representation) analysis with GO. One of the most widely used tools for this purpose is DAVID;[69] however, numerous such tools exist[70] and their operations are continuously improving.[70] The enrichment analysis carried out using these tools is not limited to GO as a functional resource, but can also integrate other functional and pathway information.[71,72]

The integration of molecular network analysis (discussed below) with GO-based enrichment analysis and expression is a common approach that has led to success, and which many tools, such as ClueGO,[73] now facilitate. These tools mainly entail building molecular networks from differentially expressed proteins and identifying sub-networks enriched in biological functions over-represented by the GO terms. A big challenge with GO analysis, however, is to overcome the lack of detailed annotated information for genes and proteins.[74] For example, many proteins in GO suffer from a lack of functional descriptions at high-resolution detail, and indeed many proteins still remain to receive annotations.[74] Bioinformatics methods and databases will need to be developed to support the creation of accurate, high-resolution annotation of detailed conditions, tissue specificity, secretome potential and context specific functions of each protein.

13.3.2 Data- and Text-mining Approaches to Prioritise Biomarkers

Many approaches for biomarker prioritisation use data-mining techniques applied to microarray gene expression profiles.[75,76] However, these approaches usually have been analysed on tissue and have a drawback in that they rely on a somewhat flawed assumption that if a gene or protein is significantly differentially expressed, it may also be so in peripheral sources, such as blood, serum or urine. Efforts to counteract this assumption and incorporate scoring strategies and likelihood measures for the protein(s) to be detected in peripheral sources would be of great value in exploiting quantitative proteomics assays. One proposed solution is the integration of databases of proteomics experiments in which candidates have been found in peripheral sources. These are becoming increasingly available in the public domain.[77]

Arguably, there is a vast source of potentially untapped biomarkers residing in the biomedical literature. One promising bioinformatics approach to extract this information from the over 22 million articles in MEDLINE is the use of text mining tools. There has been a continuous growth in biomedical text-mining leading to the emergence of a large number of services aimed at extracting and organising biomedical knowledge.[78] Methods and tools are being developed, for example, to extract disease biomarkers from MEDLINE and map their associations to pathways and disease.[79]

Inevitably, text mining is an automatic process and it remains error-prone due to the challenges of training complex algorithms to gain complete natural language understanding of text.[80] Although this is currently difficult, text mining can nevertheless be a valuable toolkit for extracting knowledge, not only for the creation of biomarker knowledge-bases, but also to assign ranked relevance scores to candidates for prioritisation. For example, using text mining, a ranked immune relevance score for all human genes was used to evaluate gene expression profiles and quantify the immunological component of tumours.[81] Further work is needed, however, to extract knowledge of candidate biomarkers from the complete corpus of the biomedical literature.

13.3.3 Network and Pathway Approaches to Rank Candidates

Biomarker prioritisation can benefit greatly from the new paradigm of "network biology",[82] which has materialised in the field of bioinformatics, genomics and proteomics in recent years. This has brought the ability to understand the function of a protein, not as a discrete unit in a linear set of biochemical events, but rather as a complex network of interdependent molecular relationships. The conventional single-protein reductionist approach is being replaced by an increased understanding of the enormous complexity inherent in large molecular networks within the cell. Consequently, rather than focusing on a one-dimensional protein list, candidate prioritisation is often better served by the analysis of the relationship and contribution of candidates towards the structure, organisation and information flow within the cell's molecular networks. These topological features may guide prioritisation strategies and also allow the possibility of studying the dynamics of the disease and normal states in these networks.[83]

Given an input set of candidate proteins, such a network analysis, begins by constructing an interaction network in which each candidate or node is connected to other nodes based on data obtained from pathway and molecular interaction databases, often requiring the introduction of additional connecting nodes not part of the original set of candidates. Two resources readily used for this purpose are STRING[84] and GeneMANIA.[85] Both have proven to be very useful in generating an output of functional networks from an input protein list. Similarly, both are comprehensive and incorporate functional associations based on observed and predicted protein–protein interactions (PPIs), co-expression and pathway associations. STRING has more comprehensive coverage of predicted functions between proteins in its many networks and includes a wider range or organisms. The advantages of GeneMANIA include additional associations, such as co-localisation, shared protein domains and genetic interactions,[86] and of having an intuitive plugin for the Cytoscape software.[87] Cytoscape is the gold standard for the visualisation of molecular networks and is widely supported with continuous development from the academic community.[86]

These two resources are among numerous alternatives for probing molecular network databases that can possibly be used for biomarker prioritisation. Pathguide[69] (www.pathguide.org) is an updated list of molecular interaction databases and currently reports 488 such resources. The availability of so many tools makes it a daunting challenge to choose the most relevant resources for various biological contexts being analysed in a proteomics experiment. However, community-wide efforts are being developed to deal with this problem. Notably, the Proteomics Standards Initiative Common Query Interface (PSICQUIC)[86] allows the query of millions of functional interactions from the entire compendium of molecular interaction databases in a non-redundant manner. Multiple software and web services have now begun to support the standard PSICQUIC molecular query service, making the retrieval of large protein interaction data feasible and reliable.

With an interaction network in place, the network analysis proceeds by applying graph theoretic methods to derive various centrality properties for each node, not dissimilar to the approach used to analyse social and computer networks. For instance, "degree centrality" simply corresponds to the number of nodes with which a candidate directly interacts, and can be used to identify network hubs that may play an important local regulatory role. "Betweenness centrality" refers to the number of shortest paths between all other nodes in the network that pass through a given node, indicating the global importance of a candidate in interconnecting different network regions and its potential role as a network bottleneck.[88] Cytoscape plugins, such as CentiScape[89] and NetworkAnalyzer,[90] can be used to readily calculate these and other measures. While many centrality properties exist, several are highly correlated, not all have a clear biological meaning and some may introduce unintended biases.[91] It should also be noted that, while topological features may point to a critical structural or informational role held by a candidate within the network, such a role may not be functionally relevant to the disease being investigated, or may not constitute the role of a biomarker in predicting disease state or therapeutic response. When applied to protein candidate prioritisation, therefore, network analysis may prove more effective when complemented by functional and contextual data, such as those derived from gene ontology and literature-mining methods discussed above, and integrated with differential expression data when available.[92]

Toolkits that use molecular networks to prioritise candidates from high-throughput clinical studies are arguably underdeveloped.[93] However, some successful approaches are emerging in the field of cancer for the prioritisation of driver genes from next generation sequencing (NGS).[93–95] These network approaches can possibly be further optimised and developed for biomarker prioritisation. For such future strategies, it is important to consider that molecular networks are not static, but rather dynamically changing in space and time in response to intra- and extra-cellular stimuli. For that reason, future developments in biomarker prioritisation may incorporate systems biology modelling of signalling networks. These toolkits may attempt to predict quantitative levels of a given protein's abundance in the context of the state of the signalling networks it interacts within. The ability to effectively predict these dynamics is not always possible due to lack of sufficient experimental data points and effective mathematical tools to analyse the data, if and when it is available. However, methods are fast evolving to deal with these challenges and deliver the next generation prioritisation tools.[96]

13.3.4 Towards Next Generation Integrative Bioinformatics Approaches

An integrated analysis of proteomic data sets incorporating different sources of biological information to prioritise candidates may open up new possibilities to discover novel biomarkers and elucidate complex mechanisms driving health and disease. For example, the integration of microarray gene expression data mining with a molecular network analysis has been used to characterise candidate biomarkers in breast cancer metastasis.[97] Similarly, the integration of tissue-specific gene and protein expression databases with a likelihood score of the protein being secreted in blood or serum has been used to identify biomarkers across several cancers.[98]

In the field of bioinformatics for personalised medicine[99] there has been a recent influx of new methods used to prioritise driver genes from NGS data.[66] These methods are adaptable and can be integrated to cater to the field of quantitative proteomics and trained on data sets for biomarker prioritisation. Ideally, these future integrated approaches will be optimised to capture proteins detectable in the peripheral sources of blood and serum in order to guide the discovery of non-invasive diagnostics. Such developments may guide the prioritisation of biomarkers in recent[66] and future high-throughput MRM assays applied to large patient cohorts.

13.4 Statistical Methods for Analysing MRM Data: Random forests and Support Vector Machines

Once peptide features for a candidate biomarker panel have been quantified and/or identified, statistical analysis must be conducted in order to calculate, firstly, if the chosen panel is predictive of the scientific question and, secondly, to check if the chosen list of biomarkers can be further filtered to identify a more parsimonious panel. This analysis can be performed in conjunction with molecular network or text-mining methods (as discussed in the previous section) to prioritise promising biomarkers, or it can be conducted without prioritisation using the raw discovery data. Further analysis on a separate cohort must also be conducted after initial discovery in order to validate the chosen biomarker panel.

There are a vast number of both statistical and machine-learning techniques that could be used to assess how well a candidate panel predicts the response variable of interest. These include random forests, regularised logistic regression, support vector machines, k-nearest neighbours, neural networks and linear and quadratic discriminant analysis among others. These and other methods have been reviewed and compared in a number of previous studies.[100–102]

The choice of which statistical method is most appropriate is case dependent and often comes down to personal preference as many of the most powerful techniques will perform comparably in the majority of cases. This section will focus on two of the most popular and widely used classification and regression methods for high-dimensional data, namely random forests and support vector machines.

Hastie *et al.*, in their seminal book "*The Elements of Statistical Learning, Data Mining, Inference, and Prediction*",[103] compared random forests to two leading ensemble methods. They found that random forests performed with a comparable accuracy to that of its major competitor, gradient boosting. One major advantage of random forests over their leading ensemble rival is that they require much less model tuning and automatically give the user access to a cross-validated error rate. Hastie *et al.* also identified support vector machines as one of the leading machine-learning techniques with respect to predictive accuracy.[103] A number of other previous studies have also identified support vector machines and random forests as leading classifier algorithms for bioinformatics data.[100,104–107]

Although the explanation in the following section assumes the outcome of the experiment is binary (for example, predicting whether a patient belongs to a disease or control group), these methods can also be used if the outcome of interest is continuous (*e.g.*, response to therapy).

13.4.1 Random Forests

Random forests were first proposed by Leo Breiman in 2001[108] and have become a very popular method in many areas of research, including bioinformatics and proteomics.[106,109] Random forests are an ensemble method, which use multiple decision trees in their model. Each decision tree consists of a list of yes/no questions that are asked until the patient can be classified into their respective groups (for example, disease/control).

An example of a decision tree can be seen in Figure 13.3. This tree model uses two proteins: haptoglobin and kininogen to predict between the disease and control groups. Using Figure 13.3 as the motivating prediction model, patients who present with haptoglobin abundance levels lower than 1.2 would be classified as belonging to the disease group. On the other hand, patients who present with haptoglobin levels greater than or equal to 1.2 and kininogen levels less than 0.85 would be classified as controls. Patients with haptoglobin levels greater than 1.2 and kininogen levels between 0.85 and 2.3 would be classified as belonging to the disease group, otherwise they would be classed as belonging to the control group. As can be seen, decision trees recursively partition the data using binary splits in order to make an overall classification.

A major disadvantage of single-decision tree models (such as that shown in Figure 13.3) is that a slight change in the training data can cause a large change in the tree model. For example, the inclusion or exclusion of a small number of samples could cause completely different proteins or completely different cut-off points to be used in the tree model. When evaluating a biomarker panel, a robust and consistent model is often more desirable. Random forests overcome this instability by using multiple decision trees (hundreds or even thousands of tree models), where each decision tree is based on a random sample with replacement of patients and a random sample of proteins. This means each tree model within the random forest is based on a different sample of patients and proteins.

To classify a new test subject, each tree in the random forest is used to classify the patient to a group, as shown in Figure 13.3. The subject is then assigned to the group for which the majority of the tree models predicted. For example, if 10 trees out of a total of 100 in the random forest model

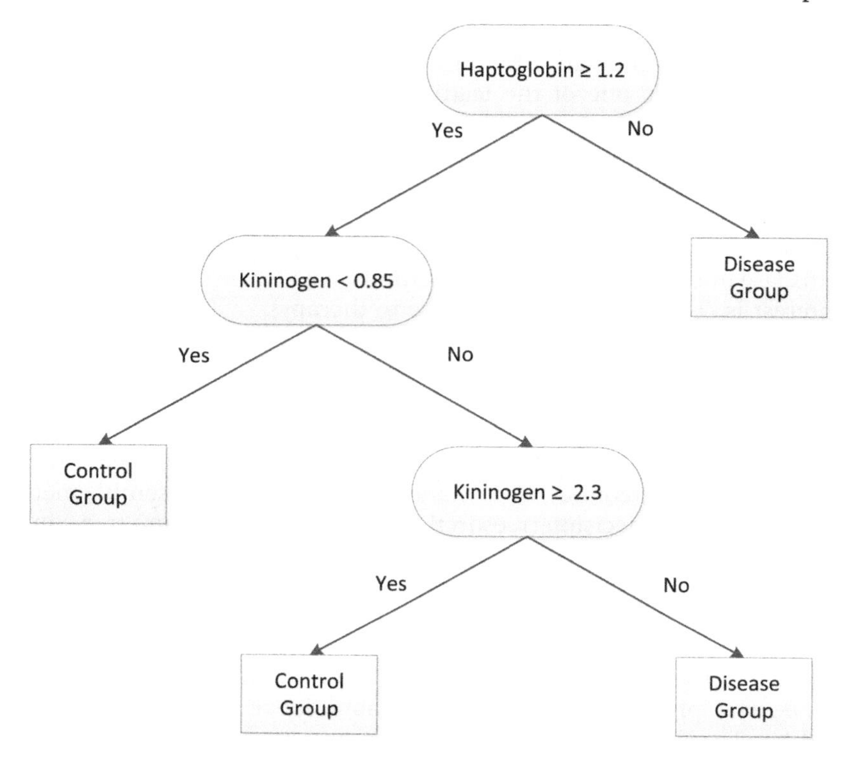

Figure 13.3 Random forest decision tree. A simple example of a decision tree that was made to predict between disease and control groups.

predict the patient as belonging to the disease group and the remaining 90 predict the patient as a control, the random forest would predict the patient as belonging to the control group overall.

One of the main advantages of random forest models are that they are easily implemented using standard statistical software packages (see Appendix). They also automatically allow for protein interactions unlike other statistical models. Although cross-validation can be performed additionally by the user, random forests give automatic access to the cross-validated error rate, which can be used to assess how well the model should perform on a separate cohort (the lower the error rate, the better the model). The AUC (area under the receiver operating characteristic (ROC) curve) can also be easily calculated from the standard output of the model. As well as giving easy access to performance metrics, such as the AUC and the cross-validated error rate, random forests also have a variable importance measure. The variable importance measure can be used to rank each of the candidate biomarkers in the panel from the most to the least important. This measure could be used to assess if a smaller, more parsimonious biomarker panel could be used for the final biomarker panel.

Although random forests can give easy access to error and AUC values, it should be noted that these predictions are point predictions only and that such methods cannot report confidence intervals for these measures. Also, in the case of very small samples sizes, random forests as with other statistical methods can tend to over-fit the data and report over optimistic AUC values.

13.4.2 Support Vector Machines (SVM)

Support vector machines (SVM) are arguably the most commonly used and popular machine-learning method in a wide array of research areas, including proteomics.[107,106] The SVM for binary classification tries to separate two groups (*e.g.*, control and disease) using a straight line called the optimal hyperplane or decision boundary, which is represented by the solid line in Figure 13.4. The objective of the SVM is to find the decision boundary that maximises the margin (represented by the dashed lines in Figure 13.4) such that all points above the decision boundary belong to one class (for example, the control) and all points below the boundary belong to the other class (for example, the disease group).

Clearly, the example given in Figure 13.4 is a simplified case as the data are linearly separable. Often, a straight line cannot separate the data groups; in this case the SVM uses non-linear functions called basis functions or kernels to map the data to a higher dimensional space that is linearly separable. A simple example of this idea can be seen in Figure 13.5. Figure 13.5 (left) shows one-dimensional data; as a motivating example, we could consider these points to be the patients' blood pressure and the two groups are the control and hypertension. As can be seen from Figure 13.5

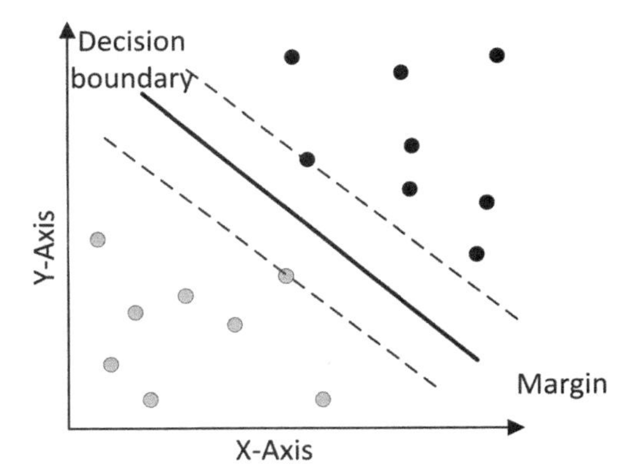

Figure 13.4 Support vector machine (linearly separable data). An example of a linearly separable SVM, where the disease and control groups can be completely separated by a single straight line.

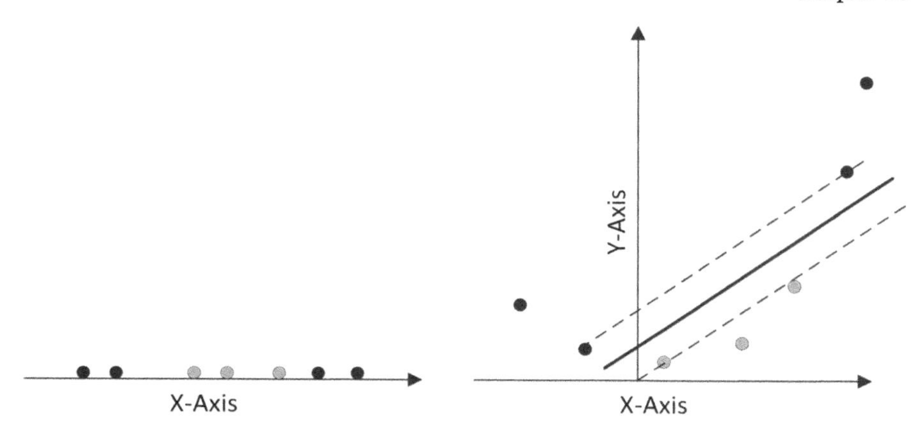

Figure 13.5 Support vector machine (non-linearly separable data). (Left) Data points in one dimension—the groups of which cannot be separated by one single line; (right) the same data mapped to two dimensions, which are now linearly separable.

(left), it is not possible to separate the control and hypertension patients using a single straight line. However, if each patient's blood pressure on the x axis is plotted against, for example, the square of their blood pressure on the y axis (Figure 13.5 (right)), it can be seen that it is now possible to separate the two groups by a single straight line. In other words the one-dimensional blood pressure data were mapped to a two-dimensional space in which the decision boundary could be calculated. In order to predict whether a new patient has hypertension their blood pressure would be mapped to Figure 13.5 (right) and they would be classified as either being control or hypertensive according to whether their data point was above or below the decision boundary depicted in Figure 13.5 (right). Although this is a very simplified example to show how SVMs work, it expresses the basic idea behind how SVMs can classify groups by mapping non-linear data to higher dimensional spaces.

SVMs cannot automatically select important candidate biomarkers, so filtering an initial list of biomarkers to a smaller panel is not quite as simply done as with random forests. However, some *ad hoc* methods for variable selection using SVM have been proposed, the most common of which is called recursive feature elimination (commonly referred to as SVM-RFE). Here, the variables (protein biomarkers) with the smallest weights are iteratively removed and the SVM is refit to the data until only significant biomarkers remain.

It should be noted that the disadvantages of random forests noted above, mainly a tendency to over fit the data in the case of small sample sizes and the inability to report confidence intervals for accuracy measures, also hold true for SVMs. These shortcomings identify the importance of validating potential biomarker panels on a separate cohort of patients.

13.5 Measurement of Protein Biomarker Panels—From Panels to Protein Signatures

13.5.1 MRM Development: Peptide Selection

The selection of appropriate proteotypic peptides is crucial to guarantee the accurate detection of specific proteins of interest. A proteotypic peptide is defined as a peptide that is unique to the protein of interest and routinely observed by mass spectrometry.[110] By measuring only selected proteotypic peptides, the presence or absence of a particular protein and its abundance can be definitively established.[111,112] Information on MS characteristics of peptides is required in order to select the best "precursor ions" for accurate detection of the desired protein. This information is best obtained from discovery liquid chromatography-tandem mass spectrometry (LC-MS/MS) experiments on biological samples of the same type as the MRM measurement will be made.[113] The same information may also be obtained by consulting publicly available spectral libraries in which data from previously run LC-MS/MS experiments are stored.[111,112] Within these spectral libraries, the intensity of each peak represents the peptide expression differences and is, therefore, the basis for selection of the most appropriate peptides for further quantitative experiments, such as MRM. Spectral library searching is currently recognised as an excellent method of peptide selection for most proteomic investigations.[114]

Synthetic peptides are often used to aid MRM assay development as they are inexpensive and can help rule out adverse matrix effects and assist in establishing suitable assay coordinates. By spiking synthetic peptides into a biological sample prior to MRM analysis, it can be ascertained whether or not the peptide of interest is detectable within a given matrix. This is particularly useful when attempting to develop MRM assays for the measurement of lower abundance proteins that are difficult to detect in biological samples.[115]

13.5.2 MRM Assays for Large Scale Analysis

For initial MRM assay development, it is preferable to use at least four transitions for each peptide. A higher number of transitions provide greater confidence that the peak detected during the analysis authentically measures the peptide of interest. As the assay is being refined, the number of transitions can be reduced to the three most intense or "highest ranking" transitions. In our view, monitoring a minimum of three transitions per peptide is critical in maintaining assay selectivity and recognising background interferences when they occur. Different peptides from the same protein can produce drastically different signal intensities when measured in a mass spectrometer, hence peptides producing the highest intensities should ideally be selected to improve assay sensitivity. Targeting multiple peptides per protein increases overall protein coverage and thereby increases

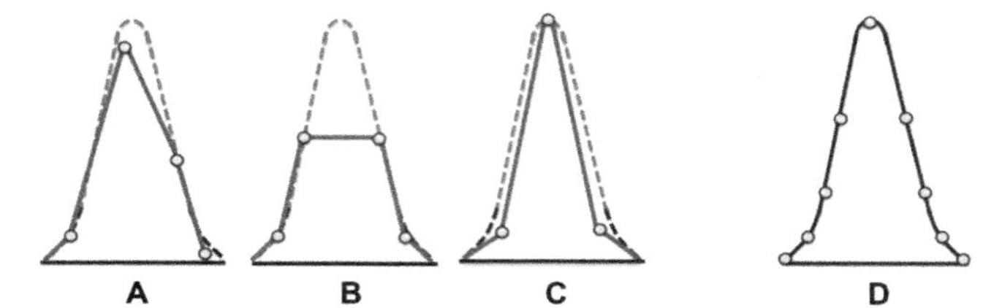

Figure 13.6 MRM assay duty cycle. A "good" peak shape (right) shows even distribution, with a narrow gap between its start and end point, and a defined peak. This requires acquisition of at least ten data points within the duty cycle time. Poor peak shapes are observed if an insufficient number of data points are acquired during the duty cycle time. Left are examples of where insufficient data points fail to reflect the true abundance/intensity of the peak (represented by the dotted line).

confidence in protein measurement.[116] However, depending on the size of the biomarker panel, multiple peptides per protein may unnecessarily limit the number of proteins that can be analysed within a single MRM method. Designing "scheduled" methods, in which each peptide is targeted to be measured within a time period that covers the known elution time of the peptide, can dramatically increase the number of peptides that can be observed and measured within a single LC run.[117]

The importance of the number of peptides and transitions included in the final method is reflected in the "duty cycle" time of the overall assay. The duty cycle is impacted by the total number of transitions and the dwell time set for each transition, as shown by eqn 13.1.

$$\text{Duty Cycle} = (\text{Dwell Time} \times \text{Number of Transitions})/1000 \qquad (13.1)$$

The duty cycle time is the time spent monitoring each analyte. The dwell time refers to the time spent acquiring the targeted MRM during each cycle. A cycle time that provides 10–15 data points across the resulting peak is optimal for accurate quantification and reproducibility, especially for low-abundant proteins. With too few or too many data points, a poor peak shape is generated. Therefore, increasing the transition list or decreasing the transition list without a concurrent increase in dwell time will have adverse effects on the quality of the results (Figure 13.6).

13.5.3 Methods for MRM Quantification

Absolute protein quantification provides a more precise and definitive measurement of the effects of a disease state on protein expression. In this approach isotopically labelled versions of proteotypic peptides are spiked into biological samples. These labelled peptides, referred to as stable-isotope-labelled internal standards (SISs), remain chemically identical to their endogenous unlabelled counterparts, differing only in molecular weight (usually 6–10 Da). The SIS and endogenous forms of the peptides co-elute during chromatographic separation in the mass spectrometer. As such, the ratios of peak areas recorded for both the light (endogenous) and heavy (SIS) version of each peptide allow for accurate calculation of protein concentration.[53,118] Another advantage associated with this technique is that the presence of internal standards within the samples can act as a means of confirming that the instrument is functioning correctly. If there is no signal detected for the endogenous peptide, it can be assumed that it is due to the fact that its concentration is below the limits of detection. Furthermore, the internal standard may allow for inter-assay comparisons and thereby aids in ensuring assay reproducibility. This technique is capable of absolute quantification of peptides across a wide dynamic range of 10^3 to 10^4.[53,57,119] The main limitation of this quantitative approach is, of course, the cost of having purified labelled peptides synthesised (see Chapter 4 for more detailed information).

For relative quantification, differential mass tags are introduced to the sample at either the protein or peptide level to allow for comparative quantification of proteins between samples. These tags are also chemically and analytically identical to their endogenous counterparts, differing only in molecular mass. Common methods of stable-isotope labelling for relative protein quantification include isotope-coated affinity tags (ICAT), stable-isotope labelling in culture (SILAC), ^{18}O incorporation, $^{13}C/^{15}N$ guanidination and, most recently, isobaric tags for relative and absolute quantification (iTRAQ)[120] (see Chapter 3 for further information).

The application of ICAT technology has been shown to be useful for both absolute and relative quantification of proteins *via* MRM analysis.[53] ICAT reagents chemically modify target peptide sequences *via* covalent binding to cysteine residues. They also contain a cleavable biotin tag for affinity purification and a linker sequence that confers either light (227 Da) or heavy (236 Da) chain status to the peptide sequence. Control and test samples are differentially labelled, pooled and subjected to digestion and LC/MS analysis. By comparing the relative peaks detected from both the heavy and light labelled samples, a relative measurement of protein concentration is obtained.[119,120] Jenkins *et al.* modified this protocol for the use of internal standard peptides labelled with a light ICAT reagent for the absolute quantification of multiple P450 enzymes in a test sample labelled with a heavy ICAT reagent (Figure 13.7). As a result, highly sensitive and specific quantitative measurements were achieved for P450 proteins across a wide dynamic range.[53] The main limitations to ICAT labelling are that only two samples can be compared simultaneously and the tags can only bind to cysteine residues, thereby limiting the choice of available target peptides to analyse.

13.5.4 Quality Control

As with any investigation, careful assay design and thorough analysis of the data are imperative to achieving accurate, high-quality results.[115] With the

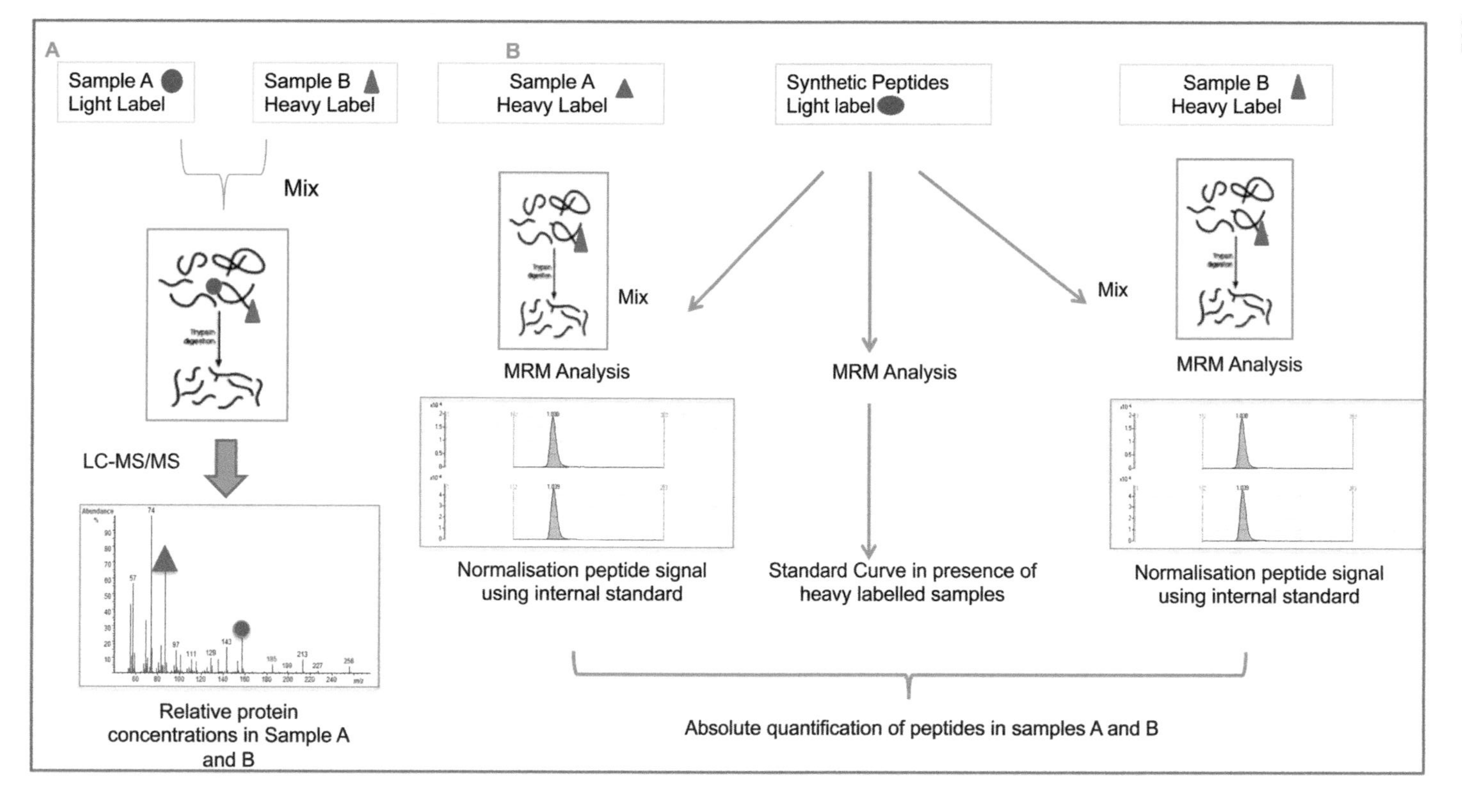

Figure 13.7 Isotope-coated affinity tags MRM workflow: the use of ICAT labelling for relative and absolute protein quantification. ICAT stable-isotope labelling may be employed for relative quantification between two samples *via* LC-MS/MS (A). When coupled to MRM analysis, a modified ICAT workflow is capable of absolute quantification of multiple peptides/proteins in different samples (B).

increasing popularity of MRM analysis, various software tools have become available, which aid in the design and subsequent analysis of the large volumes of data generated in MRM experiments. Software packages developed so far include those that aid in project management, assay development, assay validation, data export, peak integration, quality assessment and biostatistical analysis. For a more detailed overview of these software packages, see the recent review from Colangelo *et al.*[121]

13.5.5 Skyline Software

Skyline is an open-source software tool explicitly designed for the acceleration of targeted proteomics experimentation. Skyline supports all major publicly available spectral libraries from the Global Proteome Machine (GPM), National Institute of Standards (NIST), the Institute for Systems Biology (ISB) and the MacCoss Lab (http://proteome.gs.washington.edu/software/skyline).

During assay design, Skyline theoretically "digests" proteins into peptides, determines the precursor *m*/*z* value for each peptide ion and enables selection of product ions (*i.e.*, generates transitions) for MRM. Resulting transition lists can then be exported from Skyline in an Excel format for direct importation to mass spectrometer software. Skyline software is able to support the import of native output data files from the triple quadruple mass spectrometer without the need for file conversion.[122] Skyline provides an automated assessment of results from MRM analysis that allows for easy monitoring of the quality, reproducibility and robustness of an MRM assay. One of the most important outputs from Skyline is a measure of the dot product for each peptide. This value scores the similarity between MRM intensities for measured transitions against the library MS/MS spectrum, and therefore provides a very useful filter to determine authentic peptide signals from interfering "noise". Skyline will also give information on the predicted retention time for each peptide, which can be used to design "scheduled" methods, as described previously (see Figure 13.8). As well as information on the dot product and retention time for a targeted peptide, Skyline will also indicate (by colour) how well the peak intensities for each measured transition for a peptide of interest match to the library spectrum for the same peptide (see Figure 13.8). Native output files from Agilent, Applied Biosystems, Thermo Fisher Scientific and Waters triple quadrupole instruments can be imported directly to Skyline following MRM analysis without need for file conversion. This data can be loaded in under a second for an experiment with 50 injections, allowing uncomplicated sharing of even the most complex experiments.[123] Skyline can generate multiple, different graphs comparing retentions times and peak intensities between both measured peptides and sample replicates in various formats. With its unique custom report designer, Skyline generates a rich variety of calculated values and statistics for the measured proteins and peptides. These may be easily exported as a CSV file, which is perfectly

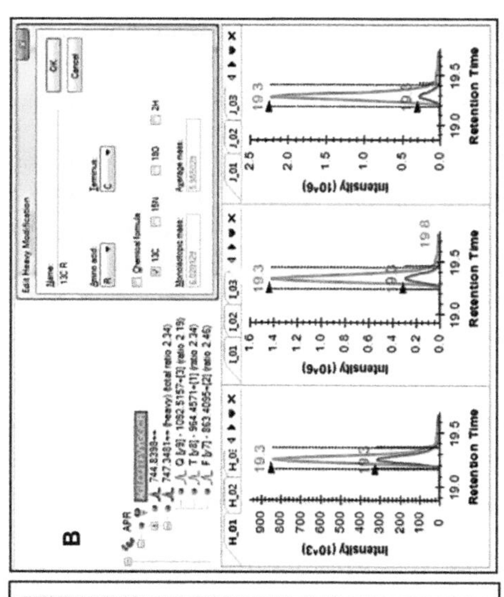
A
Colour of circle indicates quality of peak- Green = Good, Orange = ok, Red = bad
Protein
Peptide
Transitions
Background noise
Optimal retention time
Differently coloured peaks represent different transitions

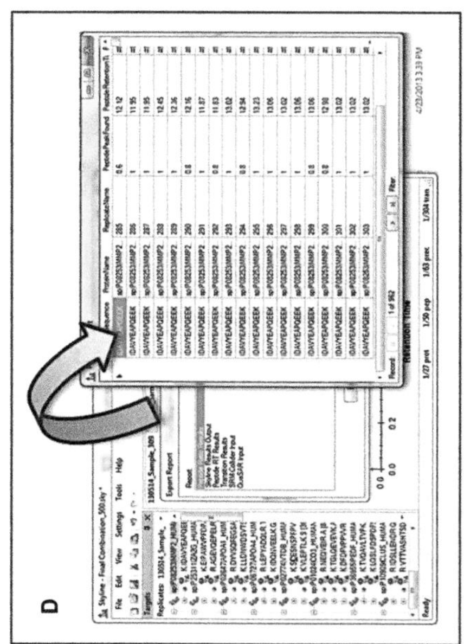
B
Edit Heavy Modification
Retention Time
Intensity (10^6)

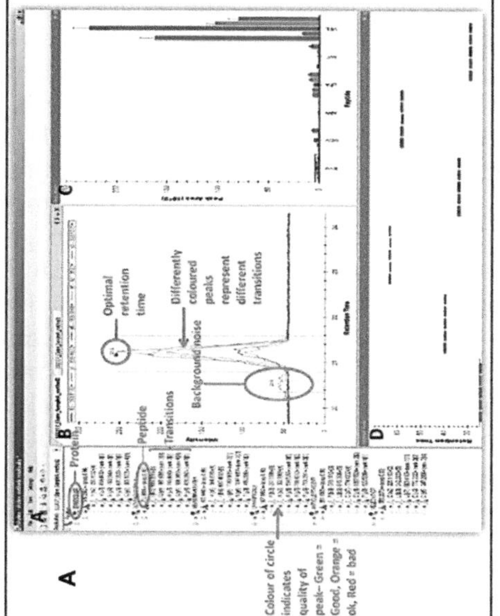
C
Retention Time
Measured Time
Score
Worm - SEATAAAEFTGK, Charge 2

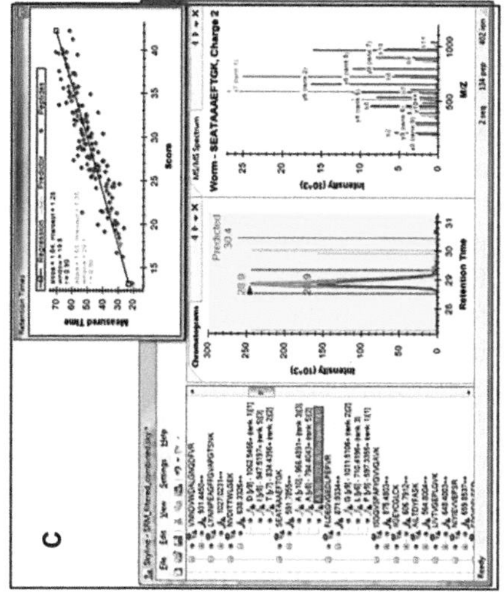
D

amenable for further analysis in any spreadsheet generator or open-source platform for continued data integration, such as Excel and R (see Figure 13.8). Skyline software will also support the design and data analysis for experiments based on isotope-labelled standards and protein quantification (see Figure 13.8). This makes MRM assay design, refinement and results analysis a very iterative process.

13.5.6 Integrated Quality Control

When undertaking larger multiplexed experiments there always exists the potential for the characteristics of the analytical system to change over the course of the analysis. Including a quality control (QC) sample within an experimental run allows monitoring of instrument performance throughout the course of the analysis. As with the incorporation of internal standards in the MRM method described above, injecting known concentrations of QC prepared from commercially available standards at various points throughout the run provides a rapid means of assessing gross performance characteristics (RT stability, peak shape, detector response and mass accuracy). This QC sample should be comprised of a limited selection of target peptides with varying masses, spanning a range of retention times. With reliable QC data, the need for technical replicates is also reduced. As an added benefit, QC data from each sample batch can be used as a tool for subsequent data normalisation following the analysis of a complete sample cohort.

Deciding on the number of samples that can potentially be analysed with optimal sensitivity and selectivity within each run is important in the early stages of experimental planning. A key consideration when making this decision is knowledge of the length of time taken to analyse each sample. When calculating the predicted length of an MRM run, inclusion of multiple QCs and "blank" samples should also be considered. Frequent injections of "blank" samples throughout the run prevents carry-over between sample injections and reduces the potential for false positives. Randomisation of the run order of samples within a batch is also important to ensure that all the experimental groups are affected to the same extent. The variability within each sample batch should be less than the variability of the entire sample set and, thus, each estimate of the treatment effect within a batch is more efficient than estimates across the entire sample.[124]

Figure 13.8 Skyline MRM development and data analysis workflow. Skyline software provides automated and integrative analysis of MRM data (A). Skyline software can be used in the design and subsequent assessment of quantitative label-based experiments (B). Various graphical outputs depicting information on retention time and peak intensities across sample replicates and for individual peptides are easily generated with Skyline software (C). Custom made reports compatible with Excel and R formatting can be generated and exported from Skyline for further analysis (D).

13.5.7 Standardisation within the Lab and Inter-lab

Substantial efforts have been made toward evaluating and controlling for potential pre-analytical and instrumental variables that might arise both within and between labs. Pre-analytical effects may be technical—sample collection, storage and treatment, *etc.*—or biological, such as a patient's age and gender, *etc.*[125] The National Cancer Institute's Clinical Proteomic Technologies for Cancer (NCI-CPTAC) initiative has been making great efforts to reduce the impact of these sources of variability. CPTAC is composed of a network of Proteome Characterisation Centre teams, a Data Coordinating Centre and a Resource Centre with the aim to improve diagnosis, treatment and the prevention of cancer by the co-ordination of research and data sharing.

One of the main priorities of NCI-CPTAC is ensuring assay reproducibility and transferability. In 2009 NCI-CPTAC conducted a series of interrelated studies that were designed to assess the reproducibility and quantitative characteristics of MRM assays across eight participating laboratories. Study I represented optimum performance using synthetic peptides as analytes as opposed to actual proteins. Variations in protein digestion were introduced in studies II and III. In study II digestion in the absence of plasma was conducted in a central location, while in study III the digestion was carried out in the presence of plasma at individual sites. It was reported that intra-laboratory coefficients of variance (CVs) remained less than 25% across all peptide concentrations, thereby demonstrating good reproducibility. It was also concluded, from the increases in CVs from study I to III, that sample preparation contributes more to assay variability than instrumental variability. Although the transferability and robustness of MRM technology within and between laboratories was demonstrated here, the only study that provided a true representation of real biomarker conditions (study III) did not perform to the standard required for clinical assays.[126] A similar study reported by Abbateillo and colleagues introduced a system suitability protocol (SSP) to assess the performance of 15 different QqQ-based nano-LC-SID-MRM-MS instrument configurations across 11 different laboratories.[127] This SSP included guidelines for peptide selection and platform-specific LC and MRM-MS optimisation. In addition the SSP included optimisation and use of Skyline for monitoring of assessment reproducibility and data quality in individual labs. Overall, this SSP was capable of facilitating rapid detection, diagnosis and correction of system faults contributing to increased CVs and reduced sensitivity over an extended time period.[127] Incorporation of standardised protocols, such as described in this study, should help ensure that quantitative MRM-based studies can be replicated with good precision both within and across labs.

13.6 Conclusion

For routine clinical use, biomarker assays must be capable of reliable indication of a disease state with high specificity and sensitivity. Many established biomarkers, such as PSA, are limited in this sense. Therefore, it is widely recognised that more effective and accurate biomarkers are needed. With the evolving concept of personalised medicine, the advantage of simultaneous analysis of panels of biomarkers or biomarker signatures is now of significant interest.[128] Although hundreds of candidate biomarkers have emerged from proteomics discovery experiments, few have progressed to use in the clinic. This is in part due to the lack of sufficiently precise methods to measure changes in the abundance of numerous candidate protein biomarkers in hundreds of patient samples.[129] Consequently, there is a growing interest in the use of multiplexed technology for the development and ultimate validation of potential candidate biomarker panels. The emergence of MRM technology has provided an ideal platform for such multiplexed assays. In comparison to the current gold standard of ELISA for biomarker verification, MRM can allow for sensitive detection of potentially thousands of biomarker proteins within a single experiment at a much lower cost. Furthermore, both absolute and relative protein quantification is now achievable with MRM technology, providing even more insight into the biological influence of disease and/or treatment strategies. This has led to increasing academic and commercial interest in the use of MRM technology for biomarker development.

Translation of an MRM assay to a "point-of-care" diagnostic test would allow for a quick, non-invasive method of disease detection and monitoring. However, in the case of complex data sets, such as protein biomarker panels, independent clinical validation is required to ensure that the new technology provides reliable analytical and clinical signals.[130] Because of the inherent biological variability in human samples, verification studies on hundreds of patients and controls are needed before accurate answers may emerge.[131]

There are many statistical and machine-learning techniques that can be used not only to assess the predictive accuracy of a biomarker panel but also to identify and remove unimportant elements. Which method is most suitable is often study specific. Random forests and support vector machines have both been shown to have high predictive accuracy in a wide range of cases and so the choice of which to use may come down to personal preference. If the candidate panel is quite large and is going to be further filtered and validated, random forests may be preferable as they automatically include a variable importance index, whereas SVM does not. Regardless of the method chosen some form of cross-validation should be performed to ensure that the model is not over fit to the data.

Subsequent clinical validation will require MRM data from a separate sample cohort, analysed in a different lab. Ultimately, the broad applicability of MRM assays in clinical research environments will depend on the method reproducibility across laboratories. Also, precision and accuracy must conform to the standards already set for other clinical assays, including immunoassays.[128] Open access to assay protocols and MRM data, as practiced by the NCI-CPTAC initiative, would be of considerable use in such efforts. In addition, with the current volume of research in this area, comprehensive meta-analysis of recent literature may enable the compilation of more robust protein biomarker panels. Panels consisting only of proteins that have previously been measured by MRM may reduce the time needed for assay development and allow for the collection of data with stronger discriminatory power. The importance of experimental design to proteomic mass spectrometry is becoming more widely recognised.[132] Although MRM technology is an effective means of rapid, sensitive and high-throughput measurement of biomarker panels, significant efforts are required to standardise the process of MRM-based analysis and bring biomarker panels forward to ultimate clinical utility.

Appendix

Software

The original Fortran code for random forests can be found at the official Random Forest site maintained by Adele Cutler (http://www.stat.berkeley.edu/~breiman/RandomForests/cc_software.htm). Random forests can also be run using the package "Random Forest" in the R statistical software. This R implementation of random forests can also be accessed through SPSS using the extension commands RANFOR to build a training model and RANPRED to predict the outcome for test samples. It is also implemented in MATLAB using either the "Random Forest" (https://code.google.com/p/randomforest-matlab/) package or the "TreeBagger" package (http://www.mathworks.co.uk/help/stats/treebagger.html) in the statistics toolbox.

The most popular implementation of support vector machines is the libsvm package written in C++ and Python (http://www.csie.ntu.edu.tw/~cjlin/libsvm/). R offers a number of packages that can run SVMs, including "e1071" (http://cran.r-project.org/web/packages/e1071/index.html), "svmpath" (http://cran.r-project.org/web/packages/svmpath/index.html) and "kernlab" (http://cran.r-project.org/web/packages/kernlab/index.html). MATHLAB also offers multiple toolboxes which can implement SVMs including the "MATLAB Support Vector Machine Toolbox" (http://theoval.cmp.uea.ac.uk/~gcc/svm/toolbox/), the "SVM and Kernel Methods MATLAB Toolbox" (http://asi.insa-rouen.fr/enseignants/~arakoto/toolbox/index.html) and the "SVM toolbox" (http://research.cs.wisc.edu/dmi/svm/). The source code to an R package, which will run recursive feature selection "mSVM-RFE", can also be downloaded (https://github.com/johncolby/SVM-RFE).

Acknowledgements

The authors wish to acknowledge the support provided by the following funding bodies and grants: Health Research Board, Science Foundation Ireland, Irish Cancer Society, Movember, SLICR. The authors also wish to thank Mr Brian Flatley for his help in editing this book chapter.

References

1. B. A. S. Atkinson, A. J. W. A. Colburn, V. G. DeGruttola, D. L. DeMets, G. J. Downing, D. F. Hoth, J. A. Oates, C. C. Peck and R. T. Schooley, *Clinical Pharmacology and Therapeutics*, 2001, **69**, 89–95.
2. J. S. Smolen, D. Aletaha, J. Grisar, K. Redlich, G. Steiner and O. Wagner, *Arthritis Research & Therapy*, 2008, **10**, 208.
3. D. J. Brennan, C. Kelly, E. Rexhepaj, P. A. Dervan, M. J. Duffy and W. M. Gallagher, *Cancer Genomics & Proteomics*, 2007, **4**, 121–134.
4. D. R. Parkinson, B. E. Johnson and G. W. Sledge, *Clinical Cancer Research: An Official Journal of the American Association for Cancer Research*, 2012, **18**, 619–624.
5. A. S. Dhamoon, E. C. Kohn and N. S. Azad, *Drug Discovery Today*, 2007, **12**, 700–708.
6. N. B. La Thangue and D. J. Kerr, *Nature Reviews Clinical Oncology*, 2011, **8**, 587–596.
7. A. J. Atherly and D. R. Camidge, *British Journal of Cancer*, 2012, **106**, 1100–1106.
8. J. A. Rindfleisch and D. Muller, *American Family Physician*, 2005, **72**, 1037–1047.
9. B. Rhodes, B. G. Fürnrohr and T. J. Vyse, *Nature Reviews Rheumatology*, 2011, **7**, 282–289.
10. R. H. Mullan, C. Matthews, B. Bresnihan, O. FitzGerald, L. King, A. R. Poole, U. Fearon and D. J. Veale, *Arthritis and Rheumatism*, 2007, **56**, 2919–2928.
11. D. V. Makarov, S. Loeb, R. H. Getzenberg and A. W. Partin, *Annual Review of Medicine*, 2009, **60**, 139–151.
12. C. Sävblom, J. Malm, A. Giwercman, J.-A. Nilsson, G. Berglund and H. Lilja, *The Prostate*, 2005, **65**, 66–72.
13. H. Lilja, D. Ulmert and A. J. Vickers, *Nature Reviews Cancer*, 2008, **8**, 268–278.
14. J. R. Prensner, M. a. Rubin, J. T. Wei and A. M. Chinnaiyan, *Science Translational Medicine*, 2012, **4**, 127rv3.
15. A. Horwich, J. Hugosson, T. de Reijke, T. Wiegel, K. Fizazi and V. Kataja, *Annals of Oncology: Official Journal of the European Society for Medical Oncology/ESMO*, 2013, 1–22.
16. J. Fransen and P. L. van Reil, *Rheumatic Diseases Clinics of North America*, 2009, **35**, 745–757.
17. N. Wild, J. Karl, V. P. Grunert, R. I. Schmitt, U. Garczarek, F. Krause, F. Hasler, P. L. van Riel, P. M. Bayer, M. Thun, D. L. Mattey, M. Sharif and W. Zolg, *Biomarkers*, 2008, **13**, 88–105.
18. M. J. Duffy, C. Sturgeon, R. Lamerz, C. Haglund, V. L. Holubec, R. Klapdor, A. Nicolini, O. Topolcan and V. Heinemann, *Annals of Oncology*, 2009, 441–447.
19. M. A. Albert, *Journal of Clinical Sleep Medicine*, 2011, **7**, S9–S11.
20. S. R. Langley, J. Dwyer, I. Drozdov, X. Yin and M. Mayr, *Cardiovascular Research*, 2013, **97**, 612–622.

21. I. Ikonomidis, C. Michalakeas, J. Lekakis, I. Paraskevaidis and D. T. Kremastinos, *Disease Markers*, 2009, **26**, 273–285.
22. B. Zethelius, L. Berglund, J. Sundström, E. Ingelsson, S. Basu, A. Larsson, P. Venge and J. Arnlöv, *The New England Journal of Medicine*, 2008, **358**, 2107–2116.
23. N. Ahluwalia, J. Blacher, F. Szabo de Edelenyi, P. Faure, C. Julia, S. Hercberg and P. Galan, *Atherosclerosis*, 2013, **228**, 478–484.
24. B. O Hartaigh, J. Bosch, D. Carroll, K. Hemming, S. Pilz, A. Loerbroks, M. E. Kleber, T. B. Grammer, J. E. Fischer, B. O. Boehm, W. März and G. N. Thomas, *European Heart Journal*, 2013, **34**, 932–941.
25. N. Rifai, M. Gillette and S. A. Carr, *Nature Biotechnology*, 2006, **24**, 971–983.
26. G. Poste, *Nature*, 2011, **469**, 156–157.
27. M. Mann, N. Nagaraj, J. R. Wisniewski, T. Geiger, J. Cox, M. Kircher, J. Kelso, S. Pa and S. Pääbo, *Molecular Systems Biology*, 2011, **60**, 1–8.
28. L. F. Waanders, K. Chwalek, M. Monetti, C. Kumar, E. Lammert and M. Mann, *Proceedings of the National Academy of Sciences of the United States of America*, 2009, **106**, 18902–18907.
29. M. Bantscheff, D. Eberhard, Y. Abraham, S. Bastuck, M. Boesche, S. Hobson, T. Mathieson, J. Perrin, M. Raida, C. Rau, V. Reader, G. Sweetman, A. Bauer, T. Bouwmeester, C. Hopf, U. Kruse, G. Neubauer, N. Ramsden, J. Rick, B. Kuster and G. Drewes, *Nature Biotechnology*, 2007, **25**, 1035–1044.
30. Z. Wu, J. B. Doondeea, A. M. Gholami, M. C. Janning, S. Lemeer, K. Kramer, S. a Eccles, S. M. Gollin, R. Grenman, A. Walch, S. M. Feller and B. Kuster, *Molecular & Cellular Proteomics: MCP*, 2011, **10**, M111.011635.
31. C. M. Dobson, *Nature*, 2003, **426**, 884–890.
32. E. Rhoades, E. Gussakovsky and G. Haran, *Proceedings of the National Academy of Sciences of the United States of America*, 2003, **100**, 3197–3202.
33. J. Lee, J. J. Han, G. Altwerger and E. C. Kohn, *Journal of Proteomics*, 2011, **74**, 2632–2641.
34. R. Aebersold and M. Mann, *Nature*, 2003, **422**, 198–207.
35. H. Mischak, J. P. A. Ioannidis, A. Argiles, T. K. Attwood, E. Bongcam-Rudloff, M. Broenstrup, A. Charonis, G. P. Chrousos, C. Delles, A. Dominiczak, T. Dylag, J. Ehrich, J. Egido, P. Findeisen, J. Jankowski, R. W. Johnson, B. a Julien, T. Lankisch, H. Y. Leung, D. Maahs, F. Magni, M. P. Manns, E. Manolis, G. Mayer, G. Navis, J. Novak, A. Ortiz, F. Persson, K. Peter, H. H. Riese, P. Rossing, N. Sattar, G. Spasovski, V. Thongboonkerd, R. Vanholder, J. P. Schanstra and A. Vlahou, *European Journal of Clinical Investigation*, 2012, **42**, 1027–1036.
36. A. G. Paulovich, J. R. Whiteaker, A. N. Hoofnagle and P. Wang, *Proteomics Clinical Applications*, 2008, **2**, 1386–1402.

37. S. F. Oon, S. R. Pennington, J. M. Fitzpatrick and R. W. G. Watson, *Nature Reviews Urology*, 2011, **8**, 131–138.
38. D. Kubota, K. Mukaihara, A. Yoshida, Y. Suehara, T. Saito, T. Okubo, M. Gotoh, H. Orita, H. Tsuda, K. Kaneko, A. Kawai, T. Kondo, K. Sato and T. Yao, *Japanese Journal of Clinical Oncology*, 2013, **43**, 669–675.
39. P. C. OLeary, S. a Penny, R. T. Dolan, C. M. Kelly, S. F. Madden, E. Rexhepaj, D. J. Brennan, A. H. McCann, F. Pontén, M. Uhlén, R. Zagozdzon, M. J. Duffy, M. R. Kell, K. Jirström and W. M. Gallagher, *BMC Cancer*, 2013, **13**, 175.
40. M. Uhlen and F. Ponten, *Molecular & Cellular Proteomics: MCP*, 2005, **4**, 384–393.
41. M. S. Pepe, Z. Feng, H. Janes, P. M. Bossuyt and J. D. Potter, *Journal of the National Cancer Institute*, 2008, **100**, 1432–1438.
42. P. M. Bossuyt, J. B. Reitsma, D. E. Bruns, C. A. Gatsonis, P. P. Glasziou, L. M. Irwig, J. G. Lijmer, D. Moher, D. Rennie and H. C. W. de Vet, *Clinical Chemistry*, 2003, **49**, 1–6.
43. A. Taguchi and S. M. Hanash, *Clinical Chemistry*, 2013, **59**, 119–126.
44. N. L. Anderson and N. G. Anderson, *Molecular & Cellular Proteomics: MCP*, 2002, **1**, 845–867.
45. S. Ray, P. J. Reddy, R. Jain, K. Gollapalli, A. Moiyadi and S. Srivastava, *Proteomics*, 2011, **11**, 2139–2161.
46. S. Pan, R. Chen, R. Aebersold and T. A. Brentnall, *Molecular & Cellular Proteomics: MCP*, 2011, **10**, R110.003251.
47. A. W. R. van Kuijk, D. M. Gerlag, K. Vos, G. Wolbink, M. de Groot, M. A. de Rie, A. H. Zwinderman, B. A. C. Dijkmans and P. P. Tak, *Annals of the Rheumatic Diseases*, 2009, **68**, 1303–1309.
48. R. B. H. Braakman, M. M. A. Tilanus-linthorst, N. Qing, C. Stingl, L. J. M. Dekker, T. M. Luider, J. W. M. Martens, J. A. Foekens and A. Umar, *Journal Proteomics*, 2012, 1–11.
49. V. Patel, B. L. Hood, a. a. Molinolo, N. H. Lee, T. P. Conrads, J. C. Braisted, D. B. Krizman, T. D. Veenstra and J. S. Gutkind, *Clinical Cancer Research*, 2008, **14**, 1002–1014.
50. T. Nishimura, M. Nomura, H. Tojo, H. Hamasaki, T. Fukuda, K. Fujii, S. Mikami, Y. Bando and H. Kato, *Journal of Proteomics*, 2010, **73**, 1100–1110.
51. A. D. Weston and L. Hood, *Journal of Proteome Research*, 2004, **3**, 179–196.
52. H. Keshishian, T. Addona, M. Burgess, E. Kuhn and S. A. Carr, *Molecular & Cellular Proteomics: MCP*, 2007, **6**, 2212–2229.
53. R. E. Jenkins, N. R. Kitteringham, C. L. Hunter, S. Webb, T. J. Hunt, R. Elsby, R. B. Watson, D. Williams, S. R. Pennington and B. K. Park, *Proteomics*, 2006, **6**, 1934–1947.
54. D. Shi Tujin, Su Dian, Liu Tao, Tang Keqi, Camp II David. G, Qian Wei-Jun and Smith Richard, *Proteomics*, 2012, **12**, 1074–1092.
55. U. Christians, J. Klepacki, T. Shokati, J. Klawitter and J. Klawitter, *Microchemical Journal, Devoted to the Application of Microtechniques in all Branches of Science*, 2012, **105**, 32–38.

56. B. Definitions and W. Group, *Clinical Pharmacology and Therapeutics*, 2001, **69**, 89–95.
57. L. Anderson and C. L. Hunter, *Molecular & Cellular Proteomics: MCP*, 2006, **5**, 573–588.
58. D. Domanski, A. J. Percy, J. Yang, A. G. Chambers, J. S. Hill, G. V. C. Freue and C. H. Borchers, *Proteomics*, 2012, **12**, 1222–1243.
59. A. J. Percy, A. G. Chambers, D. S. Smith and C. H. Borchers, *Journal of Proteome Research*, 2013, **12**, 222–233.
60. A. J. Percy, A. G. Chambers, J. Yang, D. B. Hardie and C. H. Borchers, *Biochimica et Biophysica Acta*, 2013, 4–13.
61. Z. Meng and T. D. Veenstra, *Journal of Proteomics*, 2011, **74**, 2650–2659.
62. A. Carlsson, C. Wingren, M. Kristensson, C. Rose, M. Fernö, H. Olsson, H. Jernström, S. Ek, E. Gustavsson, C. Ingvar, M. Ohlsson, C. Peterson and C. A. K. Borrebaeck, *Proceedings of the National Academy of Sciences USA*, 2011, **108**, 14252–14257.
63. V. Kaklamani, *Expert Review of Molecular Diagnostics*, 2006, **6**, 803–809.
64. B. C. Collins, C. A. Miller, A. Sposny, P. Hewitt, M. Wells, W. M. Gallagher and S. R. Pennington, *Molecular and Cellular Proteomics: MCP*, 2012, **11**, 394–410.
65. R. Malik, K. Dulla, E. a Nigg and R. Körner, *Proteomics*, 2010, **10**, 1270–1283.
66. R. Hüttenhain, M. Soste, N. Selevsek, H. Röst, A. Sethi, C. Carapito, T. Farrah, E. W. Deutsch, U. Kusebauch, R. L. Moritz, E. Niméus-Malmström, O. Rinner and R. Aebersold, *Science Translational Medicine*, 2012, **4**, 142ra94.
67. M. Harris, J. Clark, Ireland, J. Lomax, M. Ashburner, R. Foulger, K. Eilbeck, S. Lewis, B. Marshall, C. Mungall, J. Richter, G. M. Rubin, J. a Blake, C. Bult, M. Dolan, H. Drabkin, J. T. Eppig, D. P. Hill, L. Ni, M. Ringwald, R. Balakrishnan, J. M. Cherry, K. R. Christie, M. C. Costanzo, S. S. Dwight, S. Engel, D. G. Fisk, J. E. Hirschman, E. L. Hong, R. S. Nash, a Sethuraman, C. L. Theesfeld, D. Botstein, K. Dolinski, B. Feierbach, T. Berardini, S. Mundodi, S. Y. Rhee, R. Apweiler, D. Barrell, E. Camon, E. Dimmer, V. Lee, R. Chisholm, P. Gaudet, W. Kibbe, R. Kishore, E. M. Schwarz, P. Sternberg, M. Gwinn, L. Hannick, J. Wortman, M. Berriman, V. Wood, N. de la Cruz, P. Tonellato, P. Jaiswal, T. Seigfried and R. White, *Nucleic Acids Research*, 2004, **32**, D258–D261.
68. M. N. Davies, E. L. Meaburn and L. C. Schalkwyk, *Briefings in Functional Genomics*, 2010, **9**, 385–390.
69. G. Dennis, B. T. Sherman, D. a Hosack, J. Yang, W. Gao, H. C. Lane and R. a Lempicki, *Genome Biology*, 2003, **4**, P3.
70. P. Khatri, M. Sirota and A. J. Butte, *PLoS Computational Biology*, 2012, **8**, e1002375.
71. I. Vastrik, P. D'Eustachio, E. Schmidt, G. Joshi-Tope, G. Gopinath, D. Croft, B. de Bono, M. Gillespie, B. Jassal, S. Lewis, L. Matthews, G. Wu, E. Birney and L. Stein, *Genome Biology*, 2007, **8**, R39.
72. M. Kanehisa and S. Goto, *Nucleic Acids Research*, 2000, **28**, 27–30.

73. G. Bindea, B. Mlecnik, H. Hackl, P. Charoentong, M. Tosolini, A. Kirilovsky, W.-H. Fridman, F. Pagès, Z. Trajanoski and J. Galon*Bioinformatics*, Oxford, England, 2009, **25**, 1091–1093.
74. S. Y. Rhee, V. Wood, K. Dolinski and S. Draghici, *Nature Reviews Genetics*, 2008, **9**, 509–515.
75. J. T. Dudley and A. J. Butte, *Pacific Symposium on Biocomputing. Pacific Symposium on Biocomputing*, 2009, 27–38.
76. H.-C. Huang, D. Jupiter and V. VanBuren, *PloS One*, 2010, **5**, e9056.
77. H. Xue, B. Lu and M. Lai, *Journal of Translational Medicine*, 2008, **6**, 52.
78. D. Rebholz-Schuhmann, A. Oellrich and R. Hoehndorf, *Nature Reviews Genetics*, 2012, **13**, 829–839.
79. H. Li and C. Liu, *Computational and Mathematical Methods in Medicine*, 2012, **2012**, 135780.
80. F. Zhu, P. Patumcharoenpol, C. Zhang, Y. Yang, J. Chan, A. Meechai, W. Vongsangnak and B. Shen, *Journal of Biomedical Informatics*, 2013, **46**, 200–211.
81. T. Clancy, M. Pedicini, F. Castiglione, D. Santoni, V. Nygaard, T. J. Lavelle, M. Benson and E. Hovig, *BMC Medical Genomics*, 2011, **4**, 28.
82. A.-L. Barabási and Z. N. Oltvai, *Nature Reviews Genetics*, 2004, **5**, 101–113.
83. A.-L. Barabási, N. Gulbahce and J. Loscalzo, *Nature Reviews Genetics*, 2011, **12**, 56–68.
84. B. Snel, G. Lehmann, P. Bork and M. A. Huynen, *Nucleic Acids Research*, 2000, **28**, 3442–3444.
85. D. Warde-Farley, S. L. Donaldson, O. Comes, K. Zuberi, R. Badrawi, P. Chao, M. Franz, C. Grouios, F. Kazi, C. T. Lopes, A. Maitland, S. Mostafavi, J. Montojo, Q. Shao, G. Wright, G. D. Bader and Q. Morris, *Nucleic Acids Research*, 2010, **38**, W214–W220.
86. R. Saito, M. E. Smoot, K. Ono and J. Ruscheinski, *Nature Methods*, 2013, **9**, 1069–1076.
87. P. Shannon, A. Markiel, O. Ozier, N. S. Baliga, J. T. Wang, D. Ramage, N. Amin, B. Schwikowski, and T. Ideker, 2003, 2498–2504.
88. H. Yu, P. M. Kim, E. Sprecher, V. Trifonov and M. Gerstein, *PLoS Computational Biology*, 2007, **3**, e59.
89. G. Scardoni, M. Petterlini and C. Laudanna, *Bioinformatics*, 2009, **25**, 2857–2859.
90. Y. Assenov, F. Ramírez, S.-E. Schelhorn, T. Lengauer and M. Albrecht, *Bioinformatics*, 2008, **24**, 282–284.
91. J. P. Gonçalves, A. P. Francisco, Y. Moreau and S. C. Madeira, *PloS One*, 2012, 7, e49634.
92. D. Nitsch, L.-C. Tranchevent, B. Thienpont, L. Thorrez, H. Van Esch, K. Devriendt and Y. Moreau, *PloS One*, 2009, **4**, e5526.
93. A. Bashashati, G. Haffari, J. Ding, G. Ha, K. Lui, J. Rosner, D. G. Huntsman, C. Caldas, S. a Aparicio and S. P. Shah, *Genome Biology*, 2012, **13**, R124.
94. A. L. Tarca, S. Draghici, P. Khatri, S. S. Hassan, P. Mittal, J.-S. Kim, C. J. Kim, J. P. Kusanovic and R. Romero, *Bioinformatics*, 2009, **25**, 75–82.

95. C. J. Vaske, S. C. Benz, J. Z. Sanborn, D. Earl, C. Szeto, J. Zhu, D. Haussler and J. M. Stuart, *Bioinformatics*, 2010, **26**, i237–i245.
96. P. K. Kreeger and D. a Lauffenburger, *Carcinogenesis*, 2010, **31**, 2–8.
97. M. J. Jahid and J. Ruan, *BMC Genomics*, 2012, **13**(Suppl 6), S8.
98. I. Prassas, C. C. Chrystoja, S. Makawita and E. P. Diamandis, *BMC Medicine*, 2012, **10**, 39.
99. A. Valencia and M. Hidalgo, *Genome Medicine*, 2012, **4**, 61.
100. M. Hilario, A. Kalousis, C. Pellegrini and M. Müller, *Mass Spectrometry Reviews*, 2006, **25**, 409–449.
101. D. L. Sampson, T. J. Parker, Z. Upton and C. P. Hurst, *PloS One*, 2011, **6**, e24973.
102. B. Wu, T. Abbott, D. Fishman, W. McMurray, G. Mor, K. Stone, D. Ward, K. Williams and H. Zhao, *Bioinformatics*, 2003, **19**, 1636–1643.
103. T. Hastie, R. Tibshirani and J. Freidman, *The Elements of Statistical Learning Data Mining, Inference, and Prediction*, 2009, (Second Edition), Springer-Verlag, New York.
104. S. Lee, J. Park and M. & Song, *Computational Statistics & Data Analysis*, 2005, **26**, 869–885.
105. T. Abeel, T. Helleputte, Y. Van de Peer, P. Dupont and Y. Saeys, *Bioinformatics*, 2010, **26**, 392–398.
106. M. Robin Xavier, Turck Natacha, Hainard Alexandre, Lisacek Frederique, Sanchez Jean-Charles and Muller, *Expert Review of Proteomics*, 2009, **6**, 675–689.
107. Y. Saeys, I. Inza and P. Larrañaga, *Bioinformatics*, 2007, **23**, 2507–2517.
108. L. Breiman, *Machine Learning*, 2001, **45**, 5–32.
109. I. Cima, R. Schiess, P. Wild, M. Kaelin, P. Schüffler, V. Lange, P. Picotti, R. Ossola, A. Templeton, O. Schubert, T. Fuchs, T. Leippold, S. Wyler, J. Zehetner, W. Jochum, J. Buhmann, T. Cerny, H. Moch, S. Gillessen, R. Aebersold and W. Krek, *Proceedings of the National Academy of Sciences of the United States of America*, 2011, **108**, 3342–3347.
110. C.-Y. Chang, P. Picotti, R. Hüttenhain, V. Heinzelmann-Schwarz, M. Jovanovic, R. Aebersold and O. Vitek, *Molecular & Cellular Proteomics: MCP*, 2012, **11**, M111.014662.
111. J. A. Chem Mead, L. Bianco and C. Bessant, *Methods in Molecular Biology (Clifton, N.J.)*, 2010, **604**, 187–199.
112. R. Schiess, B. Wollscheid and R. Aebersold, *Molecular Oncology*, 2009, **3**, 33–44.
113. B. Domon and R. Aebersold, *Nature Biotechnology*, 2010, **28**, 710–721.
114. H. Lam, *Molecular & Cellular Proteomics: MCP*, 2011, **10**, R111.008565.
115. P. Picotti, O. Rinner, R. Stallmach, F. Dautel, T. Farrah, B. Domon, H. Wenschuh and R. Aebersold, *Nature Methods*, 2010, **7**, 43–46.
116. A. B. Stergachis, B. MacLean, K. Lee, J. A. Stamatoyannopoulos and M. J. MacCoss, *Nature Methods*, 2011, **8**, 1041–1043.
117. A. Bertsch, S. Jung, A. Zerck, N. Pfeifer, S. Nahnsen, C. Henneges, A. Nordheim and O. Kohlbacher, *Journal of Proteome Research*, 2010, **9**, 2696–2704.

118. S. A. Gerber, J. Rush, O. Stemman, M. W. Kirschner and S. P. Gygi, *Proceedings of the National Academy of Sciences of the United States of America*, 2003, **100**, 6940–6945.
119. M. H. Elliott, D. S. Smith, C. E. Parker and C. Borchers, *Journal of Mass Spectrometry: JMS*, 2009, **44**, 1637–1660.
120. R. D. Unwin, C. A. Evans and A. D. Whetton, *Trends in Biochemical Sciences*, 2006, **31**, 473–484.
121. C. M. Colangelo, L. Chung, C. Bruce and K.-H. Cheung, *Methods (San Diego, Calif.)*, 2013, **61**, 287–298.
122. B. Schilling, M. J. Rardin, B. X. Maclean, A. M. Zawadzka, B. E. Frewen, M. P. Cusack, D. J. Sorensen, M. S. Bereman, E. Jing, C. C. Wu, E. Verdin, C. R. Kahn, M. J. Maccoss, and B. W. Gibson, 2012, 202–214.
123. B. MacLean, D. M. Tomazela, N. Shulman, M. Chambers, G. L. Finney, B. Frewen, R. Kern, D. L. Tabb, D. C. Liebler and M. J. MacCoss, *Bioinformatics*, 2010, **26**, 966–968.
124. E. J. Want, I. D. Wilson, H. Gika, G. Theodoridis, R. S. Plumb, J. Shockcor, E. Holmes and J. K. Nicholson, *Nature Protocols*, 2010, **5**, 1005–1018.
125. R. E. Ferguson, D. F. Hochstrasser and R. E. Banks, *Proteomics. Clinical Applications*, 2007, **1**, 739–746.
126. T. A. Addona, S. E. Abbatiello, B. Schilling, S. J. Skates, D. R. Mani, D. M. Bunk, C. H. Spiegelman, L. J. Zimmerman, A.-J. L. Ham, H. Keshishian, S. C. Hall, S. Allen, R. K. Blackman, C. H. Borchers, C. Buck, H. L. Cardasis, M. P. Cusack, N. G. Dodder, B. W. Gibson, J. M. Held, T. Hiltke, A. Jackson, E. B. Johansen, C. R. Kinsinger, J. Li, M. Mesri, T. A. Neubert, R. K. Niles, T. C. Pulsipher, D. Ransohoff, H. Rodriguez, P. A. Rudnick, D. Smith, D. L. Tabb, T. J. Tegeler, A. M. Variyath, L. J. Vega-Montoto, A. Wahlander, S. Waldemarson, M. Wang, J. R. Whiteaker, L. Zhao, N. L. Anderson, S. J. Fisher, D. C. Liebler, A. G. Paulovich, F. E. Regnier, P. Tempst and S. A. Carr, *Nature Biotechnology*, 2009, **27**, 633–641.
127. S. E. Abbatiello, D. R. Mani, B. Schilling, B. Maclean, L. J. Zimmerman, X. Feng, M. P. Cusack, N. Sedransk, S. C. Hall, T. Addona, N. G. Dodder, M. Ghosh, J. M. Held, V. Hedrick, and H. Dorota, *Molecular & Cellular Proteomics: MCP*, 2013, **12**, 2623–2639.
128. A. Prakash, T. Rezai, B. Krastins, D. Sarracino, M. Athanas, P. Russo, M. M. Ross, H. Zhang, Y. Tian, V. Kulasingam, A. P. Drabovich, C. Smith, I. Batruch, L. Liotta, E. Petricoin, E. P. Diamandis, D. W. Chan and M. F. Lopez, *Journal of Proteome Research*, 2010, 6678–6688.
129. E. Kuhn, T. Addona, H. Keshishian, M. Burgess, D. R. Mani, T. Richard, M. S. Sabatine, R. E. Gerszten and S. A. Carr, *Clinical Chemistry*, 2010, **55**, 1108–1117.
130. J. L. Hackett and S. I. Gutman, *Journal of Proteome Research*, 2005, **4**, 1110–1113.
131. D. M. Rocke, *Seminars in Cell & Developmental Biology*, 2004, **15**, 703–713.
132. J. Hu, K. R. Coombes, J. S. Morris and K. Baggerly, *Briefings in Functional Genomics & Proteomics*, 2005, **3**, 322–331.

CHAPTER 14

MRM-based Protein Quantification with Labeled Standards for Biomarker Discovery, Verification, and Validation in Human Plasma

ANDREW J. PERCY,[a] ANDREW G. CHAMBERS,[a] CAROL E. PARKER[a] AND CHRISTOPH H. BORCHERS*[a,b]

[a] University of Victoria - Genome British Columbia Proteomics Centre, Vancouver Island Technology Park, #3101–4464 Markham Street, Victoria, BC V8Z 7X8, Canada; [b] Department of Biochemistry and Microbiology, University of Victoria, Petch Building Room 207, 3800 Finnerty Road, Victoria, BC V8P 5C2, Canada
*Email: christoph@proteincentre.com

14.1 Introduction

Non-communicable diseases (NCDs) are a global epidemic as they currently account for roughly 60% of all fatalities.[1] These diseases include cardiovascular (*e.g.*, myocardial infarction, stroke) and respiratory (*e.g.*, chronic obstructive pulmonary disease) diseases, as well as diabetes (*e.g.*, insulin-dependent), cancer (*e.g.*, breast, ovarian), and mental illness (*e.g.*, schizophrenia). To diminish morbidity and mortality, biomarker tests are conducted to help predict disease risk, categorize disease severity, and

New Developments in Mass Spectrometry No. 1
Quantitative Proteomics
Edited by Claire Eyers and Simon J Gaskell

Published by the Royal Society of Chemistry, www.rsc.org

assess efficacy of therapy. Blood plasma is one of the preferred sample types since it is minimally invasive to collect, inexpensive to assess, and is routinely collected in clinical laboratories. The proteomics community continues to devote a great deal of effort toward discovering, verifying, and validating the putative markers. At present, there is a limited number of Food and Drug Administration (FDA)-approved protein disease biomarkers,[2] and a growing list of potential protein candidates.

The verification and validation process has long been considered the bottleneck of the protein biomarker pipeline. This stems from the 100s to 1000s of clinical samples required to identify the top quality candidates to progress to the pre-clinical validation stage.[3] To accelerate the verification process, methods that overcome the limitations of traditional enzyme-linked immunosorbent assays (ELISAs) are required to make them capable of analyzing large panels of proteins. Thus, alternative quantitative proteomics methods must be developed for high degrees of multiplexing that are inexpensive to develop and run. Such methods must also be transferrable and reproducible between laboratories, while offering the sensitivity and throughput required for comprehensive protein analysis.

Biomarker discovery and verification involve different types of quantitative proteomics. Biomarker discovery can be done using relative quantification, with the output being a "fold change" or an increase or decrease in expression. For biomarker verification, particularly for a biomarker that will be used as a diagnostic tool, the output should be an "absolute" amount; for example, a concentration in terms of pg mL^{-1} of plasma. This requires comparing the peptide or protein concentrations against a known amount of standard. One analytical method that satisfies most of these requirements uses stable isotope-labeled standards (SISs) within a multiple reaction monitoring-mass spectrometry (MRM-MS)-based bottom-up proteomic workflow.[4–6] The labeled standards help normalize for ion suppression effects and variations in instrument performance.[5,7] A schematic of an MRM experiment is shown in Figure 14.1.

To improve the sensitivity for lower abundance proteins, some researchers have implemented immunoaffinity depletion[8–10] or antibody-based enrichment[11–14] at the protein- or peptide-level, prior to bottom-up MRM analysis

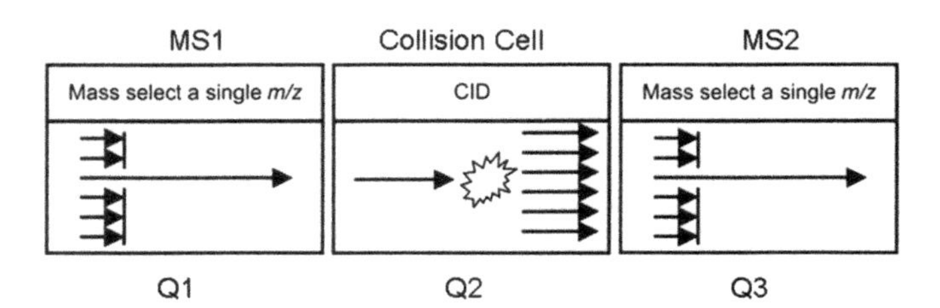

Figure 14.1 Schematic of an MRM experiment. Adapted from ref. 64.

with internal standards. Such methodological adaptations have met with increasing success. For instance, Whiteaker *et al.* quantified 15 cancer-related biomarker proteins in the low ng mL^{-1} concentration range through an immuno-MRM assay conducted on a magnetic bead-based platform.[12] Such strategies, however, are limited by increased assay costs, elevated variability,[10,15] and the increased risk of protein loss through non-covalent interactions to non-target species (*e.g.*, albumin).[16] An example of the effect of various depletion or fractionation techniques on peptide recoveries is shown in Figure 14.2.

In an attempt to reduce the problems associated with depletion and enrichment, while extending the depth of quantification, peptide fractionation through two dimensions (2D) of liquid chromatography (LC) is currently being explored. The concept of fractionation is not new to the proteomics field, having been used previously in shotgun proteomic experiments to enhance MS-based protein identification.[17,18] Its emergence in protein quantitative analyses, however, is relatively recent. Of the many LC combinations available,[19,20] an attractive option involves the use of two dimensions of reversed-phase liquid chromatography (RPLC), operated at high and low pH. This strategy was recently applied to the MRM with SIS peptide approach for the sub-ng mL^{-1} quantification of human plasma/serum proteins.[21] While this represents a significant advancement

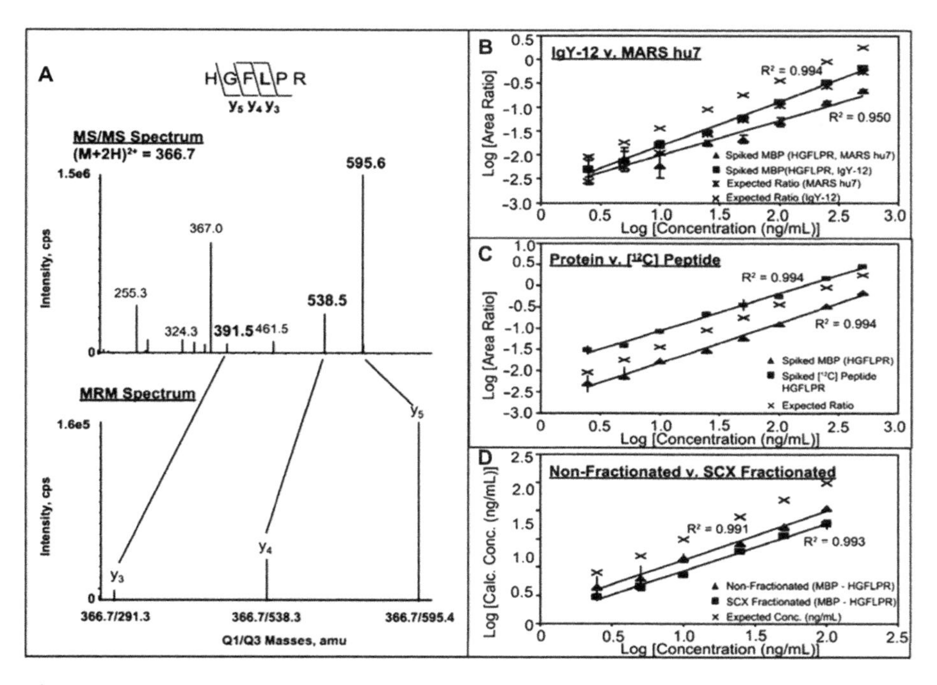

Figure 14.2 Peptide recovery after depletion or fractionation. MS/MS spectra (A) and recoveries of a peptide from myelin basic protein after antibody-based depletion (B and C) or after SCX fractionation (D). Reprinted from ref. 10 with permission.

to the quantitative proteomics field, the technique requires up-front depletion (which removes 14 potentially relevant clinical markers) and is performed under nano-flow operation, which hampers the robustness of the assay,[22] and was evaluated with just four proteins.

If depletion or enrichment are not used, this places an even greater reliance on the chromatography and mass spectrometry to ensure that only the correct transitions are being measured, particularly for a matrix as complex as plasma. Interference testing is, therefore, a key component of the development of any method, and is usually performed by a comparison of the peak shapes of the precursor and product ion chromatograms for the endogenous and the SIS peptide, as well as a comparison of the ratios of the precursor to product ion obtained in buffer and in the biological matrix.[23] An example of the effect of sample pre-treatment on interferences is shown in Figure 14.3.[10] An example of how peak height ratios can be used to find more subtle interferences[23,24] is shown in Figure 14.4.

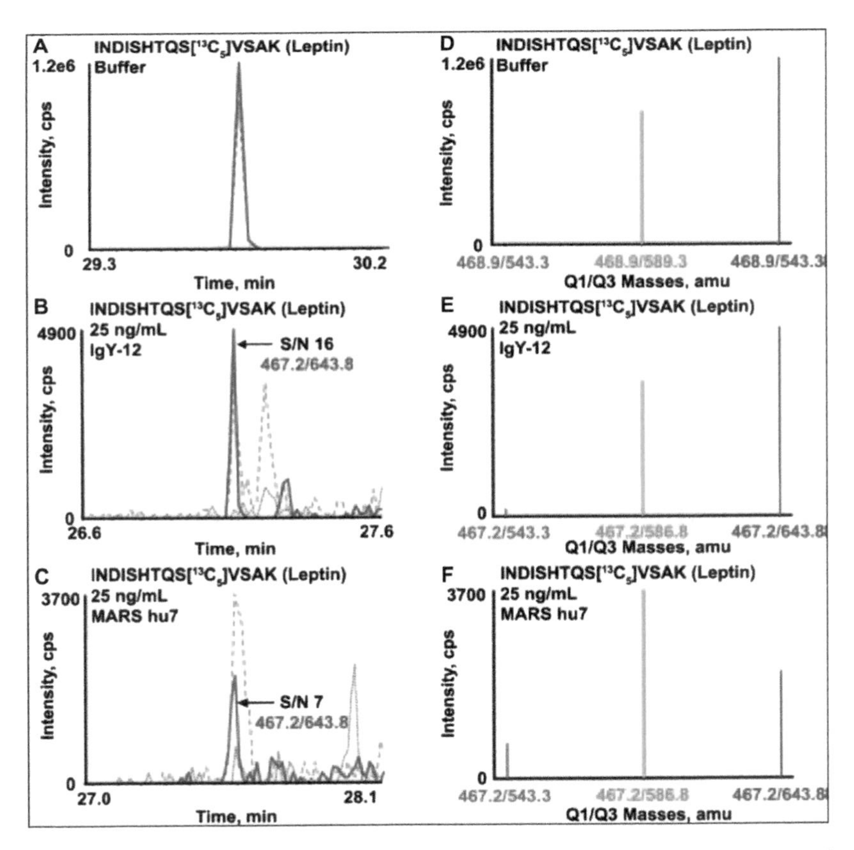

Figure 14.3 Interferences in MRM analyses. Extracted ion chromatograms (A-C) for a peptide from mouse leptin, as well as three MRM transitions (D-F) for a peptide spiked into buffer (A and D) or into various types of depleted plasma ((IgY-12 for B and E, and MARS hu7 for C and F). Reprinted from ref. 10, with permission.

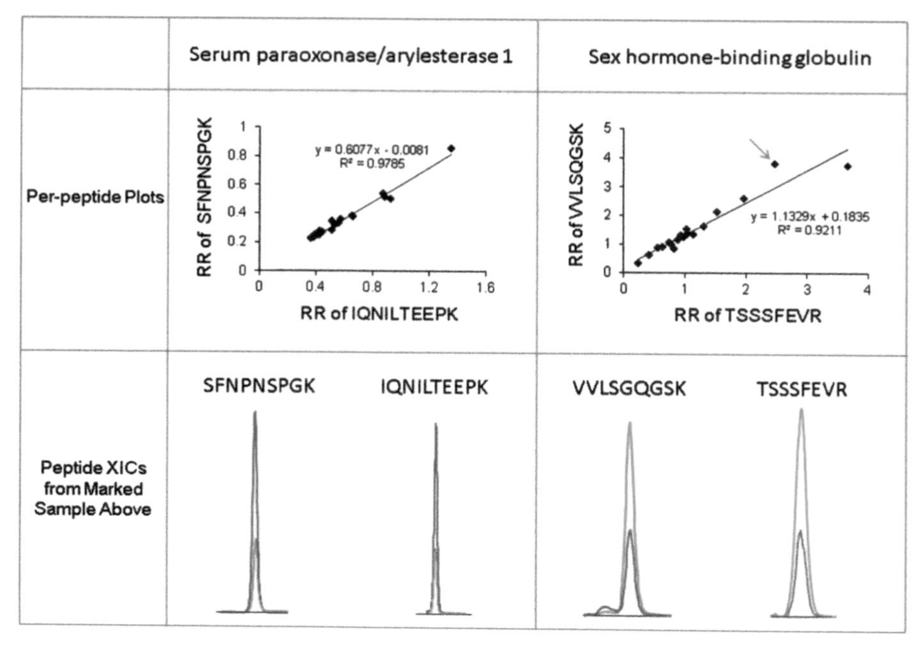

Figure 14.4 Interference detection based on relative ratios of the MRM transitions. Interferences are identified by examining the relative response (RR) for each protein, and confirming the peak shape by inspection of the peptide extracted-ion chromatograms (XICs). Due to its deviation from linearity, the sample indicated by the blue arrow contains an interference in the SIS and/or NAT MRM transitions of peptides VVLSQGSK and/or TSSSFEVR. The XIC traces for two interference-free peptides for serum paraoxonase/arylesterase 1 are shown for comparison.
Reprinted from ref. 24, with permission.

14.2 MRM for Biomarker Verification and Validation

By definition, MRM assays are targeted; in other words they require pre-selection of the target peptides and target proteins. For this reason, most MRM assays developed thus far have been used for the verification/validation of biomarkers that have been discovered by a *different* analytical technique. A wide variety of differential proteomics methods have been used for the biomarker discovery step; these include non-mass spectrometry-based methods, such as comparative 2D gel techniques (including GE Healthcare's DIGE approach, which uses different fluorescent stains), as well as label-free mass spectrometry-based approaches (such as spectral counting[25] and ion accounting[26]) and isotope-labeling techniques (including mTRAQ,[27] which uses non-isobaric tags, and iTRAQ[28] and TMT,[29] which use isobaric tags). Stable isotope labeling by amino acids (SILAC) can be used for cells grown in culture,[30] and has even been extended to mice[31] by the use of isotopically labeled feed.[32] We refer the reader to Elschenbroich and Kislinger[33] and Ohlund *et al.*[34] for excellent reviews on the advantages and disadvantages of these relative quantitative proteomic techniques, and the papers by Han and Higgs[35] and Kitteringham *et al.*[36] for a discussion on using mass spectrometry for the discovery and verification phase of biomarker development.

There are already an increasing number of studies in the literature where MRM is being used for biomarker verification (see refs 37–42 for some excellent recent reviews). In our laboratory we have used iTRAQ for discovery, followed by MRM for validation in studies on bladder cancer.[43,44] iTRAQ with MRM validation has also been used for biomarkers of Parkinson's disease.[45] In several recent studies MRM has been directly compared to ELISA (the technique that is currently most often used for biomarker validation). The study by Kuhn *et al.* used MRM of peptides from C-reactive protein (a biomarker for the diagnosis of rheumatoid arthritis), with isotopically labeled SIS peptides, and found good correlation between the ELISA and the MRM results.[46] The authors also noted that different peptides from the same protein sometimes gave different absolute protein concentrations, as we and others have also observed, possibly due to difference in digestion efficiencies.[47] In a recent study where 13 potential biomarker proteins in amniotic fluid were evaluated as a diagnostic for Down's Syndrome,[48] MRM performed at least as well as ELISA. Similarly, in our study on biomarkers of acute rejection in heart transplantation,[49] MRM also compared favourably with ELISA and immunonephelometric assays (INA)(INA). In another study where three amyloid-beta isoforms were quantified in cerebrospinal fluid (CSF)(CSF), a clearer diagnostic separation was obtained with MRM than with the ELISA approach.[50] Interestingly, in a recent study on hordein (gluten) in beer,[51] MRM results were found to be superior to ELISA, because the antibodies in the ELISA assay missed some of the hortein isoforms and gave false negative results (Figure 14.5).

	Avenin	B1	B3	D	γ3	ELISA (ppm)
	Hordein by Relative MS (% average)					
Mean	52.3	110.5	89.9	75.5	92.0	0.10
SE	22.4	37.4	47.5	23.6	21.9	0.01
Median	36.7	83.5	37.5	72.8	94.7	0.10

The mean, S.E. and median hordein composition was determined by using a representative peptide from each hordein family.
doi:10.1371/journal.pone.0056452.t003

Figure 14.5 A comparison of MRM with ELISA for hordein. MRM results for five commercial beers giving "false negative" ELISA values. Reprinted from ref. 51, with permission.

14.3 MRM for Biomarker Discovery

Although highly multiplexed assays can simply be used for the simultaneous verification of large numbers of biomarkers, importantly, they can also be used as a platform for biomarker *discovery*. A large panel of prospective biomarkers can be thought of as a "protein array", and if it is used to screen a population with several different diseases or disease stages, it is possible to discover different subsets of this protein *panel* that correlate with different diseases or disease states.[4] The smaller protein panels discovered in this way can then be validated by MRM using smaller panels of proteins, and when validated, these smaller panels can be used as diagnostic tools for specific diseases. Usually, a protein panel will provide higher diagnostic accuracy than a single biomarker. In addition, because the MRM method has already been worked out during the discovery phase, the assay can rapidly proceed to the verification without much additional development time.

Because many of the proteins in blood have been implicated as potential biomarkers of various diseases, we have been developing our large panel of plasma proteins for use as a diagnostic tool to provide this type of biomarker discovery. In a previous study on coronary artery disease (CAD), using a targeted panel of proteotypic peptides for 44 plasma proteins, we were able to find five proteins that differentiated CAD patients from those without CAD.[52] In a study on human bladder cancer patients, we used an initial MRM biomarker panel of 63 plasma proteins, and were able to refine this to a panel of only six proteins that could distinguish between bladder cancer and a urinary tract infection or hematuria with a 76.3% positive predictive value and a 77.5% negative predictive value.[43]

Several other research groups have also used MRM for discovery proteomics. The Nice group has published several studies using what they call a "hypothesis-based" MRM assay (*i.e.*, a panel of MRM assays targeting potential biomarkers found in the literature), and have applied this technique to the detection of colorectal cancer (CRC).[53] Using a combination of in-gel digestion and MRM analysis, they developed a multiplexed MRM assay for 60 proteins reported to be associated with colorectal cancer; nineteen were found in the faeces of colorectal cancer patients, and a subset consisting of five of these proteins were found in CRC patients and not in controls.[54] The Nice group used a variety of separation techniques to create a library of 108 human fecal proteins, from which they generated MRM assays for 40 non-redundant human proteins. Of these, nine proteins were found only in CRC patients.[55]

In a study on a mouse model for anaemia, similarly based on the literature, a MRM-based assay for a group of 14 plasma proteins enabled the discovery of a panel of eight proteins that could distinguish normal mice from those with iron-deficiency anemia, inflammation, and anemia/inflammation.[56] The final eight-protein biomarker panel improved the classification accuracy from 94% to 100% in the normal group, from 50% to 72% in the inflammation group, from 66% to 96% in the iron-deficiency

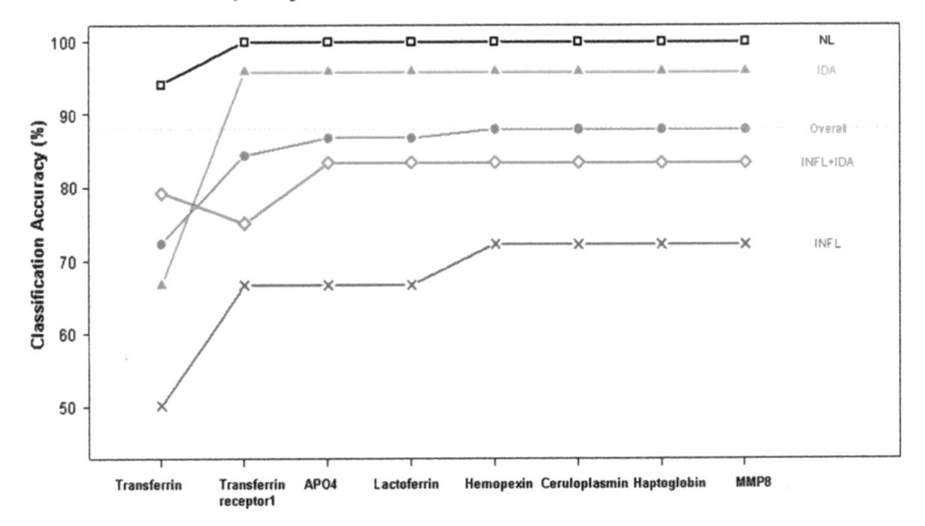

Figure 14.6 Improvement of classification accuracy with multiple-protein biomarker panels. Classification performance of a multiprotein panel for different types of anemia, showing the classification accuracy (y axis) as additional proteins are selected and incorporated into a multivariate classifier panel (x axis) by an stepwise discriminant analysis (SDA) algorithm. The classifier was built using a linear discriminant analysis (LDA)(LDA) to classify test samples from a leave-one-out cross-validation. Abbreviations used: NL = normal, IDA = Iron deficiency anaemia, INFL = inflammation.
Reprinted from ref. 56, with permission.

anemia group, and from 79% to 83% in the inflammation plus iron-deficiency anemia group, compared to the best single-protein biomarker (Figure 14.6).

A similar approach was used by Pan *et al.*,[57] where a group of five plasma proteins, previously reported as being associated with human pancreatic cancer, was monitored in a multiplexed MRM analysis to determine if they were actually biomarkers of this disease. Of the four proteins that were detectable in the plasma, three gave area-under-curve (AUC) values that were better than or equal to that of the current FDA-approved blood test (for CA19-9). At 95% specificity, a combination of two protein biomarkers showed a 5% improvement in specificity over the best single-protein biomarker. ELISA for tissue inhibitor of metalloproteinase 1 (TIMP1), performed on the same samples, consistently gave lower plasma concentration values, which the authors attributed to ELISA only measuring TIMP1 in its free form.

14.4 Conclusions

MRM with isotopically labeled standards can bridge the gap between the discovery and validation of putative disease biomarkers. This targeted quantitative approach helps overcome the limitations related to development time, cost, failure rate, and lack of multiplexing that plague immunoassays for the precise quantification of biofluid-derived proteins. Hüttenhain *et al.*, have reported quantifying 182 proteins in depleted plasma (down to 10 ng mL^{-1}), and 408 peptides in urine, which is a simpler matrix because of its narrower dynamic range of protein concentrations.[58] We have recently demonstrated the enhanced multiplexing ability of a 1D LC/MRM-MS method for the precise quantification of 142 candidate NCD-related biomarkers in undepleted human plasma.[24] The determined concentrations span six orders of magnitude (from albumin at 31 mg mL^{-1} to myeloblastin at 44 ng mL^{-1}), which covers the range of the high-to-moderate abundance plasma proteins. We feel that it should be possible to improve even further the quantitative performance of this method, and to accelerate the sample preparation steps, and we are continuing to work in this area.

These and other multiplexed methods are robust and can be used to obtain improved sensitivity and high specificity in pre-clinical studies on hundreds of patient samples. Furthermore, the high degree of analyte multiplexing already achieved should accelerate the verification of the large numbers of candidate markers that currently exist. Our current 1D method is particularly rapid, requiring less than a day of sample preparation time (includes lyophilization), followed by 43 min of instrument time for each LC/MRM-MS run. Ideally, multiple peptides per protein and multiple transitions per peptide would be monitored. However, for highly multiplexed assays, a trade-off must usually be made. Therefore, for the analysis of unknown samples, we advocate monitoring quantifier transitions only, but with multiple peptides targeted per protein. With this method, interferences can be determined on a protein-by-protein basis by plotting the relative response of one peptide against another. Only the samples that remain linear are considered to be interference-free. To determine the concentration of the protein in the plasma sample, the relative response is then applied to the regression equation (slope, y intercept, and SIS concentration are known values), as defined by the analysis of the control plasma. As a result, both interference screening and protein quantification could be performed from a single experiment. Based on this approach, the methods developed are useful for the analysis of patient samples and are transferable to the screening of large numbers of analytes in a rapid and precise manner. Moreover, if specific biomarker candidates are to be verified (*e.g.*, apolipoproteins or cancer panels), the MRM method can easily be adjusted to detect new targets. Thus, the developed methods provide flexibility for further refinement of the biomarker panel, or for routine use in screening a large number of targets.

Concurrent with the work done by our laboratory and others on improving the multiplexing capabilities, the intra-laboratory reproducibility, and the robustness and accuracy of the MRM approach, many research groups and organizations are developing libraries of MRM peptides that can be used to create large panels of MRM assays for biomarker discovery purposes. In addition to MRM Atlas/SRM Atlas[59] we are creating a library of MRM transitions for plasma proteins, while several other research groups are creating libraries or databases of MRM peptides and transitions that can be used, for example, to probe cancer-related pathways for determining therapeutic response,[60] or breast tumor proteins and peptides.[61,62] The Aebersold group has recently reported the creation of a library of MRM assays for >1000 cancer-related proteins.[58] A library of 1572 proteins from MCF-7 human breast cancer cells with five ranked MRM transitions per peptide is also being created.[63]

In the future these libraries of transitions could be used to simultaneously quantify large numbers of potential biomarker proteins, which could be used for the discovery of highly accurate biomarker panels in biological samples *via* highly multiplexed MRM. We have already developed highly multiplexed assays containing ~380 interference-free peptides per run in undepleted plasma. Once a subset of these biomarkers has been validated, these validated biomarker panels could be combined into new multiplexed assays. These new multiplexed assays would be capable of differentiating between different diseases based not only on the detection of a large number of different protein biomarkers, but also on different expression patterns of the same set of biomarkers. These highly multiplexed and accurate assays could then be used in clinical laboratories for the simultaneous diagnosis of multiple diseases.

Acknowledgements

The authors would like to thank Genome Canada, Genome BC, and the Western Economic Diversification of Canada for providing funding to the University of Victoria - Genome BC Proteomics Centre. The authors also recognize the fiscal, operational, and scientific support of the NCE CECR PROOF Centre of Excellence. The authors declare no financial/commercial conflicts of interest.

References

1. J. Olsen, R. Bertollini, C. Victora and R. Saracci, *Int. J. Epidemiol.*, 2012, **41**, 1219–1220.
2. N. L. Anderson, *Clin. Chem.*, 2010, **56**, 177–185.
3. A. G. Paulovich, J. R. Whiteaker, A. N. Hoofnagle and P. Wang, *Proteomics Clin. Appl.*, 2008, **2**, 1386–1402.
4. C. E. Parker, D. Domanski, A. J. Percy, A. G. Chambers, A. G. Camenzind, D. S. Smith and C. H. Borchers, *Chemical Diagnostics*, ed. N. L. S. Tang and T. Poon, Springer, 2014, 117–138.
5. V. Brun, C. Masselon, J. Garin and A. Dupuis, *J. Proteomics*, 2009, **72**, 740–749.
6. R. Huttenhain, J. Malmstrom, P. Picotti and R. Aebersold, *Current Opinion in Chemical Biology*, 2009, **13**, 518–525.

7. M. Elliott, D. Smith, M. Kuzyk, C. E. Parker and C. H. Borchers, *J. Mass Spectrom.*, 2009, **44**, 1637–1660.
8. T. Fortin, A. Salvador, J. P. Charrier, C. Lenz, X. Lacoux, A. Morla, G. Choquet-Kastylevsky and J. Lemoine, *Mol. Cell. Proteomics*, 2009, **8**, 1006–1015.
9. H. Keshishian, T. Addona, M. Burgess, D. R. Mani, X. Shi, E. Kuhn, M. S. Sabatine, R. E. Gerszten and S. A. Carr, *Mol. Cell. Proteomics*, 2009, **8**, 2339–2349.
10. H. Keshishian, T. Addona, M. Burgess, E. Kuhn and S. A. Carr, *Mol. Cell. Proteomics*, 2007, **6**, 2212–2229.
11. J. R. Whiteaker, L. Zhao, C. Lin, P. Yan, P. Wang and A. G. Paulovich, *Mol. Cell. Proteomics*, 2012, **11**, M111.015347.
12. J. R. Whiteaker, L. Zhao, L. Anderson and A. G. Paulovich, *Mol. Cell. Proteomics*, 2010, **9**, 184–196.
13. E. Kuhn, T. Addona, H. Keshishian, M. Burgess, D. R. Mani, R. T. Lee, M. S. Sabatine, R. E. Gerszten and S. A. Carr, *Clin. Chem.*, 2009, **55**, 1108–1117.
14. G. R. Nicol, M. Han, J. Kim, C. E. Birse, E. Brand, A. Nguyen, M. Mesri, W. FitzHugh, P. Kaminker, P. A. Moore, S. M. Ruben and T. He, *Mol. Cell. Proteomics*, 2008, **7**, 1974–1982.
15. V. Polaskova, A. Kapur, A. Khan, M. P. Molloy and M. S. Baker, *Electrophoresis*, 2010, **31**, 471–482.
16. J. Granger, J. Siddiqui, S. Copeland and D. Remick, *Proteomics*, 2005, **5**, 4713–4718.
17. D. A. Wolters, M. P. Washburn and J. R. Yates, III, *Anal. Chem.*, 2001, **73**, 5683–5690.
18. A. Motoyama, J. D. Venable, C. I. Ruse and J. R. Yates, *Anal. Chem.*, 2006, **78**, 5109–5118.
19. P. Horvatovich, B. Hoekman, N. Govorukhina and R. Bischoff, *J. Sepn. Sci.*, 2010, **33**, 1421–1437.
20. H. Malerod, E. Lundanes and T. Greibrokk, *Anal. Methods*, 2010, **2**, 110–122.
21. T. Shi, T. L. Fillmore, X. Sun, R. Zhao, A. A. Schepmoes, M. Hossain, F. Xie, S. Wu, J. S. Kim, N. Jones, R. J. Moore, L. Pasa-Tolic, J. Kagan, K. D. Rodland, T. Liu, K. Tang, D. G. Camp, R. D. Smith and W. J. Qian, *Proc. Nat. Acad. Sci. USA*, 2012, **109**, 15395–15400.
22. A. J. Percy, A. G. Chambers, J. Yang, D. Domanski and C. H. Borchers, *Anal. Bioanal. Chem.*, 2012, **404**, 1089–1101.
23. A. J. Percy, A. G. Chambers, D. S. Smith and C. H. Borchers, *J. Proteome Research*, 2013, **12**, 222–233.
24. A. J. Percy, A. G. Chambers, J. Yang, D. Hardie and C. H. Borchers, *Biochimica et Biophysica Acta*, 2013, **24**, 1338–1345.
25. C.-H. Chen, *Analytica Chimica Acta*, 2008, **624**, 16–36.
26. J. C. Silva, M. V. Gorenstein, G.-Z. Li, J. P. C. Vissers and S. J. Geromanos, *Mol. Cell. Proteomics*, 2006, **5**, 144–156.
27. Applied Biosystems, http://www.sciex.com/products/standards-and-reagents/mTRAQ-Reagents.xml, Accessed 8th November 2013.

28. P. L. Ross, Y. N. Huang, J. N. Marchese, B. Williamson, K. Parker, S. Hattan, N. Khainovski, S. Pillai, S. Dey, S. Daniels, S. Purkayastha, P. Juhasz, S. Martin, M. Bartlet-Jones, F. He, A. Jacobson and D. J. Pappin, *Mol. Cell. Proteomics*, 2004, **3**, 1154–1169.
29. Thermo Scientific, Amine-reactive 6-plex Tandem Mass Tag Reagents, http://www.piercenet.com/product/amine-reactive-6-plex-tandem-mass-tag-reagents, 2008. Accessed 8th November 2013.
30. S.-E. Ong, B. Blagoev, I. Kratchmarova, D. B. Kristensen, H. Steen, P. Akhilesh and M. Mann, *Mol. Cell. Proteomics*, 2002, **1**, 376–386.
31. S. Zanivan, M. Krueger and M. M. Meth, *Mol. Biology*, 2012, **757**, 435–450.
32. Silantes GmbH, http://www.silantes.com/silacmouse.htm, München, Germany, vol. June 17, 2013. Accessed 12th October 2013.
33. S. Elschenbroich and T. Kislinger, *Mol. Biosystems*, 2011, 7, 292–303.
34. L. B. Ohlund, D. B. Hardie, M. H. Elliott, D. S. Smith, J. D. Reid, G. V. Cohen-Freue, A. P. Bergman, M. Sasaki, L. Robertson, R. F. Balshaw, R. T. Ng, A. Mui, B. M. McManus, P. A. Keown, W. R. McMaster, C. E. Parker and C. H. Borchers, in *Sample Preparation in Biological Mass Spectrometry*, A. Ivanov and A. Lazarev (Eds.), Springer, New York, 2011, pp. 575–624.
35. B. Han and R. E. Higgs, *Briefings in Functional Genomics and Proteomics*, 2008, 7, 340–354.
36. N. R. Kitteringham, R. E. Jenkins, C. S. Lane, V. L. Elliott and B. K. Park, *J. Chromatogr. B*, 2009, **877**, 1229–1239.
37. E. S. Boja and H. Rodriguez, *Proteomics*, 2012, **12**, 1093–1110.
38. Z. Meng and T. D. Veenstra, *J. Proteomics*, 2011, **74**, 2650–2659.
39. T. Shi, D. Su, T. Liu, K. Tang, D. G. Camp, W.-J. Qian and R. D. Smith, *Proteomics*, 2012, **12**, 1074–1092.
40. D. C. Liebler and L. J. Zimmerman, *Biochemistry*, 2013, **52**, 3797–3806.
41. J. R. Whiteaker, C. Lin, J. Kennedy, L. Hou, M. Trute, I. Sokal, P. Yan, R. M. Schoenherr, L. Zhao, U. J. Voytovich, K. S. Kelly-Spratt, A. Krasnoselsky, P. R. Gafken, J. M. Hogan, L. A. Jones, P. Wang, L. Amon, L. A. Chodosh, P. S. Nelson, M. W. McIntosh, C. J. Kemp and A. G. Paulovich, *Nature Biotechnol.*, 2011, **29**, 625–634.
42. S. A. Carr and L. Anderson, *Clin. Chem.*, 2008, **54**, 1749–1752.
43. Y.-T. Chen, H.-W. Chen, D. Domanski, D. S. Smith, K.-H. Liang, C.-C. Wu, C.-L. Chen, T. Chung, M.-C. Chen, Y.-S. Chang, C. E. Parker, C. H. Borchers and J.-S. Yu, *J. Proteomics*, 2012, **75**, 3529–3545.
44. Y.-T. Chen, C. E. Parker, H.-W. Chen, C.-L. Chen, D. Domanski, D. S. Smith, C.-C. Chih- Wu, T. Chung, K.-H. Liang, M.-C. Chen, Y.-S. Chang, C. H. Borchers and J.-S. Yu, in *Comprehensive Biomarker Discovery and Validation for Clinical Application*, Péter Horvatovich, Rainer Bischoff (Eds.), Royal Society of Chemistry, Cambridge, 2013.
45. S. Lehnert, S. Jesse, W. Rist, P. Steinacker, H. Soininen, S. K. Herukka, H. Tumani, M. Lenter, P. Oeckl, B. Ferger, B. Hengerer and M. Otto, *Experimental Neurology*, 2012, **234**, 499–505.

46. E. Kuhn, J. Wu, J. Karl, H. Liao, W. Zolg and B. Guild, *Proteomics*, 2004, **4**, 1175–1186.
47. J. L. Proc, M. A. Kuzyk, D. B. Hardie, J. Yang, D. S. Smith, A. M. Jackson, C. E. Parker and C. H. Borchers, *J. Proteome Research*, 2010, **9**, 5422–5437.
48. C.-K. J. Choa, A. P. Drabovich, I. Batruch and E. P. Diamandis, *J. Proteomics*, 2011, **74**, 2052–2059.
49. G. V. Cohen Freue, A. Meredith, D. Smith, A. Bergman, M. Sasaki, K. K. Y. Lam, Z. Hollander, N. Opushneva, M. Takhar, D. Lin, J. Wilson-McManus, R. Balshaw, P. A. Keown, C. H. Borchers, B. McManus, R. T. Ng, W. R. McMaster, *PLoS Computational Biology*. 2013, **9**, e1002963.
50. J. Pannee, E. Portelius, M. Oppermann, A. Atkins, M. Hornshaw, I. Zegers, P. Höjrup, L. Minthon, O. Hansson, H. Zetterberg, K. Blennow and J. Gobom, *J. Alzheimers Dis.*, 2013, **33**, 1021–1032.
51. G. J. Tanner, M. L. Colgrave, M. J. Blundell, H. P. Goswami and C. A. Howitt, *PLoS One*, 2013, **8**, e56452.
52. G. V. Cohen Freue and C. H. Borchers, *Circ: Cardiovasc. Genet.*, 2012, **5**, 378.
53. C.-S. Ang, J. Phung and E. C. Nice, *Biomedical Chromatography*, 2011, **25**, 82–99.
54. C. S. Ang and E. C. Nice, *J. Proteome Research*, 2010, **9**, 4346–4355.
55. C. S. Ang, J. Rothacker, H. Patsiouras, P. Gibbs, A. W. Burgess and E. Nice, *Electrophoresis*, 2011, **32**, 1926–1938.
56. D. Domanski, G. Cohen Freue, L. Sojo, M. A. Kuzyk, C. E. Parker, Y. P. Goldberg and C. H. Borchers, *J. Proteomics*, 2012, **75**, 3514–3528.
57. S. Pan, R. Chen, R. E. Brand, S. Hawley, Y. Tamura, P. R. Gafken, B. P. Milless, D. R. Goodlett, J. Rush and T. A. Brentnall, *J. Proteome Research*, 2012, **11**, 1937–1948.
58. R. Hüttenhain, M. Soste, N. Selevsek, H. Röst, A. Sethi, C. Carapito, T. Farrah, E. W. Deutsch, U. Kusebauch, R. L. Moritz, E. Niméus-Malmström, O. Rinner and R. Aebersold, *Science and Translational Medicine*, 2012, **4**, 1–13.
59. SRM Atlas, www.srmatlas.org. Last accessed 8th November 2013.
60. E. R. Remily-Wood, R. Z. Liu, Y. Xiang, Y. Chen, C. E. Thomas, N. Rajyaguru, L. M. Kaufman, J. E. Ochoa, L. Hazlehurst, J. Pinilla-Ibarz, J. Lancet, G. Zhang, E. Haura, D. Shibata, T. Yeatman, K. S. M. Smalley, W. S. Dalton, E. Huang, E. Scott, G. C. Bloom, S. A. Eschrich and J. M. Koomen, *Proteomics Clinical Applications*, 2011, **5**, 383–396.
61. L. Alldridge, G. Metodieva, C. Greenwood, K. Al-Janabi, L. Thwaites, P. Sauven and M. Metodiev, *J. Proteome Research*, 2008, 7, 1458–1469.
62. G. Metodieva, C. Greenwood, L. Alldridge, P. Sauven and M. Metodiev, *Proteomics Clinical Applications*, 2009, **3**, 78–82.
63. X. Yang and I. M. Lazar, *BMC Cancer*, 2009, 96.
64. M. Kinter and N. E. Sherman, Protein sequencing and identification using tandem mass spectrometry, John Wiley & Sons, USA, 2005.

CHAPTER 15

Mass Spectrometry-based Quantification of Proteins and Peptides in Food

PHIL E. JOHNSON, JUSTIN T. MARSH AND E.N. CLARE MILLS*

Institute of Inflammation and Repair, Manchester Academic Health Science Centre, Manchester Institute of Biotechnology, University of Manchester, UK
*Email: clare.mills@manchester.ac.uk

15.1 Introduction

Proteomic analyses, together with genomic and metabolomic profiling of foods, is contributing to the development of a new field of expertise, which has been called "food 'omics". Such methodologies are giving rise to a new generation of quality assurance tools able to ensure the authenticity, quality and safety of foods. They are also making an important contribution to helping the agricultural and food industries address food security issues. For example, the application of proteomics is bringing insights into how climate change may affect food protein quality, whether it be derived from crops or livestock, and ensuring the drive for low-energy food processing procedures does not compromise the safety, flavour and nutritional quality of foods. As food prices increase, there is an emerging need for new tools for food authentication to detect food fraud and ensure effective traceability of foods in what is a complex food supply chain that can spread across several continents.

New Developments in Mass Spectrometry No. 1
Quantitative Proteomics
Edited by Claire Eyers and Simon J Gaskell

Published by the Royal Society of Chemistry, www.rsc.org

Despite such potential, the application of proteomics approaches to food has been slow, partly because of the diverse range of plant and animal species that constitute the food supply to humans. It has been estimated that around 7000 plant species have been cultivated for consumption in human history and around 35 animal species have been domesticated for use in agricultural and food production, although only about 30 crops provide 95% of human food energy needs. Until recently, sequenced genomes were only available for a few model species, data that greatly assists in annotation of a proteome. This is beginning to change, although plant genomes present many technical challenges, with physical maps having to be constructed for large genomes, such as those found in cereals.

Such a lack of basic information about plant and animal species used for food is further complicated by the fact that food proteins are structurally modified by the numerous cooking and fermentation procedures employed in food preparation. These modifications are overlaid by other chemical and biochemical changes that take place during storage. As a consequence of these processes, food chemists have to deal with both the classical post-translational modifications found in plant and animal proteins and a broad range of often poorly characterised processing-induced adducts and modified amino acids. One particularly complex processing-related reaction is that between free amino groups in proteins and sugars to form Maillard browning reaction products. These modifications often make both protein identification by mass spectrometry and the selection of targets for quantification more difficult for foods. This chapter describes some of these complexities and the strategies that are being pursued to overcome them and, hence, realise the potential of protein mass spectrometry in the food arena.

15.2 Challenges for Food Proteomics

15.2.1 Food Proteome Annotation

The widespread application of proteomics to foods, especially those of plant origin, has been hampered by the lack of genomic sequence information and protein sequences deposited in databases such as UNIPROT. For many years, the only edible crop that had a genome sequence was rice; however, the availability of plant genome sequence information is expanding rapidly, with the sequences of several important crops, including wheat, maize and soybean, having become available in recent years. There has also been an expansion in publicly available EST databases for crops, which can also be used to facilitate the annotation of proteins. However, despite these advances, a paucity of sequence information still remains for many plant species, with less than 100 UNIPROT accessions attributed to some foods, such as walnuts. The lack of genomic and protein sequence information for many animal and plant species used for food has slowed the application of proteomics to foods. There is also a lack of a systematic approach to data collection, although this is beginning to change. One initiative that is incorporating proteomic data relevant to food into more consistent formats is the Bovine Peptide Atlas.[1] This has placed 1921 proteins identified in *Bos taurus* from 107 samples from six tissues, with the aim of supporting the development of targeted quantitative proteomics studies. Such initiatives make the tools that have been available to those working with model organisms much more accessible to food researchers and will underpin efforts in the future to design quantitative proteomic approaches to profiling the protein content of cow's milk and the plethora of related dairy products, which are a major food protein source.

15.2.2 Processing-induced Modifications of Food Proteins

Numerous processing-induced modifications of proteins have been defined, although the extent and type of modification depends on the precise time–temperature combinations, water activity and pH employed during processing.[2] For example, amino acid residues, such as glutamine and asparagine, can become deamidated as a result of thermal treatments and the use of low pH processes, like pickling, or through acidification to deamidate food ingredients, such as gluten. Other reactions involving cysteine include β-elimination reactions, where a hydroxyl ion facilitates cleavage of a disulphide bond to yield dehydroalanine, which can in turn react with lysine residues to form a lysine–alanine cross-link, together with thiocysteine. This can then decompose to give a variety of products, including free cysteine. There is evidence that such modified amino acids are formed after heating gluten[3] and that cross-links, such as lysinoalanine and lanthionine, are formed during the preparation of baked goods, such as hard pretzels, where the dough is subjected to brief dipping in hot alkali (45 s, 1.0% (*w*/*v*) NaOH, 90 °C) prior to baking.[4] The application of heat, together with alkaline conditions, can also give rise to a range of other amino acids, such as methyldehydroalanine, β-aminoalanine, ornithinoalanine, histidinoalanine, phenylethylaminoalanine, lanthionine and methyl-lanthionine, and can be accompanied by racemisation of L-amino acid isomers to D-analogues.[5]

One of the most well-described processing-induced modifications are Maillard adducts, which result from a complex series of reactions that take place between amino acids and sugars. These are especially important to consider in proteomics experiments where trypsin is employed in sample workflows, since the sugars react with primary amines and, particularly, lysine. Initially, hexose sugars, such as glucose, react with protein free amino groups to form an unstable Schiff's base, which then cyclises to form more stable Amadori products (also known as early glycation products).[6] A similar reaction and rearrangement, known as the Hyne's reaction, can also take place with ketose sugars, such as fructose. These modified amino acids may subsequently undergo further reactions resulting in the formation of a complex mixture of products, which are generally poorly characterised but can form cross-links between proteins, rendering them insoluble.

The reactions involved may be quite slow, with rearrangement products forming during storage of foods rather than during the initial cooking process, especially in low water activity foods. Known as melanoidins, these advanced Maillard products are responsible for the browning of baked foods, such as bread, or roasted and fried foods and contribute to their characteristic toasted flavour. Other sugar-derived compounds, such as sugar dicarbonyls like methylglyoxal, also react with lysine, arginine and tryptophan residues and may form cross-links between proteins.

Although less well characterised, lipid oxidation products also have the potential to modify proteins, such as 2,4-decadienal. Formed as a consequence of the thermoxidation of fatty acids following roasting of foods such as hazelnuts,[7] this compound can participate in Maillard-type carbonyl–amine condensation reactions with free amino groups on residues such as lysine. The modification of lysine, an example of which is *N*ε-(carboxymethyl)lysine, can occur as a result of reactions with sugars or fatty acids[8] and is strongly dependent on the degree of unsaturation, with fatty acids, such as arachidonic acid (C20:4), being much more reactive than less saturated lipids, such as oleic acid (C18:1). Studies in model systems[8,9] have shown that peroxidation of polyunsaturated fatty acids (PUFAs) results in the formation of compounds such as malonyldialdehyde, 4-hydroxy-2-nonenal and acrolein, which can react with amino acid residues in proteins and form various cross-links. There is also evidence that *N*ε-(carboxymethyl)lysine is formed in food systems, notably cow's milk products.[10]

Certain types of processing procedures are deliberately designed to modify proteins, notably hydrolytic processes. Protein ingredients, often from soybean and the gluten fraction of wheat, are subjected to chemically induced hydrolysis or digested with microbial or fungal proteases to provide complex mixtures of peptides with particular biophysical properties. These are exploited in food formulations, particularly because of their emulsifying or foam-stabilising properties. Food proteins may also be hydrolysed in fermented foods, like yogurts, cheeses, sauces, such as soy sauce, and fermented meat products like salami, as a consequence of the action of proteases secreted by the fermentation microbes. Whilst the small peptides of 10–15 residues are tractable to mass spectrometry analysis, the larger peptides resulting from hydrolysis may not be digested in quite the same manner by endoproteases, like trypsin, when presented as intact proteins. For example, hydrolysis may result in the generation of peptides with *N*- or *C*-termini that lie adjacent to a tryptic cleavage site and that will, thus, affect the ability of trypsin to digest it further and generate, for example, a target peptide used for quantitative multiple reaction monitoring experiments.

15.2.3 Food Matrix Effects and Protein Extraction

In addition to the chemical modification of proteins induced by food processing, many of the physical conditions cause proteins to denature and aggregate. Such structural changes contribute to the formation of food structures, such as the gelled protein networks found in cooked meats, and the interfacial structures formed at either air–water (such as those found in whipped egg white) or oil–water interfaces (such as those found in emulsified sauces like mayonnaise). Protein unfolding can be triggered by changes in pH, like those occurring during the lactic acid fermentation of milk, which results in the gelled network structure of yogurt. It may also result from the application of heat, which transforms, for example, the viscous, transparent liquid of egg white into the rubbery, opaque solid of boiled egg white. The formation of the protein networks found in these foods generally involves the formation of β-sheet-rich structures, which may exhibit properties similar to those found in amyloid-type structures. Other types of aggregate network are formed from globular proteins, as is the case for the gelled structures found in foods like yogurt. Lastly, food proteins can form entangled polymer networks, with fibrous networks formed by proteins, like collagen, in gelatin gels, or viscoelastic networks formed by the gluten fraction of wheat flour that comprises the wheat seed storage prolamins linked by a combination of non-covalent interactions and intermolecular disulphide bonds.

In general the protein networks formed in foods are highly insoluble and require the use of both ionic detergents, such as sodium dodecyl sulphate (SDS) or chaotropes, like urea, and the use of reducing agents to render the constituent polypeptides soluble. The variable nature of food matrices and their complexity makes extraction problematic and poses a challenge for many classical analytical tools, including immunoassays, with different extraction conditions being used depending on the nature of the food matrix.[11] These same difficulties are encountered in the application of mass spectrometry methods to food analysis. However, the use of derivative peptides as the analytical target in a method where detection is not dependent on conformation has the potential to make workflows simpler and more adaptable to different food matrices compared to more traditional analysis.[12] However, the incompatibility of many detergents required for the extraction of complex food matrices poses a barrier to the application of mass spectrometry to food protein analysis. For example, the analysis of peanuts in cookies using only protein fractions soluble in simple buffers requires additional enrichment procedures to ensure detection of proteins known to be present in the original sample. [13] Taking a much more rigorous approach, others have explored the use of a variety of extractants and demonstrated the superior performance of combinations of chaotropes (such as urea) and detergents over simple salt solutions for extracting roasted peanut proteins.[14] One way of dealing with ionic detergents used for extraction is to remove them prior to mass spectrometry analysis.[13] This has the drawback of making sample preparation a much more lengthy and expensive process, reducing the suitability of such workflows for routine analysis. The use of acid-labile detergents (which are degraded with acidification and can be removed simply by centrifugation) is a possible solution to this problem.[15]

15.2.4 Protease Digestion in Mass Spectrometry Analysis of Proteins

The digestion of proteins from a food matrix to yield analytical peptides can also pose some challenges. Many food proteins are difficult to digest with endoproteases and this is characteristic of many animal and plant-derived food allergens.[16] Under many circumstances, proteolytic digestion of recalcitrant protein targets, such as lysozyme from egg, can be improved by relatively simple changes to digestion protocol, such as incorporating microwave treatment.[17] Whilst trypsin has been widely used for proteomic analysis, it is not necessarily the protease of choice for food analysis. Lysine modification as a consequence of food processing conditions, particularly modification through reactions with sugars to form Maillard adducts, may affect digestibility in a fashion which is target dependent.[18] In some cases it may also be necessary to remove protein modifications that affect protease digestion or detection of resulting peptide digestion products. Thus, treatment with the enzyme PNGase can be used to remove carbohydrate from proteins such as egg white ovomucoid.[19] In addition some important food proteins, such as wheat gluten, contain few lysine residues and the successful application of proteomic analyses requires use of other proteases, such as chymotrypsin.[20] Finally, the chemical and physical complexity of the matrix itself may, in some cases, hinder digestion of analytical targets either because of the presence of specific protease inhibitors or other non-specific inhibitory molecules that are co-extracted, *e.g.*, certain fatty acids and polyphenols.[21] The removal of inhibitory molecules prior to digestion (*e.g.*, by simple chromatographic methods) may need to be considered if incomplete digestion is observed. As incomplete digestion is often encountered in food proteomic analyses, especially when using classical digestion protocols, it is crucial to check whether digestion and release of analytical peptides is consistent across a range of food matrices.

15.2.5 Target Selection for Analysis of Food Proteins

The application of protein mass spectrometry to detect, for example, food adulteration, means that, rather than having a single protein species as the analytical target, a selection of targets from a range of potential candidate molecules has to be made (Figure 15.1). Ideally, candidate molecules are selected about which there is a great deal of prior knowledge to support development of effective extraction and protease digestion protocols and address issues posed by post-translational modifications (Table 15.1). One particular issue is the need for a complete protein sequence, which may not be known or only partial fragments of sequence may be available. Other aspects that need to be considered are the abundance of a target protein molecule in a foodstuff as this may have implications for the sensitivity of the final method. For example, targeting a casein molecule for the detection of milk ingredients may present problems in analysing a whey-based

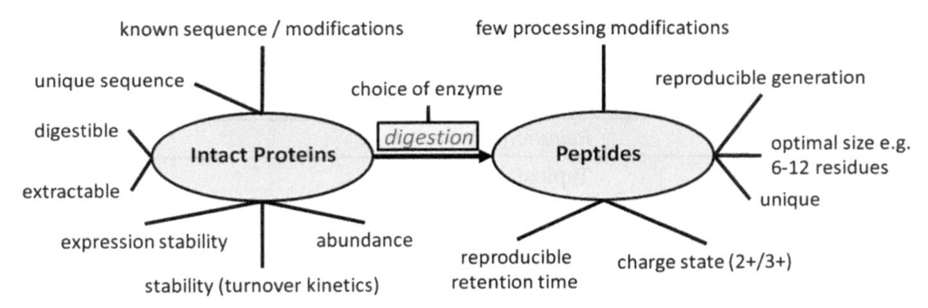

Figure 15.1 Considerations for protein and peptide target selection of food proteins.

Table 15.1 Characteristics to consider in selecting protein targets for the development of mass spectrometry-based methods for the detection of proteins in foods.

Characteristic	*Requirement*
Sequence	Full protein sequence available.
Uniqueness	Protein sequence should be unique to a food source species (plant or animal).
Post-translational modifications	The position and nature of post-translational modifications introduced when the protein is synthesised. Examples include glycosylation state, proteolylsis, disulphide bonds, phosphorylation state, prosthetic groups and ligands, including metal ions.
Processing-induced modifications	Ideally, the protein target should not be modified as a consequence of food processing procedures/matrices commonly analysed. If not possible, processing-induced modifications must be considered when selecting/identifying derived peptide targets.
Extractability	Protein should be reproducibly released from a food matrix into solution to facilitate digestion and subsequent MS analysis.
Digestibility	Protein should be reproducibly digested by the protease in the chosen sample workflow.
Abundance	Targets in an abundant protein are preferable to support development of sensitive methods with a low limit of quantification.
Stable expression	Protein expression factors (tissue specificity, species/cultivar variability, temporal variability, response to environmental/disease/stress) should be characterised and, preferably, expression variability should be minimal.

ingredient that does not contain any casein. Another factor that needs to be considered is whether a candidate molecule is subject to variations in expression depending on season, or whether it is modified during storage.

The issues affecting candidate proteins and their selection feed into the selection of resulting peptides (Figure 15.1 and Table 15.2) and also relate to issues of uniqueness of a candidate peptide, which may be difficult to

Table 15.2 Characteristics to consider in the selection of target peptides for the development of quantitative mass spectrometry-based methods for quantitative food protein analysis.

Characteristic	*Requirement*
Size	Typically 6–12 residues.
Reproducible generation of	Parent peptide and MS2 fragment produced reproducibly from various different food matrices.
Unique	Peptide/MS2 fragment produced only from digestion of protein target and not another related protein components in a food.
Robust to processing-induced modifications	Amino acid residues subjected to processing-induced modifications, notably lysine, avoided.
Retention time	Chromatographic separation step of LC-MS yields a single peak with a reproducible retention time.
Charge state	Generates 2+ or 3+ ions, which generally give the best fragmentation.

establish, especially since we lack so much sequence information about common food stuffs, as well the impact that processing-induced modifications may have on subsequent detection or fragmentation. Since, in general, the sites of such modifications have not been described, it is often advisable simply to avoid using peptides that contain residues that are likely to be modified wherever possible.

15.2.6 Calibrants and Reference Materials

As with any analytical method for the determination of proteins in foods, mass spectrometry (MS) requires relevant calibrants and reference materials. However, whilst such materials have been developed for many target analytes in foods, including mycotoxins, pesticides, and other environmental contaminants, these have not been developed for the analysis of biomacromolecules, like proteins and peptides. This is despite the fact that the analysis of protein targets, in particular, food allergens (see below), is a topic of increasing importance and increasing regulatory requirements. This lack of common calibrants and reference materials is, in part, due to the inherent difficulties in preparing food materials that are well characterised with regards to protein and peptide composition, and which have a stable protein/peptide composition over an extended period of time. There is also a need to choose calibrants that provide meaningful answers that can be used within a food arena. For example, labelling legislation for food allergens defines certain foods and "products thereof", which, for staple foods such as soybean, range from the bean itself to refined oils, soya flours, isolates, concentrates and hydrolysates, to name but a few. As yet there is no common consensus on how to select relevant calibrants for such analyses.

It is useful to consider this with respect to allergen determination, for which there is an acknowledged need for more suitable materials and increased standardisation and where two types of reference material have been identified.[22] The first is a basic foodstuff containing the allergen in question (*e.g.*, milk powder or peanut flour). An example of this type of material is NIST1549 (a skimmed milk powder), which may have been subjected to additional treatments to enhance its stability, which may have introduced modifications, such as glycation, that make it unsuitable for use as an allergen reference material.[23] Secondly, there are incurred reference materials that represent the allergen as it occurs in a foodstuff (*e.g.*, peanut flour in a baked cookie). The importance of using incurred reference materials, in part because of the tendency of protein allergens to change structure and behaviour upon processing, has been well established.[24] There are currently no incurred reference materials for allergen analysis generally available, although several are in development. The lack of such materials hinders the development of new analytical methods, as well as complicating efforts to compare different methods.

15.3 Applications of Proteomic Profiling and Quantitative Proteomics in Food Analysis

15.3.1 Detection of Allergens in Foods

Allergen regulations require that the inclusion of a food ingredient (*e.g.*, peanuts) in a recipe be labelled irrespective of the level of inclusion and the food is, therefore, ostensibly the focus of testing (*e.g.*, in Europe 2003/89/EC, 2006/142/EC and Food Information for Consumers Regulation (EU) No. 1169/2011). Such regulation has been one of the drivers for the application of proteomics techniques to food analysis, and especially the need to verify the absence of certain ingredients, such as gluten, below a certain target level in order to substantiate claims made on food labels that they are, for example, "gluten-free". There is also a need for periodic testing to support allergen control measures used to manage the unintentional presence of allergens by Hazard Analysis and Critical Control Point (HACCP).[25,26] Such testing is performed using either immunological methods, such as enzyme-linked immunosorbent assay (ELISA), or polymerase chain reaction (PCR).[27,28] The reliable detection of food allergens at a level below which they are unlikely to cause an allergic reaction is of fundamental importance to allergen analysis and already exists for gluten in relation to coeliac disease, where levels of gluten below 20 ppm are generally considered safe.[33] Although such action levels for IgE-mediated food allergies are still very much in debate, approaches such as the one proposed by EU-VITAL (www.eu-vital.org) suggest levels of an allergenic substance in food ranging from <8 mg kg^{-1} for peanut to <100 mg kg^{-1} for fish. Current methods are suitable for detecting such levels, although their use as quantitative tools for enforcement has been questioned and limits of detection vary widely according to the test kit and the food matrix being analysed.[11]

MS-based detection has the potential to provide an orthogonal detection method for proteins alongside other more traditional methods (Table 15.3).

Table 15.3 Comparison of the methods used for the detection of allergens in foods (ppm: parts per million; ppb: parts per billion).

Quality	*Immunoassay*	*Polymerase chain reaction*	*Mass spectrometry*
Analytical target	Antibodies raised to an allergen molecule, other proteins specific to the allergenic food or a mixture of allergenic food proteins.	DNA fragment that corresponds to a DNA sequence, which is specific for the allergenic food.	Peptide sequences derived from protein targets that can correspond to either an allergen molecule or other proteins specific to the allergenic food. Generally more than one peptide would be used per allergenic food.
Specificity	Whilst antibodies can have exquisite specificity, they may show cross reactions to proteins from closely related species and are not always well defined.	Highly specific probes can be developed that can give specificity to at least the species level and possibly down to cultivars or breeds.	Targets with high degree specificity can be identified at least at the species level and possibly at the cultivar/breed level.
Sensitivity	Low ppm range.	Theoretical sensitivity of ten molecules, which equates to the ppb in foods.	Low ppm range.
Quantification	At least semi-quantitative but with a narrow dynamic range meaning optimal dilution of unknown samples is crucial. Inference of protein content requires appropriate calibrators.	At least semi-quantitative, although quantification of copy numbers can be difficult. Protein content is only inferred from the DNA content and, consequently, this method is poor for foods, such as hen's egg and cow's milk, where species DNA content is low.	At least semi-quantitative with potential for absolute quantitation and wide dynamic range. Inference of protein content from derived peptides requires appropriate calibrators.
Natural variability of target	Results may vary depending on the species, plant variety or animal breed, climatic and seasonal changes.	Genotype is very stable but variation may be introduced as a consequence of harsh food processing procedures that can damage DNA and affect recoveries from foods.	Results may vary depending on species, plant variety or animal breed, climatic and seasonal changes.

Table 15.3 (*Continued*)

Quality	*Immunoassay*	*Polymerase chain reaction*	*Mass spectrometry*
Effect of matrix and processing	Most immunoassays use only simple extraction protocols but may only extract a small proportion of the sample, affecting recoveries. Optimal extraction conditions for food may not be compatible.	Inhibitors of PCR are present in food.	Potential for acid-labile detergents to improve extraction protocols and options are available that give good extraction.
Sample preparation	Easy and fast with results potentially available in hours.	Can be labour intensive.	Simple extraction protocols can be used, which are rapid. Exhaustive extraction protocols are effective but are laborious, affecting sample throughput.
Expertise required	Whilst no investment in expensive equipment is required, experience and know-how is required to allow effective selection of appropriate methods and interpretation of results.	Training in DNA extraction required and interpretation of data may be required.	Investment in expensive equipment and a lack of expertise is hampering routine applications. Specialist expertise is required to develop methods, although, once complete, analysis can be routine.

In line with effective hazard management practices, there is a general consensus that the hazard (*i.e.*, the proteinaceous allergen molecule(s) within a food) should be the analytical target whenever practicable. The detection of allergen molecules ensures the best chance of risk avoidance for the consumer, and is especially important where food has been subjected to processing or separation technologies that may result in differential effects on protein and non-protein targets.[30] In some cases the allergens within an allergenic foodstuff are poorly defined and, in most cases, there is limited information on which allergens are most problematic. In the case of peanuts the allergen that best predicts allergic response in allergic individuals is Ara h 2,[31] making this allergen an essential target to be included in any peanut detection method. However, it is evident that other allergens are also important in determining allergic reactions to foods like peanuts[31] and, for this reason, multiple allergen targets should be included in analysis.

Many allergens are represented in genomes by multiple sequences, as is illustrated by the major peanut allergens Ara h 2, Ara h 1 (Figure 15.2) and Ara h 3. For other foods, such as tree nuts, the lack of sequence information means it is not even possible to estimate the number of sequences representing allergenic proteins. In addition, for the vast majority of allergenic foods, it is not known if this sequence diversity significantly impacts on the capacity of proteins to elicit allergic reactions.[32] The presence of multiple

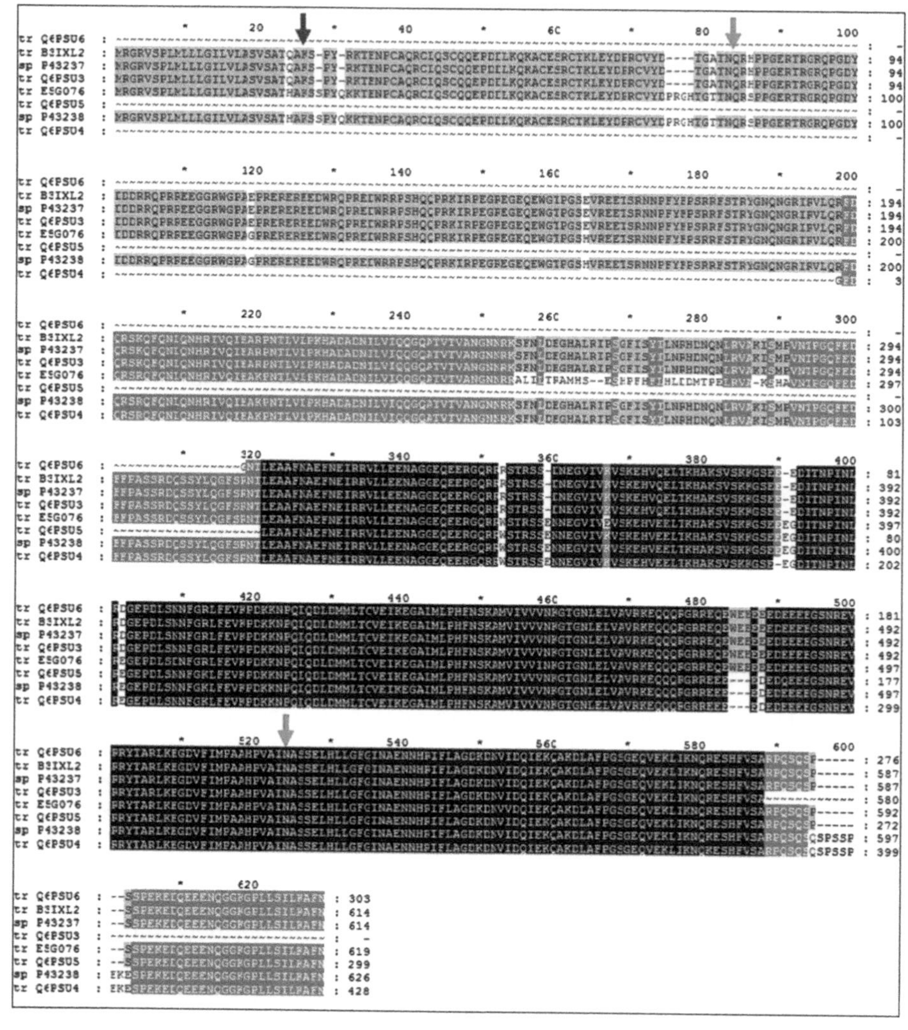

Figure 15.2 Alignment of known Ara h 1 sequences ("iso-allergens") from UniProt showing conserved regions suitable as peptide targets for detection. The signal peptide cleavage site is marked (blue arrow), as well as a known site of post-translational cleavage (green arrow) and a known *N*-glycosylation site (red arrow).

"iso-allergens" raises questions as to peptide target selection. Specifically, if the aim is to detect an allergenic protein (*e.g.*, Ara h 1), the method should detect all iso-allergens to ensure no false negative results. This favours the selection of conserved regions of the protein as peptide targets. Such conserved regions must also yield conserved peptides when digested with the enzyme of choice (usually trypsin). If such conserved regions are difficult to detect, an alternative would be to select multiple targets for each iso-allergen. However, this strategy complicates both the experimental and data analysis processes. An example of iso-allergen diversity in the peanut allergen Ara h 1 is shown in Figure 15.2. This figure also illustrates the importance of protein-level information for best candidate target peptides, showing the signal peptide cleavage site and a site of *N*-glycosylation, which should be avoided as target peptides. It should be noted that detection of all isoforms of an allergen can only be guaranteed where there is a complete genome sequence available as, in theory, unsequenced isoforms could possess sequence differences that invalidate a method to detect known isoforms. There is obviously great need for efforts to improve genome sequence coverage of commonly consumed foods, particularly tree nuts (Table 15.4). In the meantime the risk of this occurring can be minimised by selecting heavily conserved regions for peptide analytical targets.

It is likely that MS-based methods of analysis would require reliable limits of detection at a low (1–2) mg kg^{-1} level for these methods to be accepted and comfortably able to protect consumers. As a consequence of their potential to deliver high selectivity and sensitivity, multiple reaction monitoring (MRM) methods (an extension of selected reaction monitoring or SRM) using triple-quadrupole instruments have been the methods of choice.[29] However, other MS-based techniques have been employed to deliver information in specific instances where allergen contamination is an issue, especially where the matrix is relatively simple, such as liquids. Limits of detection of published MS methods for allergen detection vary hugely. For example, 0.2 μg kg^{-1} for lactosylated peptides could be detected in milk hydrosylates,[34] whilst 10 g kg^{-1} of soy globulins could be detected in milk powder.[35] However, such methods are not used routinely for allergen control and represent case-specific investigations. The only MS method described for routine allergen detection in foods is a MRM method[36] designed to simultaneously detect a range of allergens (milk, hazelnut, peanut, almond, walnut, egg and soy). The limit of detection of the most intense peptide ranged from 3 mg kg^{-1} for almond, to 70 mg kg^{-1} for walnut. Additionally, it was observed that detecting the allergen in a matrix (baked bread) resulted in far lower sensitivity than using spiked samples. It is, therefore, of paramount importance to establish the characteristics of a method on a matrix-by-matrix basis.

For a method to gain general acceptance as suitable for consumer protection, it is preferable that it has been validated against an accepted method through both international (*e.g.*, European Committee for Standardization (CEN) and AOAC) and national bodies (*e.g.*, the German Federal Office of

Table 15.4 Sequence representation of allergenic foods, for which labelling is mandatory in the EU, in the UniProt database. TaxonID and number of sequences (UniProtKB) were determined using the noted taxonomy. Number of allergens was taken from the IUIS database of allergens (www.allergen.org). Downloads were undertaken on 19/6/13 (na: not available).

Allergenic food	*Species (TaxonID)*	*Total no. sequences in UniProt*	*Number of allergens in IUIS*
Cereals containing gluten, (*i.e.*, wheat, rye, barley, oats, spelt, kamut or their hybridized strains) and products thereof	*Triticum aestivum* (4565); *Triticum durum* (4567)	*T. aestivum* (5437); *T. durum* (424)	*T. aestivum* (20); *T. durum* (0)
	Secale cereale (4550)	439	4
	Hordeum vulgare (4513)	73146	7
	Avena sativa (4498)	242	0
	Triticum spelta (58933)	97	0
	Triticum turanicum (376534)	21	0
Crustaceans and products thereof	*Crustacea* (6657)	69697	28 (Decapoda)
Eggs and products thereof	*Gallus gallus* (chicken) (9031)	29206	6
Fish and products thereof	*Teleosti* (bony fish) (32443)	414041	5
Peanuts and products thereof	*Arachis hypogaea* (3818)	920	13
Soybeans and products thereof	Glycine max (3847)	74194	7
Milk and products thereof (including lactose)	*Bos taurus* (cow) (9913)	31595	11
	Ovis aries (sheep) (9940)	4043	0
	Capra hircus (goat) (9925)	23	0

Nuts, *i.e.*, almonds, hazelnuts, walnuts, cashews, pecan nuts, Brazil nuts, pistachio nuts, macadamia nuts and Queensland nuts and products thereof	*Prunis dulcis* (almond)(3755)	316	4
	Corylus avellana (hazelnut) (13451)	383	9
	Juglans regia (English walnut) (51240) + *Juglans nigra* (black walnut) (16719)	71 + 34	4 + 2
	Anacardium occidentale (cashew) (171929)	7	3
	Carya illinoinensis (pecan) (32201)	9	2
	Bertholletia excelsa (Brazil nut) (3645)	10	2
	Pistacia vera (pistachio) (55513) + *Pistacia mexicana* (American pistachio) (246379)	15 + 10	5
	Macadamia integrifolia (macadamia nut) (60698)	13	0
	Macadamia tetraphylla (512563)	9	0
Celery and products thereof	*Apium graveolens* (4045)	76	6
Mustard and products thereof	*Sinapis alba* (white mustard) (3728); *Brassica nigra* (Black mustard) (3710); *Brassica juncea* (Indian mustard) (3707)	*S. alba* (96); *B. nigra* (115); *B. juncea* (624)	*S. Alba* (4); *B. nigra* (0); *B. juncea* (1)
Sesame seeds and products thereof	*Sesamum indicum* (4182)	191	7
Sulphur dioxide and sulphites at concentrations of more than 10 mg kg^{-1} or 10 mg L^{-1} expressed as SO_2	na	na	na
Lupin and products thereof	*Lupinus* (3869)	868	1
Molluscs and products thereof	*Mollusca* (6449)	66946	3

Consumer Protection and Food Safety (BVL)). There are currently no comparable guidelines for the validation of MS methods of allergen detection, although these do exist for ELISA methods.[37] In the absence of a reference method to perform such validation, multi-centre evaluations ("ring-trials"), in which a common material is analysed by the same method in several laboratories, are being undertaken. However, the small number of laboratories currently capable of performing such a trial is a barrier to such validation. In the interim, alternative approaches employing performance parameter-based validation of laboratory specific methods may need to be adopted by individual laboratories.

A second aspect of managing allergens in foods relates to allergenic risk assessment, especially regarding new ways of producing foods and the concerns they raise regarding the introduction of new allergenic foods into the supply chain. Quantitative mass spectrometry tools are being developed to assess varietal differences in the levels of allergens to allow effective comparison between, for example, crop varieties produced using conventional breeding and those produced using genetic modification. For example, a two-dimensional liquid chromatography-MS method has been developed for quantifying a soybean allergen that has been associated with severe reactions. Known as Gly m 4,[38] this allergen comprises between 364.5 and 608.8 $\mu g\ g^{-1}$ of soybean seed protein, with variation between samples potentially being attributed to both varietal and environmental influences. Such a complex interplay between genotype and environment has been observed in the use of MRM experiments to quantitatively profile multiple soybean allergens.[39] Using both relative and absolute quantification methods to profile twenty soybean cultivars showed little variation in allergen contents in the lines. Using the same methodology, a geographic survey demonstrated that the growing environment had a far greater influence on allergen levels than the genotype, including whether the soybean varieties had been produced using conventional breeding or through genetic modification.[40]

In addition to the growing environment, post-harvest storage conditions can affect the expression of allergens in the proteome of fresh fruits, such as apples.[41,42] MS analysis is now being extended to other fruits, such as nectarines.[43] Using this approach it has been shown that levels of key food allergens change on cold storage, notably those belonging to the lipid transfer protein allergen family (Pru p 1) and the Bet v 1 family (Pru p 2), which are pathogenesis-related proteins, expression of which may be modified in response to abiotic stress and maturation state.

15.3.2 Proteomic Profiling for Food Product Quality

There are a number of reports applying proteomics to the profiling of food products and processes in relation to product quality. One application that has received considerable attention is that of characterising the changes in the muscle proteome. In particular myofibrillar degradation that relates to fragmentation of structural proteins was observed during aging and storage of meat that affects quality characteristics, such as tenderness.[44] Other proteomics studies of meat tenderness have also shown that tenderness is associated with higher levels of glycolytic enzymes that were less phosphorylated, as well as myofibrillar degradation.[45] Proteomic profiling has also been used to identify potential stress-related effects of slaughter processes on the muscle proteome of fish in order to ensure the process is undertaken in as humane a manner as possible.[46]

Techniques, such as isobaric tags for relative and absolute quantitation (iTRAQ), used in combination with two-dimensional gel electrophoresis have also been used to monitor changes in protein profiles during the ripening of cheese. Using this approach, quantification of both microbial proteins and those from the curd-fraction of milk could be quantified.[47] A similar approach has been taken to monitor the malting process,[48] which confirmed that the quantities of the hordein storage proteins decreased during malting, with components such as C hordein being reduced by around 65%. Analysis of beer proteins using untargeted analysis also showed the presence of serpin-Z4, LTP1, α-amylase trypsin inhibitors, defensins, heat shock proteins and primarily B-, C- and D-hordeins.[49] Evidence of amino acid modifications, such as methionine oxidation, was also found. Glycation of proteins abundant in beer, such as non-specific lipid transfer proteins LTP1 and LTP2, the α-amylase trypsin inhibitor CMb and a single peptide from D-hordein, was also observed. In addition MS analysis revealed the presence of wheat proteins and peptides, even though wheat was not necessarily declared on the ingredients label; traces of gluten proteins, including α-gliadin, γ-gliadins and low molecular weight subunits of glutenin, were detected, as well as defensins, α-amylase trypsin inhibitors and LTPs. Protein MS has also been applied to the analysis of residual grape proteins in wine[50] with the pathogenesis related proteins, including allergens such as LTPs, dominating. This is probably a reflection of their thermostability and resistance to proteolysis, which allows them to survive the vinification process. However, this was dependent on grape variety with the proteins being observed in Portugieser and Dornfelder red wines, but not in wines made from grape varieties such as Cabernet Sauvignon, Shiraz and Pinot Noir.

Protein MS is also beginning to be used to profile the complex mixtures of peptides generated in the preparation of hydrolysates.[51] Although these data are giving insights into protease activity and specificity in complex food systems, a more quantitative approach to protein MS analysis has yet to be explored for such complex mixtures. Targeted analysis of the formation of peptides with particular biological activities, such as those with antihypertensive properties, is also being undertaken.[52] Thus, methods have been developed for two such peptides derived from αs_1-casein corresponding to residues 90–94 (RYLGY) and 143–149 (AYFYPEL), and have been used to optimise the generation of these peptides from powdered milk. This investigation also demonstrated the value of having rigorous methods for the direct determination of bioactive peptides rather than simply relying on biological measures, such as angiotensin-converting enzyme (ACE)-inhibitory activity.

15.4 Conclusions

Whilst many hurdles remain to be overcome before protein MS can be routinely applied to the analysis of foods, it has great potential both with regards to profiling complex ingredients and providing alternative means of quantitative analysis of proteins in foods. Some of these challenges are common to all methods of analysis for proteins in foods, such as the lack of common calibrants and reference materials and the need for effective extraction protocols. Others are unique to MS-based analysis with the lack of sequence information for the sources of food protein ingredients being a major gap, which hampers its application to food analysis. This is further complicated by the poor knowledge-base with regards to the processing-induced modifications of food proteins and concerns over the reproducibility and robustness of protease digestion steps used in proteomics sample preparation. Despite these issues, the MS-based analysis of allergens in foods represents a significant step forward in terms of a protein-based methodology for confirmation of ELISA methods, and additionally, has great potential for facilitating routine allergen analysis in support of allergen control in the long term. MS techniques also have much to offer for other types of food analysis relating to sourcing and control of food ingredients and the impact of processing procedures on product quality. This is especially pertinent to the analysis of specialist ingredients, such as hydrolysates, used in the production of specialist foods, such as infant formula.

References

1. S. L. Bislev, E. W. Deutsch, Z. Sun, T. Farrah, R. Aebersold, R. L. Moritz, E. Bendixen and M. C. Codrea, *Proteomics*, 2012, **18**, 2895.
2. P. J. Davis and S. C. Williams, *Allergy*, 1998, **53**, 102.
3. B. Lagrain, I. Rombouts, K. Brijs and J. A. Delcour, *J. Agric. Food Chem.*, 2011, **59**, 2034.
4. I. Rombouts, B. Lagrain, K. Brijs and J. A. Delcour, *Amino Acids*, 2012, **42**, 2429.
5. M. Friedman, *J. Agric. Food Chem.*, 1999, **47**, 1295.
6. C. M. Oliver, L. D. Melton and R. A. Stanley, *Crit. Rev. Food. Sci. Nutr.*, 2006, **46**, 337.
7. Y. Karademir, N. Göncüoğlu and V. Gökmen, *Food Funct.*, 2013, **4**, 1061.
8. M. Lima, S. H. Assar and J. M. Ames, *J. Agric. Food Chem.*, 2010, **58**, 1954.
9. H. H. Refsgaard, L. Tsai and E. R. Stadtman, *Proc. Natl. Acad. Sci. U.S.A.*, 1999, **97**, 611.
10. S. Drusch, V. Faist and H. F. Erbersdobler, *Food Chem.*, 1999, **65**, 547.
11. M. L. Downs and S. L. Taylor, *J. Agric. Food Chem.*, 2010, **58**, 10085.
12. J. Heick, M. Fischer and B. Popping, *J. AOAC Int.*, 2012, **95**, 388.

13. R. Pedreschi, J. Nørgaard and A. Maquet, *Nutrients*, 2012, **4**, 132.
14. C. M. Hebling, M. A. McFarland, J. H. Callahan and M. M. Ross, *J. Agric. Food Chem.*, 2013, **61**, 5638.
15. Y. Q. Yu, M. Gilar, P. J. Lee, E. S. Bouvier and J. C. Gebler, *Anal. Chem.*, 2003, **75**, 6023.
16. J. D. Astwood, J. N. Leach and R. L. Fuchs, *Nat. Biotechnol.*, 1996, **14**, 1269.
17. B. N. Pramanik, U. A. Mirza, Y. H. Ing, Y. H. Liu, P. L. Bartner, P. C. Weber and A. K. Bose, *Protein Sci.*, 2002, **11**, 2676.
18. I. V. Gmoshinskiĭ, V. K. Mazo and N. F. Samenkova, *Vopr Pitan.*, 1981, **6**, 19.
19. M. Offengenden, M. A. Fentabil and J. Wu, *Glycoconj. J.*, 2011, **28**, 113.
20. G. Mamone, S. De Caro, A. Di Luccia, F. Addeo and P. Ferranti, *J. Mass Spectrom.*, 2009, **44**, 1709.
21. B. H. Doell, C. J. Ebden and C. A. Smith, *Plant Foods Hum. Nutr.*, 1981, **31**, 139.
22. M. Lacorn and U. Immer, *Accred. Qual. Assur.*, 2010, **15**, 207.
23. P. Johnson, M. Philo, A. Watson and E. N. Mills, *J. Agric. Food Chem.*, 2012, **14**, 12420.
24. S. L. Taylor, J. A. Nordlee, L. M. Niemann and D. M. Lambrecht, *Anal. Bioanal. Chem.*, 2009, **395**, 83.
25. R. Ward, R. Crevel, I. Bell, N. Khandke, C. Ramsay and S. Paine, *Trends Food Sci. Tech.*, 2010, **21**, 619.
26. C. Diaz-Amigo and B. Popping, *J. AOAC Int.*, 2010, **93**, 434.
27. R. E. Poms, C. L. Lein and E. Anklam, *Food Addit. Contam.*, 2004, **21**, 1.
28. S. Kirsch, S. Fourdrilis, R. Dobson, M. L. Scippo, G. Maghuin-Rogister and E. De Pauw, *Anal. Bioanal. Chem.*, 2009, **395**, 57.
29. P. E. Johnson, T. Aldick, C. V. L. Giosafatto, A. Watson, E. N. C. Mills, S. Baumgartner, C. Bessant, J. Heick, G. Mamone, G. O'Connor, R. Poms, B. Popping, A. Reuter, F. Ulberth and L. Monaci, *J. AOAC Int.*, 2011, **94**, 1026.
30. E. Scaravelli, M. Brohée, R. Marchelli and A. J. van Hengel, *Anal. Bioanal. Chem.*, 2009, **395**, 127.
31. N. Nicolaou, C. Murray, D. Belgrave, M. Poorafshar, A. Simpson and A. Custovic, *J. Allergy Clin. Immunol.*, 2011, **127**, 684.
32. L. H. Christensen, E. Riise, L. Bang, C. Zhang and K. Lund, *J. Immunol.*, 2010, **184**, 4966.
33. C.-H. Chan and S. Hattersley, *BMJ*, 2011, **343**, 7208.
34. D. Molle, F. Morgan, S. Bouhallab and J. Leonil, *Anal. Biochem.*, 1998, **259**, 152.
35. D. M. Luykx, J. H. Cordewener, P. Ferranti, R. Frankhuizen, M. G. Bremer, H. Hooijerink and A. H. America, *J. Chromatogr. A*, 2007, **1164**, 189.
36. J. Heick, M. Fischer and B. Pöpping, *J. Chromatogr. A*, 2011, **1218**, 938.

37. M. Abbott, S. Hayward, W. Ross, S. B. Godefroy, F. Ulberth, A. J. Van Hengel, J. Roberts, H. Akiyama, B. Popping, J. M. Yeung, P. Wehling, S. L. Taylor, R. E. Poms and P. Delahaut, *J. AOAC Int.*, 2010, **93**, 442.
38. S. Julka, K. Kuppannan, A. Karnoup, D. Dielman, B. Schafer and S. A. Young, *Anal. Chem.*, 2012, **84**, 10019.
39. N. L. Houston, D. G. Lee, S. E. Stevenson, G. S. Ladics, G. A. Bannon, S. McClain, L. Privalle, N. Stagg, C. Herouet-Guicheney, S. C. MacIntosh and J. J. Thelen, *J. Proteome Res.*, 2011, **10**, 763.
40. S. E. Stevenson, C. A. Woods, B. Hong, X. Kong, J. J. Thelen and G. S. Ladics, *Front. Plant Sci.*, 2012, **3**, 196.
41. A. I. Sancho, R. Foxall, T. Browne, R. Dey, L. Zuidmeer, G. Marzban, K. W. Waldron, R. van Ree, K. Hoffmann-Sommergruber, M. Laimer and E. N. Mills, *J. Agric. Food Chem.*, 2006, **54**, 5917.
42. L. Zuidmeer, W. A. van Leeuwen, I. Kleine Budde, H. Breiteneder, Y. Ma, C. Mills, A. I. Sancho, E. J. Meulenbroek, E. van de Weg, L. Gilissen, F. Ferreira, K. Hoffmann-Sommergruber and R. van Ree, *Int. Arch. Allergy Immunol.*, 2006, **141**, 1018.
43. E. Giraldo, A. Díaz, J. M. Corral and A. García, *J. Proteomics*, 2012, **75**, 5774.
44. T. Sayd, C. Chambon, E. Laville, B. Lebret, H. Gilbert and P. Gatellier, *Food Chem.*, 2012, **15**, 2238.
45. A. D'Alessandro, S. Rinalducci, C. Marrocco, V. Zolla, F. Napolitano and L. Zolla, *J. Proteomics*, 2012, **75**, 4360.
46. T. S. Silva, O. D. Cordeiro, E. D. Matos, T. Wulff, J. P. Dias, F. Jessen and P. M. Rodrigues, *J. Agric. Food Chem.*, 2012, **60**, 9443.
47. J. Jardin, D. Mollé, M. Piot, S. Lortal and V. Gagnaire, *Int. J. Food Microbiol.*, 2012, **155**, 19.
48. D. Flodrová, J. Ralplachta, D. Benkovská and J. Bobálová, *Eur. J. Mass Spectrom.*, 2012, **18**, 323.
49. M. L. Colgrave, H. Goswami, C. A. Howitt and G. J. Tanner, *J. Proteome Res.*, 2012, **11**, 386.
50. P. Wigand, S. Tenzer, H. Schild and H. Decker, *J. Agric. Food Chem.*, 2009, **57**, 4328.
51. W. Margatan, K. Ruud, Q. Wang, T. Markowski and B. Ismail, *J. Agric. Food Chem.*, 2013, **61**, 3460.
52. M. Contreras Mdel, B. Gómez-Sala, P. J. Martín-Alvarez, L. Amigo, M. Ramos and I. Recio, *Anal. Bioanal. Chem.*, 2010, **397**, 2825.

Subject Index

3T3-L1 adipocytes 221

β-elimination reactions 331
Δ*r* (desolvation distance) 9

a priori knowledge, SRM assays 85–7
AAA (amino acid analysis) 112
AAL (*Aleuria aurantia* lectin) 164, 165, 166–7
absolute protein abundance index (APEX) 95
absolute protein quantification
  applications 81–2
  challenges
    complete protein extraction 96–7
    optimal peptide selection 98–100
    specific/complete protein digestion 97–8
    translation of label-free MS intensities into absolute protein quantities 100–1
  ICP-MS method 118–25
  LC–MS/MS workflows 4, 5–6
  need for 80–3
  phosphorus-quantified phosphopeptide standards 113–16
    peptide/phosphopeptide ratio standards 118
    phosphorus-free 116–17
  selected reaction monitoring 80–103
  selenium-quantified but selenium-free protein standards 123–5
  targeted mass spectrometry 83–8
  technologies 82–3
  tyrosine phosphorylation 223, 224
absolute quantification (AQUA) peptides 90
abundances of proteins, LC–MS/MS 16
accuracy
  isobaric tag methods 59–62, 71, 243
  LOPIT 202
  MS^E data-independent acquisition 182
  spatial dynamic proteomics 203, 205
  synthetic peptide standards 91
accurate mass and retention time alignment (AMRT) 139, 140, 141
acetylation, primary amines 54
acetyllysine-containing peptides 157
actin cytoskeletons 65
adherens junctions 188
adipocytes 221
ADP-ribosylated PTMs enrichment 248
agarose beads 215
alcohol dehydrogenase (ADH) 149
Aleuria aurantia lectin (AAL) 164, 165, 166–7
allergens detection in food 337–44

Amadori products 331
amine-specific tandem mass tags 58
amines, primary, acetylation 54
amino acid analysis (AAA) 112
AMRT (accurate mass and retention time alignment) 139, 140, 141
analyte concentrations/ionisation efficiency 9
analytical fractionation method 198
anchor proteins 100, 101
animal models
  mouse
    anemia 322–3
    leptin 319
    liver 157, 261
    whole brain lysis 182
  SILAC labelling 197
annotation
  food proteome 330–1
  gene ontology annotation 196, 205–6, 227, 290–1
antibodies
  absolute protein quantification 82–3
  biomarker discovery/ development 285
  immunoaffinity depletion 317, 318
  immunocytochemical methods 189
  immunodepletion QC metrics 147, 148
  immunonephelometric assays 321
  immunoprecipitation 191–2, 239
  phosphotyrosine-specific as enrichment tools 214–15
  spatial dynamic proteomics 189, 191–2
APEX (absolute protein abundance index) 95
Apex3D algorithm 178
apoptosis 261
AQUA (absolute quantification) peptides 90, 111, 112
Arabidopsis thaliana SYP61 compartment 191
Arah 1/2 allergen (peanut) 337, 339, 340, 341
area under curve (AUC) 138, 142, 273, 323
arginine 241, 259
arthritis biomarkers 286–7
Aspergillus niger 68
axin 188

balance groups 55, 60, 61
barocycling 96
basal breast cancer cell lines 169, 170
basicity and ESI response 9–10
basis functions 297
Bcr-Abl fusion tyrosine kinase 213
bead proteome 239
beer proteins 345
best practice, SOPs 145–8
beta-elimination reactions 331
'betweenness' centrality 293
bias 8–10, 116
BiNGO Cytoscape plug-in 205
bioactive peptides 345
biochemical assays 58
biochemical reactivity 122
bio-engineering using isobaric tagging 67–8
biofluids
  see also urine
  blood
    biomarker discovery/ validation 286, 291, 316–25
    blood pressure 297–8, 345
    breast cancer biomarkers 164, 166–8, 169
    protein species 69
  chemical labelling 53–5
biofuels production 67–8
bioinformatics
  see also data analysis; software packages/programs

biomarkers 293–4
label-free LC-MS identification/quantification 133, 134
phosphotyrosine cellular signalling 227
spatial dynamic proteomics 189–90
biomarkers
approved cancer biomarkers list 282
breast cancer 164, 166–8, 169, 280, 288, 294
subtype-specific glycoprotein 157
cancer driver gene sequencing 293
data/text-mining 291
definition 279–80
discovery/development 81, 82, 279–308
human blood/plasma 286, 291, 316–25
methods used 320–1
gene ontology as tool 290–1
ideal attributes 280
integrative bioinformatics 293–4
for prioritisation 289–94
multiple reaction monitoring 294–8, 299, 300–1, 302
data analysis statistical methods 294–8
network biology paradigm 291–3
panels/signature 281, 283, 288–9
assembly 288–9
human blood/plasma 322, 323
measurement 299–306
prostate cancer 69, 281, 282
quality control 301, 303, 305–6
Skyline software 303–5
validation 279–308
biomedical industry 81, 82
biosensors for wastewater treatment 68
biotechnological iTRAQ/TMT applications 67–8
biotinylation 192
bis(4-nitrophenyl)phosphate (BNPP) 113, 114, 115
bladder cancer 282, 321
blood
biomarkers
breast cancer 164, 166–8, 169
discovery/validation 286, 291, 316–25
erythrocytes 143, 261
PBMCs 225
plasma 286, 291, 316–25
pressure 297–8, 345
red blood cell membranes 143
BNPP (bis(4-nitrophenyl)phosphate) 113, 114, 115
bone marrow, healthy human 225
Bovine Peptide Atlas 331
breast cancer
approved biomarkers 282
biomarker panels 288
HER2-driven 157, 163–4, 165–8, 169, 214, 226, 228
MRM studies 280, 282, 288, 294
MS1 label-free studies 157
subtype-specific glycoprotein 157
subtype-specificity determination 164, 166–8, 169, 170
buffers 55, 236, 237
BVL (German Federal Office of Consumer Protection and Food Safety) 341, 344

C-traps 33, 34, 35, 39, 40, 217
CAD (coronary artery disease) 322
calibrants for food proteomics 336–7

calmodulin-like protein 3 (CALML3) 120
Campylobacter 67
cancer 69–70, 293, 294
- bladder cancer 282, 321
- breast cancer 228
  - MRM studies 280, 282, 288, 294
  - subtype-specific glycoprotein 157
  - subtype-specificity determination 164, 166–8, 169, 170
- chronic myelogenous leukemia 213
- glyoblastoma 228
- hepatocellular 66
- human colon carcinoma 67, 322
- non-small cell lung carcinoma 214, 224
- pancreatic 70
- phosphotyrosine cellular signalling 213–14, 222, 224
- prostate cancer 69, 281, 282
- tyrosine kinase inhibitors 228

candidate biomarkers 280, 283–7, 294–8
carbon-13
- LC-MS identification/ quantification 132
- MRM quantification of biomarkers 301
- stable isotope labelling 88, 90, 92
- tyrosine phosphorylation quantification 219

cardiovascular disease (CVD) 281–2
β-catenin 188
cathepsin C 166, 167, 168, 169
Ccl1-Kin28-Tfb3 complex 65
CD74-ROS-driven malignant invasion 222
cell compartments 236, 237
- *see also* organelles; spatial dynamic proteomics

cell culture *see* stable isotope-labelling of amino acids in cell culture (SILAC)
cell death 261
cell lines
- breast cancer 164, 166–8, 169, 170
- iTRAQ/TMT 68, 69
- organelles for spatial dynamic proteomics 197

cell lysis 53–5, 102
cell signalling 118, 188, 211–28, 247
cell walls, rigid 96
CentiScape software 293
centrifugation 190–1, 236, 237
charge residue model (CRM) 8
chemical labelling 53–5, 241–3, 245
- *see also* isobaric tags for relative and absolute quantification (iTRAQ); tandem mass tags (TMTs)

chi squared tests 198
Chico insulin receptor substrate homologue complex 65
Chlamydomonas reinhardtii starch-less mutants 67–8
chloroplasts 193, 195, 196
chronic myelogenous leukemias (CML) 213
chymotrypsin digestion 334
CID (collision-induced dissociation) 62, 63, 84, 216, 243
classification/regression methods 294–8
classifier creation/uncertainty 205
cleanup, LC-MS label-free studies 146
cleavage sites 60, 61, 97, 98
clinical biomarkers *see* biomarkers
clinical samples 284, 286–7, 305, 317
- *see also* biofluids

clinical trials 283
ClueGO software tool 290
clustering methods 227, 271
collision energy optimisation 43, 44

collision-induced dissociation (CID) 62, 63, 84, 216, 243
colon/colorectal carcinoma 67, 322
colorimetric methods 112
companion biomarkers 280
complete protein extraction 96–7
computer-based methods 189–90, 226–7
 see also bioinformatics
concatenated peptides 91–2
confounding variables 16, 18–19
 see also accuracy; contamination; experimental error
contamination
 ions, DDA 176
 organelles for spatial dynamic proteomics 193, 194, 197
 PTMs enrichment 239
 spatial dynamic proteomics 192
coomassie blue stain 147, 148
copies per cell 96
coronary artery disease (CAD) 322
correction factors 133, 136
corroboration rates 206
CRM (charge residue model) 8
crop genome sequences 330
cross-comparison of data 81, 82
cross-talk, PTMs 64
crystallin, bovine 67
cycle times 14–15, 85, 300
cysteine residues 54, 242
 SPARC/osteonectin 166, 167, 168, 169, 170
cytometry 224–5
Cytoscape software 292

data analysis
 see also machine learning; software packages/programs
 alignment strategies 140, 141
 data mining 154–72, 291
 datasets integration 82, 250
 iTRAQ/TMT 63–4
 machine learning 189–90, 203, 204, 294–8, 307
 outliers determination 150
 protein turnover studies 262–75
 spatial dynamic proteomics 202–5, 250–1
 spectral counting for label-free quantification 134
data-dependent acquisition (DDA)
 absolute protein quantification 83–8
 definition 175, 176
 HR/AM instrumentation 32–3
 iTRAQ/TMT 57
 LC-MS-based 27–31
 LC-MS/MS absolute workflows 4, 5–6
 serendipitous 140
 spectral counting for label-free quantification 134
 SWATH-MS 87–8, 171, 176, 207
data-independent acquisition (DIA)
 DDA comparison 175–6
 ion intensity-based label-free quantification 142
 MS1 filtering label-free quantification folow-up 170, 171
 MS^E acquisition
  fundamentals 177
  ion mobility 179–80
  performance evaluation 180–3
  protein identification 178
  protein quantification 178–9
DAVID software tool 290
de novo amino acid synthesis 258, 259
deamidation reactions 331
decision boundary, SVM 297, 298
decision trees, random forests 295–7
de-glycosylated peptides 164, 165, 166–8, 169, 170

degree centrality 293
density gradient ultra-centrifugation 191
dephosphorylation, enzymatic 116, 117
depletion 178, 317, 318, 319
Design of Experiments (DoE) methods 6, 16–19
desolvation distance (Δr) 9
detergents 333
deuterium 54, 240, 259
DIA *see* data-independent acquisition (DIA)
cis 1,2-dichloroethylene (DCE) 68
differential centrifugation 190–1
differential permeabilisation 191
digestion
  in-gel 238
  receptor tyrosine protein kinase 163–4, 165–8
  specific/complete 97–8
  trypsin catalysed 216, 241, 259, 334
  urine proteins 41–3
dimethyl labelling 219, 221, 242
dipeptidyl peptidase 1 166, 167, 168, 169
directed peak picking 159, 160
Directives/Regulations 337
discovery shotgun proteomics 4–5, 8, 9, 26–7
disease states
  see also cancer; *individual disease states*
  biomarkers validation 279–308
    human blood/plasma 316–25
  phosphotyrosine cellular signalling 211–28
double labelling of standards 111, 112, 119, 120, 122, 123, 124
Down's syndrome 321
droplet size (r) 9
Drosophila melanogaster 204
dual detection methods 63
duty cycle 14–15, 85, 300
dwell time 85, 216, 300
dynamic protein quantification *see* spatial dynamic proteomics; temporal dynamic proteomics
dynamic range 16, 234–5, 236
dynamic SILAC 264

early glycation products 331
EGF (epidermal growth factor) 221, 226
EGFR (epidermal growth factor receptor) 214, 222, 228, 280
electron transfer dissociation (ETD) 59, 63
electrophoresis 192, 238
electrospray ionisation (ESI) 4, 6–12, 84, 110–25
β-elimination reactions 331
Elucidator® (Rosetta) 156
elution times 87
emitter-to-capillary distance 6, 7, 10, 11
emPAI (exponentially modified protein abundance index) 136, 144
end user driven biomarker discovery 284, 285
enrichment
  GO-based for biomarker discovery 290
  MRM-MS-based workflow 317, 318, 319
  organelles for spatial dynamic proteomics 193, 194, 195, 196
  phosphotyrosine profiling 214–18
  precursors for turnover studies 260–1
  PTMs for spacial/temporal proteomics 246–50
ensemble methods 189–90, 203, 204, 294–8, 307
enzymatic dephosphorylation 116, 117
enzymatic labelling 54

enzyme-linked immunosorbent assay (ELISA)
biomarkers discovery/ validation 287–8, 307
food proteins 337, 338, 339
human blood/plasma biomarkers discovery 317, 321, 323
ephrin-B1 cell-specific signalling network 222
epidermal growth factor (EGF) 221, 226
epidermal growth factor receptor (EGFR) 214, 222, 228, 280
equaliser peptides 89, 92, 117
equimolarity through equaliser peptide (EtEP) 89, 92
ErbB2 receptor tyrosine-protein kinase (HER2)
MS1 label-free quantification 157, 163–4, 165–8, 169
phosphotyrosine signalling 214, 226, 228
Erk pathway 226–7
error *see* experimental error
erythrocytes 143, 261
see also blood
Escherichia coli
auxotrophic strain to generate full-length labelled proteins 93
calmodulin recombinant expression 121
Campylobacter glycosylation machinary overexpression 67
toluene *ortho*-monooxygenase 68
tryptic digest for MS^E data-independent acquisition 180, 181
ESI (electrospray ionisation) 4, 6–12, 84, 110–25
ETD (electron transfer dissociation) 59, 63
EtEP (equimolarity through equaliser peptide) 89, 92
eukaryotic translation elongation factor 1 69
European food legislation 337
experimental design 6, 16–19, 145–50
experimental error
false classification rates 197
false discovery rates 162
false positive rates 178
iTRAQ/TMT 59–62, 71
label-free methods 244
LOPIT 202
proline 259
spatial dynamic proteomics 203, 205
synthetic peptide standards 91
experimental workflows
biomarker discovery/ development 284
ICATs MRM 301–2
ion intensity-based label-free quantification 137–9
iTRAQ/TMT 55–8
label-free Skyline MS1 filtering 155–6
LC-MS
identification/ quantification 131–2
label-free study guidelines 145–8
LC–MS/MS relative/absolute workflows 4–6
MRM experiment 317
protein turnover studies 263
data analysis 270–1
quantitative bottom-up proteomics 111, 112
reproducibility 162–5
Skyline MRM development/ data analysis 303, 304–5
spatial dynamic proteomics 193–202, 251
exponentially modified protein abundance index (emPAI) 136, 144
extracted ion chromatograms (XIC) 138–43, 154–72, 320
extraction, complete 96–7, 102

factorial design, LC–MS/MS 17–19
false classification rates 197
false discovery rate (FDR) 162
false positive rates 178
fathead minnow livers 143
Fe-IMAC phosphopeptide enrichment 65
FFE (free flow electrophoresis) 192
fibroblast growth factor receptors 214
file conversion, Skyline 303
Fisher's exact test 135
FLEXIQuant (full-length expressed stable isotope-labelled proteins for quantification) 93
flow cytometry 224–5
flow rate stability, RP-HPLC 12–13
fluorescence 82, 83, 189, 224–5
flux *see* spatial dynamic proteomics; temporal dynamic proteomics
fold changes 135, 181
follistatin-related protein 1 (FSTL1) 164–9
food 'omics' 329–46
    applications 337–46
        allergens detection 337–44
        quality control 344–6
    challenges 330–7
        calibrants/reference materials 336–7
        matrix effects on extraction 332–3, 339
        processing-induced modifications 331–2, 334, 335
        protease digestion 334
        proteome annotation 330–1
        target selection 334–6
    proteins/peptides MS-based quantification 329–46
formaldehyde 221, 242
Fourier transform mass spectrometry (FT-MS) 234
fractional factorial design (FracFD) 17, 18, 19
fractionation 317, 318
    see also pre-fractionation methods
free flow electrophoresis (FFE) 192
FSTL1 (follistatin-related protein 1) 164–9
FT-MS (Fourier transform mass spectrometry) 234
full factorial design (FullFD) 17–19
full scan mode 33, 34, 35, 36
full-length expressed stable isotope-labelled proteins for quantification (FLEXIQuant) 93
full-length protein standards 89, 92–3
FullFD (full factorial design) 17–19
functional analyses 226–7
fusion proteins 189, 213
fusion tyrosine kinase Bcr-Abl 213

GDN (glia-derived nexin) 166, 167, 168, 169, 170
gelled protein networks in food 333
gene ontology (GO) annotation 196, 205–6, 227, 290–1
GeneMANIA software 292
genome sequences 330, 341, 342–3
German Federal Office of Consumer Protection and Food Safety (BVL) 341, 344
glia-derived nexin (GDN) 166, 167, 168, 169, 170
GLILVGGYGTR isotopically-labelled peptide 40
glioblastoma cell lines 228
global protein expression/shotgun proteomics 4–5, 8, 9, 26–7
glycans 66–7
glyceraldehyde-3-phosphate dehydrogenase 188
glycopeptide analysis 163–4
glycoproteomic analysis 66–7
glycosylation motifs 98
grape varieties 345
green fluorescent protein (GFP) 83

growth factors 212, 213
see also epidermal growth factor receptor (EGFR); ErbB2 receptor tyrosine-protein kinase
guanidination of peptides 242, 301
guidelines
see also standards
sample handling 145–8

haptoglobin proteins 295, 296
HASTA (high-amplitude short-time excitation) 62
Hazard Analysis and Critical Control Point (HACCP) 337
HCD *see* higher energy C-trap dissociation (HCD)
heavy isotopes 52, 224, 225
heavy–light peptide pairs 262, 263, 265, 269, 270, 273
HER2 *see* ErbB2 receptor tyrosine-protein kinase
heregulin (HRG) 221
high-amplitude short-time excitation (HASTA) 62
high-performance liquid chromatography (HPLC) 84
see also reverse phase high-performance liquid chromatography (RP-HPLC); reverse phase liquid chromatography–tandem mass spectrometry (LC–MS/MS)
HPLC-ESI-MS/MS analysis 164
LC-MS label-free studies 131–50
LC-MS^E acquisition, data structure 177
micro-LC-ICP-MS 113–18
size-exclusion chromatography 238
splitless nanoflow HPLC systems 13
ultra-high pressure LC systems 13
high-performance tandem mass spectrometers 13–14
see also linear ion trap/quadrupole Orbitraps (LTQ-Orbitrap/Q-Exactive); quadrupole time-of-flight (Q-TOF)
high-resolution/accurate mass (HR/AM) hybrid mass spectrometry 27–46
ion intensity-based label-free quantification 140, 141
LC–MS-based targeted proteomics 27–31
precursor ion-based quantification 33–9
product ion-based quantification 39–45
spatial/temporal dynamic proteomics 234, 240–6
targeted proteomic experiments 32–3
high-throughput proteomics 195, 226, 233–53
higher energy C-trap dissociation (HCD)
iTRAQ/TMT-labelled peptides 62–3
parallel reaction monitoring 40, 43
phosphotyrosine signalling 217
PTMs in space/time 243
quadrupole–orbitrap instrument 33, 39, 40
HILIC (hydrophilic interaction chromatography) 57, 248, 249, 250
histidine, phosphorylation 247
HMEC (human mammary epithelial cells) 221
hordein 321, 345
HPLC *see* high-performance liquid chromatography (HPLC)
HR/AM *see* high-resolution/accurate mass (HR/AM) hybrid mass spectrometry

human epidermal growth factor *see* ErbB2 receptor tyrosine-protein kinase
human mammary epithelial cells (HMEC) 221
Human Protein Atlas (HPA) Project 93, 189
HUPO Proteomics Standards Initiative 155
hybrid quadrupole–orbitrap mass spectrometer (*Q-Exactive*) 33, 34, 246
hydrazide-functionalised resins 66
hydrolytic processes, food 332
hydrophilic interaction chromatography (HILIC) 57, 248, 249, 250
hydrophobicity 9–10
Hyne's reaction 331

IBAQ (intensity-based absolute quantification) 89
ICAM5 (intercellular adhesion molecule 5) 166, 167, 168, 169
ICATs (isotope-coded affinity tags) 54, 132, 241–2, 301, 302
ICP-MS (inductively-coupled plasma mass spectrometry) 110–25
IDPicker software 144
IEM (ion evaporation model) 7, 8
IGEL (isotopic glycosidase elution and labelling on lectin-column chromatography) 66
IMAC (immobilised metal affinity chromatography) 248
immobilised metal affinity chromatography (IMAC) 248
immune system *see* antibodies
α-importin 190
in vitro chemical labelling 194, 199–202, 206, 207
in vitro metabolic labelling 4, 5
in-gel digestion 238
INA (immunonephelometric assays) 321
inductively-coupled plasma mass spectrometry (ICP-MS) 110–25
inflammation biomarkers validation 280–1, 286–7
infrared multiphoton dissociation (IRMPD) 63
inlet capillaries 7, 10–12
insoluble material removal 146
instrumentation
    high resolution/accurate mass (HR/AM) hybrid mass spectrometry 32–3
    ion intensity-based label-free quantification 140
    iTRAQ/TMT 57
    LC-MS performance qualification 148–50
    LC–MS/MS 3–19
        contemporary 12–16
        control problems 12, 17
        high-performance 13–14
insulin receptor substrate homologue Chico complex 65
insulin signalling pathway 221
intensity-based absolute quantification (IBAQ) 89
interaction partners 239
intercellular adhesion molecule 5 (ICAM5) 166, 167, 168, 169
interference effects
    ion intensity-based label-free quantification 138–9
    iTRAQ/TMT 59, 61
    MRM-MS-based workflow 319
    quadrupole–orbitrap instruments
        HR/AM product ions measurements 42
        PRM optimisation 44–5
        sensitivity 39
internal standards 146, 199, 305
iodine 113
iodixanol 200, 201
Ion Accounting 178
ion evaporation model (IEM) 7, 8
ion funnel technology 11, 12

ion intensity-based label-free
  quantification 137–43
ion mobility (IM) 179–80, 181
ion transmission efficiency 7, 10–12
ion trapping capabilities 35–6, 243
ionisation efficiency ratios 117
ionisation mechanisms 7–10
IPTL (isobaric peptide termini
  labelling) 71
IRMPD (infrared multiphoton
  dissociation) 63
isobaric tags for relative and
  absolute quantification
  (iTRAQ) 51–71, 245
    bio-engineering 67–8
    cheese ripening 345
    contamination 243
    data analysis 63–4
    glycoproteomic analysis 66–7
    labelling approaches 53–9
    LC-MS identification/
      quantification 132, 133
    LC–MS/MS 16
        workflows 4, 5
    LOPIT 194, 195, 200
    medical research
      applications 68–70
    MRM-MS-based workflow 321
    MS platforms 62–3
    MS1 filtering label-free
      quantification 156, 157
    phosphoproteomic analysis
      65–6
    post-translational
      modifications 64–7
    reagents 60, 61
    technical limitations 59–62
    tyrosine phosphorylation
      quantification 219, 220, 221
    usefulness 51–71
    variants 58–9
    workflows 55–8
isotope-coded affinity tags (ICATs)
  54, 132, 133, 241–2, 301, 302
isotopes *see* label-based methods;
  label-free methods
isotopic glycosidase elution and
  labelling on lectin-column
  chromatography (IGEL) 66
iTRAQ *see* isobaric tags for relative
  and absolute quantification
  (iTRAQ)

kininogen proteins 295, 296

label-based methods
    *see also individual label-based
      methods*
    ICP-MS/ESI-MS 110–25
    isotopically-labelled
      peptides 41–3
    iTRAQ/TMT isobaric
      tagging 55–71
    metabolic labelling 240–1,
      245
    parallel reaction monitoring
      41–3
    selected reaction monitoring
      80–103
    stable isotopes used 218, 219
    standards 110–25
label-free methods
    absolute protein
      quantification 89, 90, 94–5
    current interest in 133
    dynamic range 16
    ion intensity-based 137–43
    LC-MS 131–50
    LC-MS^E data-independent
      acquisition 175–83
    MacQuant 156
    method 245
    MS1 quantification using ion
      intensity chromatograms in
      Skyline 154–72
    PTMs in space/time 243–6
    quantification units 93–4
    requirements 52
    Rosetta Elucidator® 140, 141,
      143, 156
    software packages 143–5
    strategy overview 134–43

label-free methods (*continued*)
tyrosine phosphorylation 223–4
workflow 5
labelled standards for MRM 316–25
labelling strategies 258–61, 262–6
lanthionine 331
lapatinib-resistance 228
LC-MS/MS *see* reverse phase liquid chromatography–tandem mass spectrometry (LC-MS/MS)
lectins 66–7, 163, 164
legislation, food allergens 337
leptin, murine 319
light peptides 262, 263, 265, 269, 270, 273, 301
linear dynamic range *see* dynamic range
linear ion trap/quadrupole Orbitraps (LTQ-Orbitrap/Q-Exactive) 13, 14
linear regression 100, 101
lipid oxidation products 332
lipophilicity 121, 122
liquid chromatography *see* high-performance liquid chromatography (HPLC)
literature studies 53, 291
liver 66, 143, 157, 261, 282
localisation of organelle proteins by isotope tagging (LOPIT) 194, 195, 199–202, 206, 207
see also spatial dynamic proteomics
low-copy number protein detection 234–5, 236
LVALVR isotopologous peptides mixture 35–6, 38–9
lysine 241, 259, 332
lysinoalanine 331

MAbs (monoclonal antibodies) 215
machine learning 189–90, 203, 204, 294–8, 307
macromolecular assemblies *see* spatial dynamic proteomics
Maillard browning 330, 331, 332
MALDI-MS (matrix-assisted laser desorption ionisation) 242
manual data analysis 266, 267
Mascot software 57, 58, 156, 250
mass cytometry 103
mass spectrometry (MS)
see also individual MS methods
absolute protein quantification 82, 83
chemical labelling 241–2, 245
food sample preparation 334
run-to-run variation 197, 244
mathematical modelling 81
matrix effects, food 332–3, 339
matrix-assisted laser desorption ionisation (MALDI)-MS 242
MaxQuant label-free method 156
meat tenderness 345
medical research applications 68–70
see also cancer; disease states
metabolic fate 258, 259
metabolic labelling
iTRAQ/TMT/SILAC 53
LC-MS identification/ quantification 132–3
LC–MS/MS workflows 4, 5
method 240–1, 245
metalloproteinase 1 (TIMP1) 323
methylglyoxal 332
micro-LC-ICP-MS (liquid chromatography–inductively-coupled plasma–mass spectrometry) 113–18
microscopy-driven analysis 188–9
mitochondria 157, 193, 195, 196
molecular network biology 291–3
molecular weight 9–10
monoclonal antibodies (MAbs) 215
Motif-X algorithm 227
mouse model
anemia 322–3
leptin 319
liver 157, 261
whole brain lysis 182
MRM *see* multiple reaction monitoring (MRM)

MS1 filtering label-free
quantification
ion intensity chromatograms
in Skyline 154–72
lab research pipelines 168–71
MS1Probe Python-encoded
bioinformatics tool 161, 162
research applications 162–71
statistical tools development
161–2
subtype-specific breast cancer
164, 166–8, 169, 170
MS^E data-independent
acquisition 176–83
fundamentals 177
ion mobility 179–80
performance evaluation 180–3
protein identification 178
protein quantification 178–9
MSight program 144
MSInspect program 144
MSnbase package 203–4
MSQuant program 144
mTRAQ reagents 58
MuDPIT (multidimensional peptide
identification technology) 176
multiple datasets integration 250
multiple reaction monitoring (MRM)
see also selected reaction
monitoring (SRM)
biomarkers 307, 308
discovery in human
plasma 316–25
measurement of panels/
signatures 299–306
multiple candidate
biomarkers 284, 285–6
prioritisation strategies
290
statistical methods for
data analysis 294–8
validation 279–308
verification 288, 289, 290
food analysis 341
MRM-HR 171
multiple samples 242–3
multiplexing capabilities 54–5, 71
multivariate statistical approaches
201
Mycobacterium tuberculosis 95
myoglobin 138

N-hydroxysuccinimide ester
(NHS-ester) 55, 60, 61, 220
N-linked sialylated
glycoproteins 248, 250
nano-electrospray ionisation (nESI)
6, 7, 8
National Cancer Institute, Clinical
Proteomic Technologies for
Cancer initiative 306, 307
nESI (nano-electrospray ionisation)
6, 7, 8
network biology paradigm 291–3
NetworkAnalyzer software 293
networks of gelled proteins in
food 333
new biomarker entities 280, 283–7
Next Generation proteomics *see*
spatial dynamic proteomics;
temporal dynamic proteomics
next generation sequence (NGS) 293
NHS (*N*-hydroxysuccinimide)-
ester 55, 60, 61, 220
nitrogen-15
LC-MS identification/
quantification 132
MRM quantification of
biomarkers 301
protein turnover studies 259
stable isotope labelling 88,
90, 92
NLS Mapper software 190
non-communicable diseases
(NCDs) 316–25
non-isobaric tagging 54
see also isotope-coded affinity
tags (ICATs); stable isotope-
labelling of amino acids in
cell culture (SILAC)
non-small cell lung carcinoma
(NSCLC) 214, 224

normalisation strategies 142, 244, 246
normalisation to selected proteins (NSP) 135
normalised spectral abundance factor (NSAF) 95, 135, 136
normaliser groups 55, 60, 61
NSAF (normalised spectral abundance factor) 95, 135, 136
NSP (normalisation to selected proteins) 135
nucleus 193, 195, 196

oa-ToF MS (orthogonal acceleration time-of-flight mass spectrometry) 179
one shot total cell lysate method 234, 235
one-third rule 62
open-source programs 144, 271–3
 see also Skyline open source program
optimising strategies 16–19, 43–5, 273–5
orbitrap instruments *see* quadrupole–orbitrap LC-MS/MS spectrometers
organelles
 see also spatial dynamic proteomics
 organelle discovery approach 238
 organelle-specific gene ontology annotation 196
 pre-fractionation for spatial/temporal studies 236–8
 purification 193, 194, 195, 196
orthogonal acceleration time-of-flight (oa-ToF) mass spectrometry 179
osteonectin/SPARC 166, 167, 168, 169, 170
outliers determination 150
oxygen-18 labelling 5, 54, 301

PAI protein abundance index) 136
pan-specific high affinity anti-phosphotyrosine monoclonal antibodies 215
Pan1 actin cytoskeleton regulatory complex protein 65
pancreatic cancer 70
PANORAMICS software 144
parallel reaction monitoring (PRM) 33, 34, 39–45, 88
parent droplets formation 8
partial least squares regression (PLSR) 226
PASTA (phosphorus-based absolutely quantified standard) 111, 112
pathway approaches, biomarkers 291–3
PC-IDMS (protein cleavage–isotope dilution mass spectrometry) 3–6, 14
PCA (principal component analysis) 204
PCP *see* protein correlation profiling (PCP)
peaks
 alignment, LC-MS label-free data 140–2
 capacity, LC–MS/MS 15–16
 measurement, PTMs in space/time 244
 picking, MS1 filtering label-free quantification 159–60
peanut allergen Arah 1/2 337, 339, 340, 341
PepC software 144
PEPPeR program 144
Pep_Prob algorithm 135
peptide mass tags (PMT) 140
peptide-N-glycosidase (PNGase) 334
PeptideAtlas database 86
peptides
 concatenated 91–2
 elution times, selected reaction monitoring 87
 equaliser peptides 89, 92
 ionisation response/bias 8–10
 proteotypic 98–100
 selection 85–6, 98–100

standards 90–2, 110–25
synthetic 90–1
PepTracker® 250
performance evaluation 148–9, 150, 180–3
peripheral blood mononuclear cells (PBMCs) 225
Perl statistical programming tool 270, 271
personalised medicine 280, 307
PFL (Protein Frequency Library) 239
phenotype discovery algorithms 203–4
phosphatase inhibitors 247
phosphoester bonds 247–50
phosphopeptides 65–6, 110–25, 147, 148
phosphorus-based absolutely quantified standard (PASTA) 111, 112
phosphorus-quantified but phosphorus-free standards 116–17
phosphorus-quantified phosphopeptide standards 113–16
phosphoryl groups enzymatic removal 116, 117
phosphorylation 211–28, 246–50
phosphotyrosine binding domains (PTB) 213, 215
phosphotyrosine cellular signalling 211–28
bioinformatics tools 227
computational modelling 226–7
deregulation in disease 213–14
enrichment for phosphotyrosine profiling 214–18
functional analyses 226–7
process summary 212–13
quantification of tyrosine phorphorylation 218–26
SH2 domains as phosphotyrosine profiling tools 215–16
single-cell level quantification 224–6
tandem mass spectrometry 216–18
Plasma Atlas 168
plasma/serum samples 286, 291, 316–25
PLGS (ProteinLynx Global Server) 178, 179
PMT (peptide mass tags) 140
PNGase (peptide-N-glycosidase) 334
polymerase chain reaction (PCR) 337, 338, 339
polyunsaturated fatty acids (PUFAs) 332
post-metabolic chemical labelling approaches *see* isobaric tags for relative and absolute quantification (iTRAQ); stable isotope-labelling of amino acids in cell culture (SILAC); tandem mass tags (TMT)
post-translational modifications (PTMs)
enrichment 247–50
ETD fragmentation 63
exclusion from proteotypicity 98
food proteins 335
iTRAQ/TMT 64–7
label-free absolute quantification 102
phosphotyrosine cellular signalling 211–28
spacial/temporal proteomics 233–53
challenges 234–9
data analysis 250–1
enrichment 246–50
quantitative proteomics 240–6
synthetic peptide standards 90
PQD (pulsed Q dissociation) 243
precision *see* accuracy

precursor ions
DDA/DIA 176, 177
label-free quantification 89, 93, 95
LOPIT 202
quantification on quadrupole–orbitrap instruments 33–9
tandem mass spectrometry 216, 217
precursor peptide pools
enrichment 260–1
protein turnover studies 258, 260–1
relative-isotope abundance 260, 261–74
predictive *in silico* methods 189–90, 203, 204, 294–8, 307
pre-fractionation methods 234–9, 248, 249, 250
PrEST (protein epitope signature tags) 89, 93–4
PRIDE proteomic repository 86
primary amines labelling by acetylation 54
principal component analysis (PCA) 204
prioritisation biomarker studies 289–94
PRM (parallel reaction monitoring) 33, 34, 39–45, 88
processed food protein modifications 331–2, 334, 335
product ion spectra
DDA/DIA 176, 177
label-free absolute quantification 93, 96
quadrupole–orbitrap instruments 39–45
tandem mass spectrometry 216–18
tryptic peptides 13
Progenesis & PPP/R software 268, 270, 271
Progenesis LC-MS (Nonlinear Dynamics) program 144
proline and experimental error 259
pRoloc package 203, 205
prostate cancer 69, 281, 282
protease digestion 162–4, 216, 241, 259, 334
protein abundance index (PAI) 136
Protein Atlas project 285
protein cleavage–isotope dilution mass spectrometry (PC-IDMS) 3–6, 14
protein copies per cell 96
protein correlation profiling (PCP)
LOPIT comparison 202
organelles for spatial dynamic proteomics 194, 198–9, 202, 206, 207
protein epitope signature tags (PrESTs) 89, 93, 94
Protein Frequency Library (PFL) 239
protein localisation *see* spatial dynamic proteomics
protein standard absolute quantification (PSAQ) 92–3
absolute quantification 110–25
phosphorus-quantified phosphopeptide standards 113–16
definition/use 111, 112
selected reaction monitoring 89, 92–4
types/summary 111
protein turnover studies
data analysis 262–75
labelling strategies 258–61, 262–6
manual data analysis 266, 267
precursor peptide enrichment 260–1
precursor peptide intensities acquisition 265–6
sampling times/frequencies 261
ProteinLynx Global Server (PLGS) 178, 179
Proteomics Standards Initiative Common Query Interface (PSICQUIC) 292

proteostasis 257
proteotypic peptides 98–100
ProteoWizard library 155, 159, 171
proton transfer ion–ion reactions (PTR) 61
PSAQ *see* protein standard absolute quantification (PSAQ)
pseudo SRM 171
PSICQUIC (Proteomics Standards Initiative Common Query Interface) 292
PTMScout Web resource 227
PTR (proton transfer ion–ion reactions) 61
pulsed Q dissociation (PQD) 243

Q-Exactive hybrid quadrupole–orbitrap mass spectrometer 33, 34, 246
QC (quality control) 146–50, 305–6, 329, 344–6
QconCAT (quantification concatamer) strategy 5, 89, 91–2, 111, 112
Qi value correction factor 136
QQQ (triple quadrupole mass spectrometers) 14, 84, 341
quadrupole time-of-flight (Q-TOF) 13, 14, 32–3, 220
quadrupole–orbitrap LC-MS/MS spectrometry 27–46
  HR/AM
    instrument characteristics 32–3, 34
    precursor ion quantification 33–9
    trapping capabilities 35–6
  iTRAQ/TMT 63
  phosphotyrosine cellular signalling 217
quality control (QC) 146–50, 305–6, 329, 344–6
quantotypic peptides selection 99–100

racemisation, food proteins 331
radioactive labelling 97
random forests 294, 295–7, 307
Rayleigh limit 7, 8
reagents for iTRAQ/TMT 55, 58, 60, 61
rearrangement products, food 332
receptor tyrosine kinases (RTKs) 212, 213, 226
  see also ErbB2 receptor tyrosine-protein kinase
receptor tyrosine protein kinase (ErbB2)
  digestion using different proteases 163–4, 165–8
  MS1 label-free studies 157, 164, 166–8, 169, 170
recombinant isotope-labelled and quantified (RIQ) standard 123, 124, 125
recombinant isotope-labelled and selenium quantified (RISQ) protein standard 111, 112, 119, 120, 122, 123, 124
recombinant and selenium quantified (RSQ) standard 123, 124, 125
recursive feature elimination, SVM 298
recycling of labelled amino acids 259, 260
red blood cells 143, 261
reference standards 81–2, 241, 336–7
relative quantification methods
  common scenario 110–11
  iTRAQ/TMT 51–71
  LC–MS/MS workflows 4–5
  tyrosine phosphorylation 218–22
relative transition intensities 86
relative-isotope abundance (RIA) 260, 261–74
replicate experiments for validation 206
reporter groups, iTRAQ/TMT 55, 60, 61

reporter ions
fewer with ETD 59
iTRAQ/TMT methods 59, 242–3
LOPIT 200, 201
tandem mass spectrometry 217, 220, 243
reporting standards 286
reproducibility
chromatography 244, 246
MS^E data-independent acquisition 182
multiple reaction monitoring 324
RP-HPLC 12–13
workflows 162–5
restrospective studies, biomarkers 283
retention times
LC-ESI-MS 114
MS1 filtering label-free quantification 159, 160
RP-HPLC 13
selenium-containing peptides 120, 121, 122
re-utilisation, labelled amino acids 259, 260
reverse phase high-performance liquid chromatography (RP-HPLC) 12–13, 57, 244, 318
reverse phase liquid chromatography–tandem mass spectrometry (LC-MS/MS)
see also quadrupole–orbitrap LC-MS/MS spectrometry
absolute protein quantification 83–103
characteristics 3–6, 14–16
contemporary instrumentation 12–16
DoE optimising strategies 16–19
electrospray ionisation 6–12
ICATs multiple reaction monitoring 301–2
phosphotyrosine cellular signalling 216–18
protein correlation profiling 194, 198–9, 202, 206, 207
protein turnover studies 257–75
size fractionation of PTMs 238
rheumatic disease biomarkers 280–1
RIA (relative-isotope abundance) 260, 261–74
RIBAR approach 136
rigid cell walls 96
RIQ (recombinant isotope-labelled and quantified standard) 123, 124, 125
risk assessment, food allergy 344
RISQ (recombinant isotope-labelled and selenium quantified) protein standard 111, 112, 119, 120, 122, 123, 124
Rosetta Elucidator® 140, 141, 143, 156
RP-HPLC *see* reverse phase high-performance liquid chromatography (RP-HPLC)
RSQ (recombinant and selenium quantified) standard 123, 124, 125
RTKs *see* receptor tyrosine kinases (RTKs)
run-to-run variation 197, 244

Saccharomyces cerevisiae 86, 135
SAF (spectral abundance factor) 135
safety of food 329, 337–46
samples
clinical for biomarker discovery 284, 286–7, 305
human blood/plasma 317
food proteins 332–3, 339
label-free absolute quantification 102
LC–MS/MS workflows 4–6
MS1 label-free quantification tools 162–4

preparation, food allergen detection methods 339
sampling times/frequencies, protein turnover studies 261
spatial/temporal dynamic proteomics 234–9
saturation effects, spectral counting 137
Scaffold (Proteome Software) 144
Scansite algorithm 227
scheduled selected reaction monitoring 84, 85
scoring systems 281, 285
SCX (strong-cation exchange) 134, 135, 248
SDLAVPSELALLK peptide dilution series 41–3
SDS-PAGE gels 147, 148, 238
see also high-performance liquid chromatography (HPLC)
SEC (size exclusion chromatography) 238
selected ion monitoring (SIM) mode 33, 34, 35–41
selected reaction monitoring (SRM)
absolute protein quantification 80–103
food analysis 341
principles 84–5
quadrupole–orbitrap LC–MS/MS instruments 36–8, 41–3
SRM assay 85–7
suitability for absolute quantification 87
tyrosine phosphorylation 223, 224
selection of peptides 85–6, 98–100, 137, 140
selection of targets 334–6, 338
selenium ICP-tags 118–25
selenium-quantified but selenium-free protein standards 123–5
separative centrifugation 190–1
serendipitous precursor selection 137, 140
serotransferrin 120, 121, 124
SH2 (src homology) domains 215–16
shared peptides problem, iTRAQ/TMT 64
shotgun proteomics 4–5, 8, 9, 26–7
sickle cell anemia 143
SIEVE (Thermo Scientific) 144
signal-to-noise ratio (S/N) 64, 85
signalling proteins 118, 188, 211–28, 247
SILAC *see* stable isotope-labelling of amino acids in cell culture (SILAC)
SIM (selected ion monitoring) mode 33, 34, 35–41
single-cell level quantification 103, 224–6
single-stage MS acquisition 33, 34
sirtuin 3 (SIRT3) 157
SIS (stable isotope-labelled standard) peptides 89, 90–1, 300–1, 317
site-specific phosphorylation 118
size exclusion chromatography (SEC) 238
size fractionation 238
skeletal muscle 261
skimmer cone/region, ESI sources 11, 12
Skyline open source program
biomarkers 303–5
improved/extended features 158–61
LC-MS system qualification 148
MS1 filtering label-free quantification 154–72
advantages 157–8
applications 157–62
overview 144
SkylineRunner 162
slaughter method effects on meat 345

software packages/programs
see also individual packages/programs
computational modelling 227
gene ontology for biomarker discovery 290
ion intensity-based label-free quantification 138–9
label-free datasets 138–9, 143–5
LC-MS^E acquisition 178, 179
network biology paradigm 292, 293
protein turnover data 262–75
optimal software 273–5
spatial dynamic proteomics data 189–90, 203, 204, 205, 250–1
SPARC/osteonectin 166, 167, 168, 169, 170
spatial dynamic proteomics 187–207
data analysis 202–5
definition 188
in silico methods 189–90
microscopy-driven analysis 188–9
MS analysis 190, 193–202
analytical fractionation 198
localisation of organelle proteins by isotope tagging 194, 199–202, 206, 207
organelle purification/cataloguing 193, 194, 195, 196
protein correlation profiling 194, 198–9, 202, 206, 207
subtractive proteomics 194, 195, 197–8
post-translational modifications in space/time 233–53
subcellular fractionation 190–3
differential permeabilisation 191
immunocapture/affinity purification 191–2
separative centrifugation 190–1
zone electrophoresis 192
validation of localisation studies 205–7
SpC *see* spectral count (SpC)
specific protein digestion 97–8
specific/nonspecific binding 239
specificity of food allergen detection 338
spectral abundance factor (SAF) 135
spectral count (SpC)
definition 134
label-free absolute quantification 89, 94, 95, 100, 101, 134–7
phosphotyrosine signalling studies 223
PTMs in space/time 244
spectral indices in label-free quantification 136
spectral libraries 158, 159
spermatozoa 261
splitless nanoflow HPLC systems 13
Src homology (SH2) domains 215–16
Src kinases 222, 228
SRM *see* selected reaction monitoring (SRM)
SSP (system suitability protocol) 306
sta6 *Chlamydomonas reinhardtii* starch-less mutant 67–8
stable isotope labelling 88–94, 110–25, 132–3
see also stable isotope-labelling of amino acids in cell culture (SILAC)
stable isotope-labelled standard (SIS) peptides 89, 90–1, 300–1, 317

stable isotope-labelling of amino acids in cell culture (SILAC)
chemical labelling alternative 241–3, 245
dynamic SILAC 264
full-length labelled proteins 93
immunoprecipitation/affinity selection 239
LC-MS identification/ quantification 132, 133
LC–MS/MS 16
workflows 4, 5–6
method 240, 241, 245
MRM quantification of biomarkers 301
MS1 filtering label-free quantification 156, 157
PCP-SILAC approach 199, 206, 207
phosphotyrosine cellular signalling 218, 219, 220
protein turnover studies 259
SILACanalyser software 268, 271
SILACtor software 268, 269
triplex SILAC for spatial dynamic proteomics 197–8
usefulness 52, 53
standard operating procedures (SOPs) 145–8
standards
biomarker discovery/ development 306
human blood/plasma 316–25
chemical labelling 241
food proteomics 336–7
HUPO Proteomics Standards Initiative 155
label-based methods 52
peptide standards 90–2
protein correlation profiling 199
reference standards 81–2, 241, 336–7
reporting 286
RIQ 123, 124, 125
RISQ 12, 111, 112, 119, 120, 122, 123
RSQ 123, 124, 125
selenium-quantified but selenium-free 123–5
stable isotope labelling 88–94, 110–25
ICP/ESI-MS 110–25
Standards for Reporting of Diagnostic Accuracy 286
STAT tyrosine kinases 225
STAT5 signaling protein 115, 117
statistical tests/methods 58, 63–4, 161–2
see also data analysis
stoichiometries 81
storage of food 344, 345
streptavidin affinity purification 192
STRING software 292
strong-cation exchange (SCX) 134, 135, 248
Student's t-test 162
subcellular fractionation 190–3, 236–8
subcellular protein localisation *see* spatial dynamic proteomics
subtractive proteomics 194, 195, 197–8
Super SILAC approach 133
supervised data analysis methods 203
support vector machines (SVMs) 294, 295, 297–8, 307
SVM-RFE (support vector machine-recursive feature elimination) 298
SWATH-MS data-independent acquisition 87–8, 171, 176, 207
synovial tissue samples 286, 287
synthetic peptides 90–1
SYP61 *Arabidopsis thaliana* compartment 191
system suitability protocol (SSP) 306

TAILS (terminal amino isotope labelling of substrates) 70

tandem mass spectrometers (MS/MS) 13–14, 216–18, 242–3, 245
see also quadrupole–orbitrap LC–MS/MS spectrometry; reverse phase liquid chromatography tandem mass spectrometry (LC–MS/MS)
tandem mass tags (TMTs)
LC-MS identification/ quantification 132
LOPIT 200
method 245
MS1 filtering label-free methods 156, 157
tyrosine phosphorylation methods 219, 220, 221
usefulness 51–71
tandem peptides concept 117
target selection 334–6, 338
targeted methods *see* data-dependent acquisition (DDA); multiple reaction monitoring (MRM); selected reaction monitoring (SRM)
TaxonID 342–3
Taylor cone 7
temperature and ionisation efficiency 9, 10–11
temporal dynamic proteomics
phosphotyrosine cellular signalling in disease 211–28
post-translational modifications in space/time 233–53
protein turnover on proteome-wide scale 257–75
subcellular localisation studies 187–207
tyrosine phosphorylation relative quantification 221–2
terminal amino isotope labelling of substrates (TAILS) 70
testicular cancer biomarkers 282
text-mining 291
TIC (total ion current) 144
TIMP1 (metalloproteinase 1) 323
tissues, chemical labelling 53–5
titanium dioxide enrichment 248, 249, 250
TMT *see* tandem mass tags (TMTs)
toluene *ortho*-monooxygenase 68
Topograph software 268, 269
total ion current (TIC) 144
total spectral count normalisation (TSpC) 135
training data 203, 205, 295
transferrin, endogenous human 37
transitions
libraries 325
multiple reaction monitoring 299, 324, 325
selected reaction monitoring 84, 85, 86, 87
Skyline 303
transmission of ions 7, 10–12
transplant rejection 321
triple quadrupole mass spectrometers (QQQ) 14, 84, 341
triplex SILAC 197–8
truncated protein standards 89, 93–4
trypsin digestion 216, 241, 259, 334
TSKgel Amide-80 columns 248
TSpC (total spectral count normalisation) 135
turnover studies *see* protein turnover studies
two-level full factorial design 17–19
tyrosine kinase inhibitors (TKIs) 228
tyrosine kinases (TKs) 211–28
see also ErbB2 receptor tyrosine-protein kinase (HER2)
tyrosine phosphorylation 211–28
tyrosine-containing peptides 113

ultra-high pressure (UHP) LC systems 13
underestimation in iTRAQ/TMT studies 59, 60, 61, 62, 64, 71

UNIPROT database 330, 342–3
units of expression 81, 93–4
unlabelling strategies 258, 261, 263, 264–6
urine
biomarker discovery/ validation 291, 324
endogenous human transferrin 37
human, XIC-based label-free LC-MS 143
protein digests, SDLAVPSELALLK peptide dilution series 41–3

validation
biomarkers using MRM 279–308
clinical smaples 286–7
discovery/development 283–6
human plasma 316–25
integrative bioinformatics 289–94
panels/signatures 281–3, 288–9, 299–306
quality control 305–6
Skyline software 303–5
standardisation 306
statistical methods 294–8
verification 287–9
protein localisation studies 205–7
variability 16, 18–19
variance-stabilising transforms 64
vendor-independent software solutions *see* Skyline open source program

wastewater biosensors 68
Western blot analyses 83
wine proteins 345
Wnt signalling pathway 188
WoLF PSORT software 190
workflows *see* experimental workflows

XIC (extracted ion chromatograms) 138–43, 154–72, 320
xRIBAR method 136
X!Tandem 156

Y-split μLC[ICP/ESI]MS system 114, 115
YhcN/YchH stress proteins 68

zebrafish 215
zone electrophoresis (ZE) 192